Fusion Energy

Small Non-Expensive Electric Inertial Fusion Reactors (v.2)

Alexander Bolonkin

USA, LULU, 2019

Small Non-Expensive Electric Inertial Thermonuclear Reactors (v2)

Alexander Bolonkin <abolonkin@gmail.com>

ISBN 978-0-359-52770-0

Book is collection of independent articles Small inertial Fusion Reactors.

The author offers the new, small cheap electric impulse and cumulative thermonuclear reactors, which increases the temperature and pressure of its nuclear fuel by millions of times, reaches the required ignition stage and, ultimately, constant contained thermonuclear reaction. Electric Impulse and Cumulative AB Reactors contain several innovations to achieve its product.

Chief among them in version one the electric thermonuclear reactors are using electric field voltage $50 \div 1000$ kV (an electric condenser discharge), which allows to heat the primary compressed fuel in special pellet by electric impulse up hundreds millions degrees of temperature.

Author offers a new capacitor as driver Inertial Thermonuclear Reactors.

The additional compressing and combustion time the fuel nucleus may have from electric pinch-effect and heavy nucleus of the fuel cartridge cover. The main advantages of the offered method are very small electric fuel cartridge (11-18 mm) and small of the full reactor installation (reactor has the spherical diameter 0.3 - 3 m), using the many thermonuclear fuels at room temperature and possibility of using the offered thermonuclear reactor for transportation (ships, trains, aircrafts, rockets, etc.). Author gives theory and estimations of the suggested reactors.

Author also is discussing the problems of converting the received thermonuclear energy into mechanical (electrical) energy and into rocket thrust.

Offered small micro-reactors may be used as heaves (ignition, fuse) for small artillery nuclear projectiles and bombs.

Publisher: USA, LULU, www.lulu.com, 2/28/2019

3

Contents:

Abstract

In the last sixty years, the government spent the tens billion dollars attempting to develop useful thermonuclear energy. However, they cannot yet reach a stable thermonuclear reaction. They still are promising publically, after another 15 – 20 years, and more tens of billions of US dollars to finally design the expensive workable industrial installation, which possibly will produce electric energy more expensive than current heat, wind and hydroelectric stations can in 2015.

The author offers the new, small cheap electric impulse and cumulative thermonuclear reactors, which increases the temperature and pressure of its nuclear fuel by millions of times, reaches the required ignition stage and, ultimately, constant contained thermonuclear reaction. Electric Impulse and Cumulative AB Reactors contain several innovations to achieve its product.

Chief among them in version one the electric thermonuclear reactors are using electric field voltage $50 \div 1000$ kV (an electric condenser discharge), which allows to heat the primary compressed fuel in special pellet by electric impulse up hundreds millions degrees of temperature.

Author offers a new capacitor as driver Inertial Thermonuclear Reactors.

The additional compressing and combustion time the fuel nucleus may have from electric pinch-effect and heavy nucleus of the fuel cartridge cover. The main advantages of the offered method are very small electric fuel cartridge (11-18 mm) and small of the full reactor installation (reactor has the spherical diameter 0.3 - 3 m), using the many thermonuclear fuels at room temperature and possibility of using the offered thermonuclear reactor for transportation (ships, trains, aircrafts, rockets, etc.). Author gives theory and estimations of the suggested reactors.

Author also is discussing the problems of converting the received thermonuclear energy into mechanical (electrical) energy and into rocket thrust.

Offered small micro-reactors may be used as heaves (ignition, fuse) for small artillery nuclear projectiles and bombs.

Keywords: *Small Inertial Fusion reactor, Impulse thermonuclear reactor, electric thermonuclear reactor, cumulative thermonuclear reactor, transport thermonuclear reactor, aerospace thermonuclear propulsion , nuclei fuse, thermonuclear rocket.*

INTRODUCTION IN CURENT THERMONUCLÉAIRE REACTORS

Main difference between the current and offered AB Thermonuclear Reactors

As it is well-known the thermonuclear reaction occurs when Lawson criterion
$$L = nT\tau > c,$$
where n is matter (fuel) density, [1/m^3]; T is temperature, [KeV], 1 eV = 1.16×10^4 K; τ is reaction time, [s]; c is constant for given nuclear fuel. For tritium-deuterium fuel (T + D) $c \approx 10^{20} \div 10^{21}$. The current Inertial Confinement Fusion (ICF) uses the laser compression method (high matter density n) and low temperature T and low reaction time τ. With a compression by 10^3, the compressed density will be 200 g/cm^3 (T+D), and the compressed radius can be as small as 0.05 mm. For this density ($n = 4.8 \cdot 10^{31}$ m^{-3}) and $\tau = 10^{-9}$ s the Lawson criterion gives a need temperature $T = 2$ eV ≈ 8300 K. It is very few for nuclear reaction. But laser ICF cannot reach this temperature. Trying to warm up the fuel capsule by additional X-rays and other particles have been unsuccessful.

   The other method of the current thermonuclear reaction is Magnetic Confinement Fusion (MCF). It uses the other idea – high time reaction (τ = seconds, up minutes) and low plasma density ($n = 10^{21}$m^{-3}) and low temperature. That method has a lot of technical problems, is very expensive and also not reaches the stable ignition.

   Both current main methods (ICF and MCF) are developed more 60 years by the thousands scientists in all main countries. The governments spent the billions of dollars for their R&D (Research and Development) and are spending hundreds millions dollars every years. But optimist sciences only promise to reach the useful stable nuclear reaction throw 10 – 15 years (after 2016) and build the industrial electric station after the additional 5 – 10 years. The other scientists show: the price of the nuclear energy used tritium fuel (main fuel for current reactors is T+D) will be cost ten times more than in present electric stations using the natural fuel (tritium costs 30,000 \$/gram, trend up 100,000 \$/g)).

   The author offers the new method (reactor)(see Chapter 2). Main idea is getting a high temperature by high intensity electric field. Reactor can easy to get the very high temperature up 300 keV (1.3 billion K), has enough compression (up 600 - 1000 atm) and conformation (10^{-6} sec). One has Lawson criterion in thousands times more than need, can work on cheap D+D nuclear fuel (1 gram of deuterium cost only 1\$), is very cheap and has a small installation. The main test (getting the thermonuclear reaction) costs only same thousands dollars. If test will be successful, we can immediately design the engines for ships, trains, submarines, electric stations and propulsions for rockets.

Brief Information about Current Thermonuclear Reactors

Fusion power is useful energy generated by nuclear fusion reactions. In this kind of reaction two light atomic nuclei fuse together to form a heavier nucleus and release energy. The largest current nuclear fusion experiment, JET, has resulted in fusion power production somewhat larger than the power put into the plasma, maintained for a few seconds. In June 2005, the construction of the experimental reactor ITER, designed to produce several times more fusion power than the power into it generating the plasma over many minutes, was announced. The unrealized production of net electrical power from fusion machines is planned for the next generation experiment after ITER.

   Unfortunately, this task is not easy, as scientists thought early on. Fusion reactions require a very large amount of energy to initiate in order to overcome the so-called *Coulomb barrier* or *fusion barrier energy*. The key to practical fusion power is to select a fuel that requires the minimum amount of energy to start, that is, the lowest barrier energy. The best fuel from this standpoint is a one-to-one mix of deuterium and tritium; both are heavy isotopes of hydrogen. The D-T (Deuterium and Tritium) mix has suitable low barrier energy. In order to create the required conditions, the fuel must be heated to tens of millions of degrees, and/or compressed to immense pressures.

At present, D-T is used by two main methods of fusion: inertial confinement fusion (ICF) and magnetic confinement fusion (MCF)--for example, tokomak device.

In inertial confinement fusion (ICF), nuclear fusion reactions are initiated by heating and compressing a target. The target is a pellet that most often contains deuterium and tritium (often only micro or milligrams). Intense focused laser or ion beams are used for compression of pellets. The beams explosively detonate the outer material layers of the target pellet. That accelerates the underlying target layers inward, sending a shockwave into the center of each pellet's mass. If the shockwave is powerful enough, and if high enough density at the center is achieved, some of the fuel will be heated enough to cause pellet fusion reactions. In a target which has been heated and compressed to the point of thermonuclear ignition, energy can then heat surrounding fuel to cause it to fuse as well, potentially releasing tremendous amounts of energy.

Magnetic confinement fusion (MCF). Since plasmas are very good electrical conductors, magnetic fields can also be configured to safely confine fusion fuel. A variety of magnetic configurations can be used, the basic distinction being between magnetic mirror confinement and toroidal confinement, especially tokomaks and stellarators.

Lawson criterion. In nuclear fusion research, the Lawson criterion, first derived by John D. Lawson in 1957, is an important general measure of a system that defines the conditions needed for a fusion reactor to reach *ignition* stage, that is the heating of the plasma by the products of the fusion reactions is sufficient to maintain the temperature of the plasma against all losses without external power input. As originally formulated the Lawson criterion gives a minimum required value for the product of the plasma (electron) density n_e and the "energy confinement time" τ. Later analyses suggested that a more useful figure of merit is the "triple product" of density, confinement time, and plasma temperature T. The triple product also has a minimum required value, and the name "Lawson criterion" often refers to this important inequality.

The key to practical fusion power is to select a fuel that requires the minimum amount of energy to start, that is, the lowest barrier energy. The best known fuel from this standpoint is a one-to-one mix of deuterium and tritium; both are heavy isotopes of hydrogen. The D-T (Deuterium and Tritium) mix has a low barrier.

In order to create the required conditions, the fuel must be heated to tens of millions of degrees, and/or compressed to immense pressures. The temperature and pressure required for any particular fuel to fuse is known as the Lawson criterion. For the D-T reaction, the physical value is about

$$L = n_e T \tau > (10^{14} \div 10^{15}) \quad \text{in} \quad \text{"cgs"} \quad \text{units}$$
$$\text{or} \quad L = nT\tau > (10^{20} \div 10^{21}) \quad \text{in} \quad \text{CI} \quad \text{units},$$

where T is temperature, [KeV], 1 eV = 1.16×10^4 K; n_e is matter density, [1/cm^3]; n is matter density, [1/m^3]; τ is time, [s]. Last equation is in metric system. The thermonuclear reaction of $^2H + ^3D$ realizes if $L > 10^{20}$ in CI (meter, kilogram, second) units or $L > 10^{14}$ in 'cgs' (centimeter, gram, second) units.

This number has not yet been achieved in any fusion reactor, although the latest generations of fusion-making machines have come significantly close to doing so. For instance, the reactor TFTR has achieved the densities and energy lifetimes needed to achieve Lawson at the temperatures it can create, but it cannot create those temperatures at the same time. Future ITER aims to do both.

The Lawson criterion applies to inertial confinement fusion as well as to magnetic confinement fusion but is more usefully expressed in a different form. Whereas the energy confinement time in a magnetic system is very difficult to predict or even to establish empirically, in an inertial system it must be on the order of the time it takes sound waves to travel across the plasma:

$$\tau \approx \frac{R}{\sqrt{kT/m_i}}$$

where τ is time, s; R is distance, m; k is Boltzmann constant; T is temperature, K; m_i is mass of ion, kg.

Following the above derivation of the limit on $n_e\tau_E$, we see that the product of the density and the radius must be greater than a value related to the minimum of $T^{3/2}/\langle\sigma v\rangle$ (here σ is Boltzmann constant, v is ion speed). This condition is traditionally expressed in terms of the mass density ρ:

$$\rho R > 1 \text{ g/cm}^2 .$$

To satisfy this criterion at the density of solid D+T (0.2 g/cm³) would require implausibly large laser pulse energy. Assuming the energy required scales with the mass of the fusion plasma ($E_{\text{laser}} \sim \rho R^3 \sim \rho^{-2}$), compressing the fuel to 10^3 or 10^4 times solid density would reduce the energy required by a factor of 10^6 or 10^8, bringing it into a realistic range. With a compression by 10^3, the compressed density will be 200 g/cm³, and the compressed radius can be as small as 0.05 mm. The radius of the fuel before compression would be 0.5 mm. The initial pellet will be perhaps twice as large since most of the mass will be ablated during the compression stage by a symmetrical energy input bath.

The fusion power density is a good figure of merit to determine the optimum temperature for magnetic confinement, but for inertial confinement the fractional burn-up of the fuel is probably more useful. The burn-up should be proportional to the specific reaction rate ($n^2\langle\sigma v\rangle$) times the confinement time (which scales as $T^{1/2}$) divided by the particle density n: burn-up fraction $\sim n^2\langle\sigma v\rangle T^{-1/2} / n \sim (nT) (\langle\sigma v\rangle/T^{3/2})$

Thus the optimum temperature for inertial confinement fusion is that which maximizes $\langle\sigma v\rangle/T^{3/2}$, which is slightly higher than the optimum temperature for magnetic confinement.

Short history of thermonuclear fusion. One of the earliest (in the late 1970's and early 1980's) serious attempts at an ICF design was *Shiva*, a 20-armed neodymium laser system built at the Lawrence Livermore National Laboratory (LLNL) in California that started operation in 1978. Shiva was a "proof of concept" design, followed by the *NOVA* design with 10 times the power. Funding for fusion research was severely constrained in the 80's, but NOVA nevertheless successfully gathered enough information for a next generation machine whose goal was ignition. Although net energy can be released even without ignition (the breakeven point), ignition is considered necessary for a *practical* power system.

The resulting design, now known as the National Ignition Facility, commenced being constructed at LLNL in 1997. Originally intended to start construction in the early 1990s, the NIF is now six years behind schedule and over-budget by some $3.5 billion. Nevertheless many of the problems appear to be due to the "Big Science Laboratory" mentality and shifting the focus from pure ICF research to the nuclear stewardship program, LLNLs traditional nuclear weapons-making role. NIF "burned" in 2010, when the remaining lasers in the 192-beam array were finally installed. Like those earlier experiments, however, NIF has failed to reach ignition and is, as of 2015, generating only about 1/3rd of the required energy levels needed to reach full fusion stage of operation.

Laser physicists in Europe have put forward plans to build a £500m facility, called HiPER, to study a new approach to laser fusion. A panel of scientists from seven European Union countries believes that a "fast ignition" laser facility could make a significant contribution to fusion research, as well as supporting experiments in other areas of physics. The facility would be designed to achieve high-energy gains, providing the critical intermediate step between ignition and a demonstration reactor. It would consist of a long-pulse laser with energy of 200 kJ to compress the fuel and a short-pulse laser with energy of 70 kJ to heat it.

Confinement refers to all the conditions necessary to keep plasma dense and hot long enough to undergo fusion:

- *Equilibrium:* There must be no net forces on any part of the plasma, otherwise it will rapidly disassemble. The exception, of course, is inertial confinement, where the relevant physics must occur faster than the disassembly time.
- *Stability:* The plasma must be so constructed that small deviations are restored to the initial state, otherwise some unavoidable disturbance will occur and grow exponentially until the plasma is destroyed.
- *Transport:* The loss of particles and heat in all channels must be sufficiently slow. The word "confinement" is often used in the restricted sense of "energy confinement".

To produce self-sustaining fusion, the energy released by the reaction (or at least a fraction of it) must be used to heat new reactant nuclei and keep them hot long enough that they also undergo fusion reactions. Retaining the heat generated is called energy *confinement* and may be accomplished in a number of ways.

Hydrogen bomb weapons require no confinement at all. The fuel is simply allowed to fly apart, but it takes a certain length of time to do this, and during this time fusion can occur. This approach is called *inertial confinement* (Figure 1). If more than about a milligram of fuel is used, the explosion would destroy the machine, so controlled thermonuclear fusion using inertial confinement causes tiny pellets of fuel to explode several times a second. To induce the explosion, the pellet must be compressed to about 30 times solid density with energetic beams. If the beams are focused directly on the pellet, it is called *direct drive*, which can in principle be very efficient, but in practice it is difficult to obtain the needed uniformity. An alternative approach is *indirect drive*, in which the beams heat a shell, and the shell radiates x-rays, which then implode the pellet. The beams are commonly laser beams, but heavy and light ion beams and electron beams have all been investigated and tried to one degree or another.

They rely on fuel pellets with a "perfect" globular shape in order to generate a symmetrical inward shock wave to produce the high-density plasma, and in practice these have proven difficult to produce. A recent development in the field of laser-induced ICF is the use of ultra-short pulse multi-petawatt lasers to heat the plasma of an imploding pellet at exactly the moment of greatest density after it is imploded conventionally using terawatt-scale lasers. This research will be carried out on the (currently being built) OMEGA EP petawatt and OMEGA lasers at the University of Rochester in New York and at the GEKKO XII laser at the Institute for Laser Engineering in Osaka, Japan which, if fruitful, may have the effect of greatly reducing the cost of a laser fusion-based power source.

Fig.1. One laser installation of NIF

At the temperatures required for fusion, the fuel is in the form of plasma with very good electrical conductivity. This opens the possibility to confine the fuel and the energy with magnetic fields, an idea known as *magnetic confinement* (Figure 2).

Much of this progress has been achieved with a particular emphasis on tokomaks (Figure 2).

In fusion research, achieving a fusion energy gain factor $Q = 1$ is called *breakeven* and is considered a significant although somewhat artificial milestone. *Ignition* refers to an infinite Q, that is, a self-sustaining plasma where the losses are made up for by fusion power without any external input. In a practical fusion reactor, some external power will always be required for things like current drive, refueling, profile control, and burn control. A value on the order of $Q = 20$ will be required if the plant is to deliver much more energy than it uses internally.

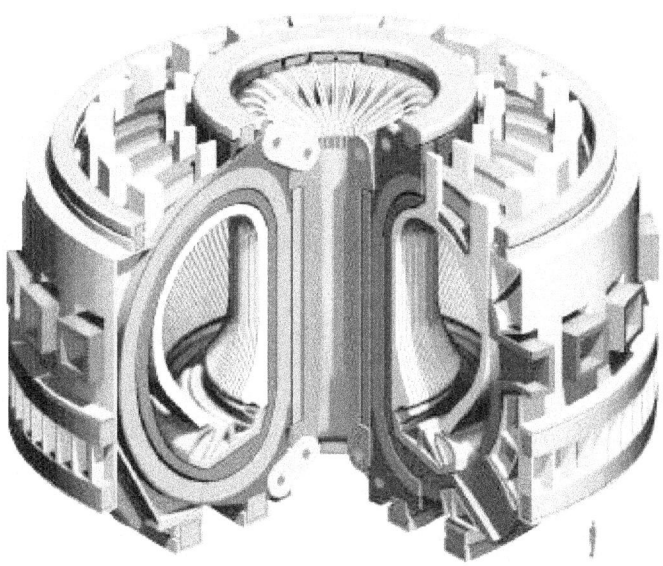

Fig. 2. Magnetic thermonuclear reactor. The size of the installation is obvious if you compare it with the "Little Blue Man" inside the machine at the bottom. Cost is some tens of billions of dollars.

In a fusion power plant, the nuclear island has a *plasma chamber* with an associated vacuum system, surrounded by a plasma-facing components (first wall and diverter) maintaining the vacuum boundary and absorbing the thermal radiation coming from the plasma, surrounded in turn by a blanket where the neutrons are absorbed to breed tritium and heat a working fluid that transfers the power to the balance of plant. If magnetic confinement is used, a *magnet* system, using primarily cryogenic superconducting magnets, is needed, and usually systems for heating and refueling the plasma and for driving current. In inertial confinement, a *driver*(laser or accelerator) and a focusing system are needed, as well as a means for forming and positioning the *pellets.*

The magnetic fusion energy (MFE) program seeks to establish the conditions to sustain a nuclear fusion reaction in plasma that is contained by magnetic fields to allow the successful production of fusion power.

In thirty years, scientists have increased the Lawson criterion of the ICF and tokomak installations by tens of times. Unfortunately, all current and some new installations (ICF and tokomak) have a Lawrence criterion that is tens of times lower than is necessary (Figure 3).

Data of same current inertial laser installations:
1. NOVA uses laser NIF (USA), has 192 beams, impulse energy up 120 kJ. One reach density 20 g/cm^3, speed of cover is up 300 km/s. NIF has failed to reach ignition and is, as of 2013, generating about 1/3rd of the required energy levels. NIF cost is about $3.5B.
2. YiPER (EU) has impulse energy up 70 kJ.
2. OMEGA (USA) has impulse energy up 60 kJ.
3. Gekko-XII (Japan) has impulse energy up 20 kJ. One reaches density 120 g/cm^3.
4. Febus (France) has impulse energy up 20 kJ.
5. Iskra-5 (Russia) has impulse energy up 30 kJ.

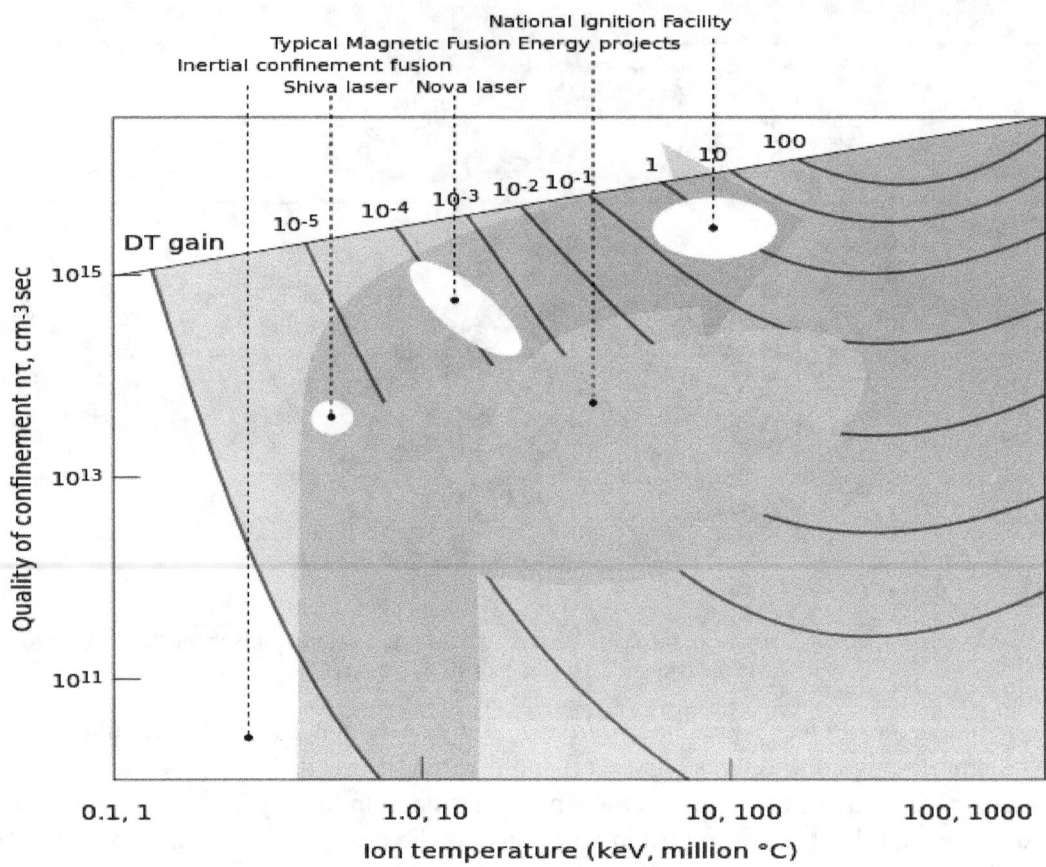

Fig. 3. Parameter space occupied by inertial fusion energy and magnetic fusion energy devices. The regime allowing thermonuclear ignition with high gain lies near the upper right corner of the plot.

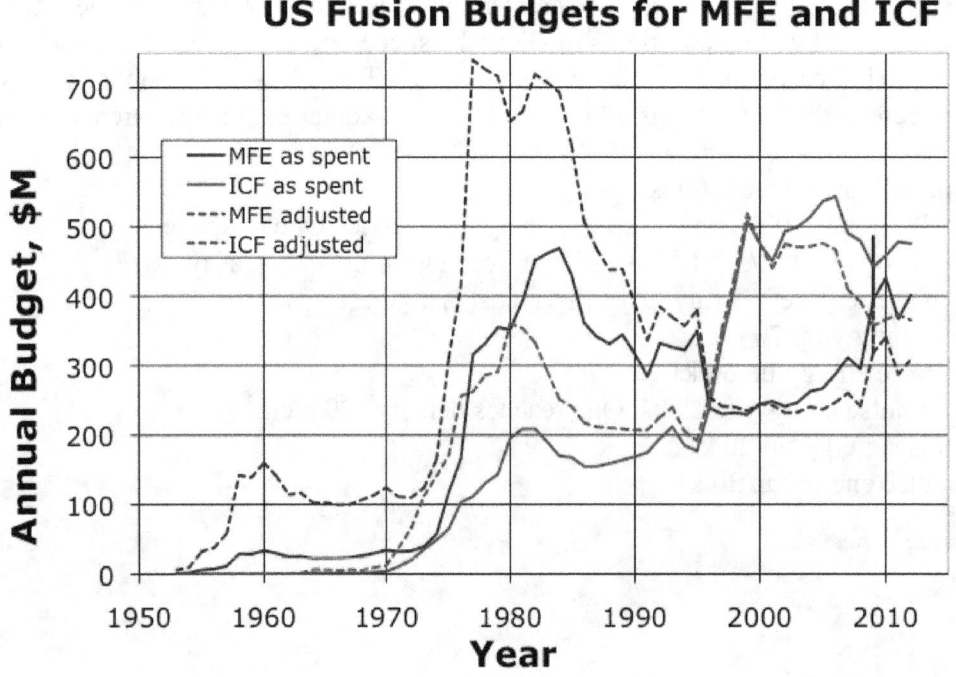

Chapter 1.
Theory of the Thermonuclear Reactors

1. The following reactions are suitable for thermonuclear fusion:

Table 1. Suitable reactions for thermonuclear fusion

#	Syntezis	Result (received Energy, MeV)	%
1	D+T→	^{4}He(3.5)+n(14.1)	
2a	D+D→	T(1.01)+p(3.02)	50%
2b	D+D→	^{3}He(0.82)+n(2.45)	50%
3	D+^{3}He→	^{4}He(3.06)+p(14.7)	
4	T+T→	^{4}He+2n(+11.3)	
5	^{3}He+^{3}He→	^{4}He+2p(+12.9)	
6a	^{3}He+T→	^{4}He+p+n(+12,1)	51%
6b	^{3}He+T→	^{4}He(4.8)+D(9.5)	43%
6c	^{3}He+T→	^{5}He(2.4)+p(+11.9)	6%
7	p+^{6}Li →	^{4}He(1.7)+^{3}He(2.3)	
8a	p+^{7}Li →	2^{4}He(17.3)	20%
8b	p+^{7}Li →	^{7}Be+n(1.6)	80%
9	D+^{6}Li→	2^{4}He(22.4)	
10	p+^{11}B→	3^{4}He(+8.7)	
11	n+^{6}Li→	^{4}He(2.1)+T(2.7)	
12	^{3}He+^{6}Li→	2^{4}He+p(+16.9)	

Here are: p = ^{1}H (protium), D = ^{2}H (deuterium), and T = ^{3}H (tritium) are shorthand notation for the main three isotopes of hydrogen. ^{4}He = ά –alpha particle.

Very important value is the **cross section of thermonuclear reaction**. The nuclear cross section of a nucleus is used to characterize the **probability** that a nuclear reaction will occur. The concept of a nuclear cross section can be quantified physically in terms of "characteristic area" where a larger area means a larger probability of interaction. The standard unit for measuring a nuclear cross section (denoted as σ) is the **barn**, which is equal to 10^{-28} m² or 10^{-24} cm². Nuclear cross section very strong depent from kinetic energy of particles. Typical thermonuclear cross section main fuel particles are shown in fig.4.

For reactions with two products, the energy is divided between them in inverse proportion to their masses, as shown. In most reactions with three products, the distribution of energy varies. For reactions that can result in more than one set of products, the branching ratios are given.

Some reaction candidates can be eliminated at once. The D+^{6}Li reaction has no advantage compared to p+^{11}B because it is roughly as difficult to burn but produces substantially more neutrons through D+D side reactions. There is also a p+^{7}Li reaction, but the cross-section is far too low accepted possible for $T_i > 1$ MeV, but at such high temperatures, an endothermic, direct neutron-producing reaction also becomes very significant. Finally, there is also a p+^{9}Be reaction, which is not only difficult to burn, but ^{9}Be can be easily induced to split into two alphas and a neutron.

Typical nuclear radii are of the order 10^{-14} m. Assuming spherical shape, we therefore expect the cross sections for nuclear reactions to be of the order of πr^2 or 10^{-28} m² (i.e. 1 barn). Observed cross

sections vary enormously - for example, **slow neutrons** absorbed by the (n,   ) reaction show a cross section much higher than 1,000 barns in some cases (boron-10, cadmium-113, and xenon-135), while the cross sections for **transmutations** by **gamma-ray** absorption are in the region of 0.001 barn.

The some cross sections and corresponding energy are shown in Table 2. In Table 2 we use the shortly notation the nuclear reaction. For example D+T → ^{4}He + n as ^{2}H(t,n)^{4}He. If result has many products canals, we do not write them.

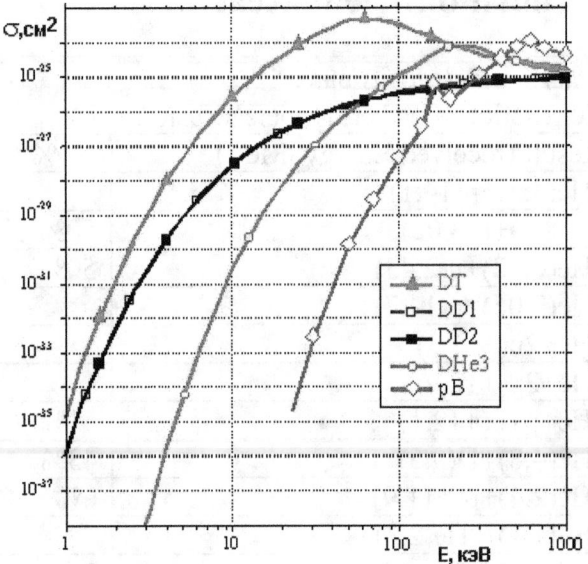

Fig.1. Thermonuclear cross section reaction D+T, D+D (2a), D+D (2b), D+^{3}He, and p+B vs kinetic energy E [keV] of the light particles.

Table 2. Cross section and corresponding energy of some thermonuclear reaction

Fig. #	Reaction	Reaction Energy MeV	Temperat. ≈10 kV and σ	Temperat. ≈10² kV and σ	Temperat. ≈10³ kV and σ	Temperat. ≈10⁴ kV and σ	Temperat. and σ=max
43-1(2)	^{2}H(d,n)^{3}H	4.033	10kV, σ=10^{-29}	10²kV, σ=2h·10^{-26}	10³kV, σ=7·10^{-26}	10⁴kV, σ=7·10^{-26}	10³kV, σ=7·10^{-26}
43-1(7)	^{6}Li(p,á)^{3}He	4.021	-	10²kV, σ=8·10^{-27}	10³kV, σ=10^{-26}	10⁴kV, σ=2·10^{-22}	-
43-1(8)	^{6}Li(t,n)···	16	-	-	10³kV, σ=3·10^{-25}	-	-
43-1(3)	^{2}H(t,d)^{4}H	4.321	-	10²kV, σ=1.1·10^{-26}	10³kV, σ=7·10^{-26}	10⁴kV, σ=7·10^{-26}	10³kV, σ=7·10^{-26}
43-2(2)	^{3}H(d,n)^{4}He D+T→^{4}He+n	17,6	10kV, σ=2·10^{-27}	10²kV, σ=5·10^{-24}	10³kV, σ=2·10^{-25}	10⁴kV, σ=6·10^{-26}	10²kV, σ=5·10^{-24}
43-2(1)	^{2}H(d,n)^{3}He	7.3	10¹kV, σ=1.1·10^{-29}	10²kV, σ=5·10^{-26}	10³kV, σ=0.9·10^{-25}	10²kV, σ=0.8·10^{-25}	10²kV, σ=0.9·10^{-25}
43-2(4)	^{3}H(d,p)^{4}He	18.35	-	10²kV, σ=3·10^{-27}	10³kV, σ=3·10^{-26}	10⁴kV, σ=5·10^{-27}	4·10²kV, σ=9·10^{-26}
43-2(5)	^{3}H(t,pn)^{4}He	12	-	10²kV, σ=7·10^{-28}	10³kV, σ=6·10^{-26}	-	-
43-2(6)	^{6}Li(d,p)^{7}Li	5.028	-	10²kV, σ=6·10^{-28}	10³kV, σ=10^{-25}	-	-
43-3(4)	^{6}Li(d,á)^{4}He	22.375	-	10²kV, σ=7·10^{-28}	7·10²kV, σ=7·10^{-26}	-	-
43-4(6)	^{9}Be(p,á)^{6}Li	2.126	-	5·10²kV, σ=3·10^{-26}	1.1·10³kV, σ=3·10^{-25}	-	-

Source: [12] pp. 947-950.

In addition to the fusion reactions, the following reactions with neutrons are important in order to "breed" tritium in "dry" fusion bombs and some proposed fusion reactors:

$$n + {}^6Li \rightarrow T + {}^4He + 4.5 \text{ MeV} , \quad n + {}^7Li + 2.5 \text{ MeV} \rightarrow T + {}^4He + n' .$$

To evaluate the usefulness of these reactions, in addition to the reactants, the products, and the energy released, one needs to know something about the cross section. Any given fusion device will have a maximum plasma pressure that it can sustain, and an economical device will always operate near this maximum. Given this pressure, the largest fusion output is obtained when the temperature is selected so that $\langle\sigma v\rangle/T^2$ is a maximum. This is also the temperature at which the value of the triple product $nT\tau$ required for ignition is a minimum. This chosen optimum temperature and the value of $\langle\sigma v\rangle/T^2$ at that temperature is given for a few of these reactions in the following table.

Table 3. Optimum temperature and the value of $\langle\sigma v\rangle/T^2$ at that temperature

fuel	T [keV]	$\langle\sigma v\rangle/T^2$ [m³/s/keV²]
D-T	13.6	1.24×10^{-24}
D-D	15	1.28×10^{-26}
D-^{3}He	58	2.24×10^{-26}
p-^{6}Li	66	1.46×10^{-27}
p-^{11}B	123	3.01×10^{-27}

Note: that many of the reactions form chains. For instance, a reactor fueled with T and ^{3}He will create some D, which is then possible to use in the D + ^{3}He reaction if the energies are "right". An elegant idea is to combine the reactions (7) and (12). The ^{3}He from reaction (7) can react with ^{6}Li in reaction (12) before completely thermalizing. This produces an energetic proton which in turn undergoes reaction (7) before thermalizing. A detailed analysis shows that this idea will not really work well, but it is a good example of a case where the usual assumption of a Maxwellian plasma is not appropriate.

Any of the reactions above can, in principle, be the basis of fusion power production. In addition to the temperature and cross section discussed above, we must consider the total energy of the fusion products E_{fus}, the energy of the charged fusion products E_{ch}, and the atomic number Z of the non-hydrogenic reactant.

Specification of the D-D reaction entails some difficulties, though. To begin with, one must average over the two branches (2) and (3). More difficult is to decide how to treat the T and ^{3}He products. T burns so well in a deuterium plasma that it is almost impossible to extract from the plasma. The D-^{3}He reaction is optimized at a much higher temperature, so the burn-up at the optimum D-D temperature may be low, so it seems reasonable to assume the T but not the ^{3}He gets burned up and adds its energy to the net reaction.

Thus we will count the D-D fusion energy as $E_{fus} = (4.03+17.6+3.27)/2 = 12.5$ MeV and the energy in charged particles as $E_{ch} = (4.03+3.5+0.82)/2 = 4.2$ MeV.

Another unique aspect of the D-D reaction is that there is only one reactant, which must be taken into account when calculating the reaction rate.

With this choice, we tabulate parameters for four of the most important reactions.

Table 4. Parameters of the most important reactions

Fuel	Z	E_{fus} [MeV]	E_{ch} [MeV]	neutronicity
D-T	1	17.6	3.5	0.80
D-D	1	12.5	4.2	0.66
D-^{3}He	2	18.3	18.3	~0.05
p-^{11}B	5	8.7	8.7	~0.001

The last column is the *neutronicity* of the reaction, the fraction of the fusion energy released as neutrons. This is an important indicator of the magnitude of the problems associated with neutrons like radiation damage, biological shielding, remote handling, and safety. For the first two reactions it is calculated as $(E_{fus}-E_{ch})/E_{fus}$. For the last two reactions, where this calculation would give zero, the values quoted are rough estimates based on side reactions that produce neutrons in a plasma in thermal equilibrium.

Table 5. Aneutronic reactions.

High nuclear cross section aneutronic reactions[2]		
Isotopes	**Reaction**	
Deuterium–helium-3	^2D + ^3He → ^4He + ^1p	+ 18.3 MeV
Deuterium–lithium-6	^2D + ^6Li → 2 ^4He	+ 22.4 MeV
Proton–lithium-6	^1p + ^6Li → ^4He + ^3He	+ 4.0 MeV
Helium-3–lithium-6	^3He + ^6Li → 2 ^4He + ^1p	+ 16.9 MeV
Helium-3-helium-3	^3He + ^3He → ^4He + 2 ^1p	+ 12.86 MeV
Proton–lithium-7	^1p + ^7Li → 2 ^4He	+ 17.2 MeV
Proton–boron	^1p + ^{11}B → 3 ^4He	+ 8.7 MeV
Proton–nitrogen	^1p + ^{15}N → ^{12}C + ^4He	+ 5.0 MeV

Of course, the reactants should also be mixed in the optimal proportions. This is the case when each reactant ion plus its associated electrons accounts for half the pressure. Assuming that the total pressure is fixed, this means that density of the non-hydrogenic ion is smaller than that of the hydrogenic ion by a factor $2/(Z+1)$. Therefore, the rate for these reactions is reduced by the same factor, on top of any differences in the values of $<\sigma v>/T^2$. On the other hand, because the D-D reaction has only one reactant, the rate is twice as high as if the fuel were divided between two hydrogenic species.

Thus, there is a "penalty" of $(2/(Z+1))$ for non-hydrogenic fuels arising from the fact that they require more electrons, which take up pressure without participating in the fusion reaction. There is, at the same time, a "bonus" of a factor 2 for D-D due to the fact that each ion can react with any of the other ions, not just a fraction of them.

The maximum value of $<\sigma v>/T^2$ is taken from a previous table. The "penalty/bonus" factor is that related to a non-hydrogenic reactant or a single-species reaction. The values in the column "reactivity" are found by dividing (1.24×10^{-24} by the product of the second and third columns. It indicates the factor by which the other reactions occur more slowly than the D-T reaction under comparable conditions. The column "Lawson criterion" weights these results with E_{ch} and gives an indication of how much more difficult it is to achieve ignition with these reactions, relative to the difficulty for the D-T reaction. The last column is labeled "power density" and weights the practical reactivity with E_{fus}. It indicates how much lower the fusion power density of the other reactions is compared to the D-T reaction and can be considered a measure of the economic potential.

Bremsstrahlung (Brake) Losses

1. *Bremsstrahlung*, (from the German *bremsen*, to brake and *Strahlung*, radiation, thus, "braking radiation"), is electromagnetic radiation produced by the acceleration of a charged particle, such as an electron, when deflected by another charged particle, such as an atomic nucleus. The term is also used to refer to the process of producing the radiation. Bremsstrahlung has a continuous spectrum. The phenomenon was discovered by Nikola Tesla (1856-1943) during high frequency research he conducted between 1888 and 1897.

Bremsstrahlung may also be referred to as free-free radiation. This refers to the radiation that arises as a result of a charged particle that is free both before and after the deflection (acceleration) that causes the emission. Strictly speaking, bremsstrahlung refers to any radiation due to the acceleration of a charged particle, which includes synchrotron radiation; however, it is frequently used (even when not speaking German) in the more literal and narrow sense of radiation from electrons stopping in matter.

The ions undergoing fusion will essentially never occur alone but will be mixed with electrons that neutralize the ions' electrical charge and form a plasma. The electrons will generally have a temperature comparable to or greater than that of the ions, so they will collide with the ions and emit Bremsstrahlung. The Sun and stars are opaque to Bremsstrahlung, but essentially any terrestrial fusion reactor will be optically thin at relevant wavelengths. Bremsstrahlung is also difficult to reflect and difficult to convert directly to electricity, so the ratio of fusion power produced to Bremsstrahlung radiation lost is an important figure of merit. This ratio is generally maximized at a much higher temperature than that which maximizes the power density (see the previous subsection). The following table shows the rough optimum temperature and the power ratio at that temperature for several reactions.

Table 6. Rough optimum temperature and the power ratio of fusion and Bremsstrahlung radiation lost

Fuel	T_i (keV)	$P_{fusion}/P_{Bremsstrahlung}$
D-T	50	140
D-D	500	2.9
D-^{3}He	100	5.3
^{3}He-^{3}He	1000	0.72
p-^{6}Li	800	0.21
p-^{11}B	300	0.57

The actual ratios of fusion to Bremsstrahlung power will likely be significantly lower for several reasons. For one, the calculation assumes that the energy of the fusion products is transmitted completely to the fuel ions, which then lose energy to the electrons by collisions, which in turn lose energy by Bremsstrahlung. However because the fusion products move much faster than the fuel ions, they will give up a significant fraction of their energy directly to the electrons. Secondly, the plasma is assumed to be composed purely of fuel ions. In practice, there will be a significant proportion of impurity ions, which will lower the ratio. In particular, the fusion products themselves *must* remain in the plasma until they have given up their energy, and *will* remain some time after that in any proposed confinement scheme. Finally, all channels of energy loss other than Bremsstrahlung have been neglected. The last two factors are related. On theoretical and experimental grounds, particle and energy confinement seem to be closely related. In a confinement scheme that does a good job of retaining energy, fusion products will build up. If the fusion products are efficiently ejected, then energy confinement will be poor, too.

The temperatures maximizing the fusion power compared to the Bremsstrahlung are in every case higher than the temperature that maximizes the power density and minimizes the required value of the fusion triple product (Lawson criterion). This will not change the optimum operating point for D-T very much because the Bremsstrahlung fraction is low, but it will push the other fuels into regimes where the power density relative to D-T is even lower and the required confinement even more difficult to achieve. For D-D and D-^{3}He, Bremsstrahlung losses will be a serious, possibly prohibitive problem. For ^{3}He-^{3}He, p-^{6}Li and p-^{11}B the Bremsstrahlung losses appear to make a fusion reactor using these fuels impossible.

In a plasma, the free electrons are constantly producing Bremsstrahlung in collisions with the ions. The power density of the Bremsstrahlung radiated is given by

$$P_{Br} = \frac{16\alpha^3 h^2}{\sqrt{3} \, m_e^{3/2}} n_e^2 T_e^{1/2} Z_{eff}$$

T_e is the electron temperature, α is the fine structure constant, h is Planck's constant, and the "effective" ion charge state Z_{eff} is given by an average over the charge states of the ions:

$$Z_{eff} = \Sigma \, (Z^2 n_Z) \, / \, n_e$$

This formula is derived in "Basic Principles of Plasmas Physics: A Statistical Approach" by S. Ichimaru, p. 228. It applies for high enough T_e that the electron deBroglie wavelength is longer than the classical Coulomb distance of closest approach. In practical units, this formula gives

$$P_{Br} = (1.69 \times 10^{-32} \, / \text{W cm}^{-3}) \, (n_e/\text{cm}^{-3})^2 \, (T_e/\text{eV})^{1/2} \, Z_{eff}$$
$$= (5.34 \times 10^{-37} \, / \text{W m}^{-3}) \, (n_e \, /\text{m}^{-3})^2 \, (T_e \, /\text{keV})^{1/2} \, Z_{eff}$$

where Wcm^{-3}, cm^{-3}, eV, Wm^{-3}, m^{-3}, keV are units of corresponding magnitudes. For very high temperatures there are relativistic corrections to this formula, that is, additional terms of order $T_e/m_e c^2$.

Below are some equations useful for computation:

2. The Deep of Penetration of outer radiation into plasma is

$$d_p = \frac{c}{\omega_{pe}} = 5.31 \cdot 10^5 n_e^{-1/2} \; . \; [\text{cm}] \tag{1}$$

For plasma density $n_e = 10^{22}$ 1/cm^3 $d_p = 5.31 \times 10^{-6}$ cm. That means the most of brake radiation will heat the fuel.

3. The Gas (Plasma) Dynamic Pressure, p_k, is

$$p_k = nk(T_e + T_i) \quad \text{if} \quad T_e = T_k \quad \text{then} \quad p_k = 2nkT , \tag{2}$$

where $k = 1.38 \times 10^{-23}$ is Boltzmann constant; T_e is temperature of electrons, °K; T_i is temperature of ions, ʌ°K. These temperatures may be different; n is plasma density, 1/m^3; p_k is plasma pressure, N/m^2.

4a. The gas (plasma) ion pressure, p, is

$$p = \frac{2}{3} nkT \tag{3a}$$

Here n is plasma density in 1/m^3.

4b. Ion collision rate

$$\nu_i = 4.80 \times 10^{-8} z^4 \mu^{-1/2} n_i \ln\Lambda \, T_i^{-3/2} \quad [1/\text{s}]; \tag{3b}$$

Here $\ln\Lambda$ is Columbus logarithm (~10); T_i is in eV. For $T_i = 10^4$ eV $\nu_i = 3.4 \cdot 10^6$, for $T_i = 10^6$ eV $\nu_i = 3.4 \cdot 10^3$.

4c. Loss energy in collisions of balls

$$W_{k1}/W_{k1,0} = [(m_1 - m_2)/(m_1 + m_2)]^2; \quad \text{For } m_1 = 100 m_2 \text{ Loss energy is 4\%,} \tag{3c}$$

5. The magnetic p_m and electrostatic pressure, p_s, are

$$p_m = \frac{B^2}{2\mu_0}, \quad p_s = \frac{1}{2} \varepsilon_0 E_s^2 , \tag{4}$$

where B is electromagnetic induction, Tesla; $\mu_0 = 4\pi \times 10^{-7}$ electromagnetic constant; $\varepsilon_0 = 8.85 \times 10^{-12}$, F/m, is electrostatic constant; E_s is electrostatic intensity, V/m.

6. Ion thermal velocity is

$$v_{Ti} = \left(\frac{kT_i}{m_i} \right)^{1/2} = 9.79 \times 10^5 \mu^{-1/2} T_i^{1/2} \quad \text{cm/s} , \tag{5}$$

where $\mu = m_i/m_p$, m_i is mass of ion, kg; $m_p = 1.67 \times 10^{-27}$ is mass of proton, kg.

7. Transverse Spitzer plasma resistivity

$$\eta_\perp = 1.03 \times 10^{-2} Z \ln \Lambda T^{-3/2}, \quad \Omega \, \text{cm} \quad \text{or} \quad \rho \approx \frac{0.1 Z}{T^{3/2}} \quad \Omega \, \text{cm} , \tag{6}$$

where $\ln \Lambda = 5 \div 15 \approx 10$ is Coulomb logarithm, Z is charge state.

8. *Reaction rates* <σv> (in cm^3 s^{-1}) averaged over Maxwellian distributions for low energy (T<25 keV) may be represent by

$$(\overline{\sigma v})_{DD} = 2.33 \times 10^{-14} T^{-2/3} \exp(-18.76 T^{-1/3}) \quad \text{cm}^3\text{s}^{-1}, \qquad (7)$$

$$(\overline{\sigma v})_{DT} = 3.68 \times 10^{-12} T^{-2/3} \exp(-19.94 T^{-1/3}) \quad \text{cm}^3\text{s}^{-1},$$

where *T* is measured in keV.

Reaction rates are presented in Table 7 below:

Table 7. Reaction rates <σv> (in cm^3 s^{-1}) averaged over Maxwellian distributions

Temperature, keV	D+D, (2a + 2b)	D+T, (1)	D+^{3}He, (3)	T+T, (4)	T+^{3}He, (6a-c)
1.0	1.5×10^{-22}	5.5×10^{-21}	10^{-26}	3.3×10^{-22}	10^{-28}
2.0	5.4×10^{-21}	2.6×10^{-19}	1.4×10^{-23}	7.1×10^{-21}	10^{-25}
5.0	1.8×10^{-19}	1.3×10^{-17}	6.7×10^{-21}	1.4×10^{-19}	2.1×10^{-22}
10.0	1.2×10^{-18}	1.1×10^{-16}	2.3×10^{-19}	7.2×10^{-19}	1.2×10^{-20}
20.0	5.2×10^{-18}	4.2×10^{-16}	3.8×10^{-18}	2.5×10^{-18}	2.6×10^{-19}
50.0	2.1×10^{-17}	8.7×10^{-16}	5.4×10^{-17}	8.7×10^{-18}	5.3×10^{-18}
100.0	4.5×10^{-17}	8.5×10^{-16}	1.6×10^{-16}	1.9×10^{-17}	7.7×10^{-17}
200.0	8.8×10^{-17}	6.3×10^{-16}	2.4×10^{-16}	4.2×10^{-17}	9.2×10^{-17}
500.0	1.8×10^{-16}	3.7×10^{-16}	2.3×10^{-16}	8.4×10^{-17}	2.9×10^{-16}
1000.0	2.2×10^{-16}	2.7×10^{-16}	1.8×10^{-16}	8.0×10^{-17}	5.2×10^{-16}

Source: AIP, Desk Reference, Third Edition, p. 644.

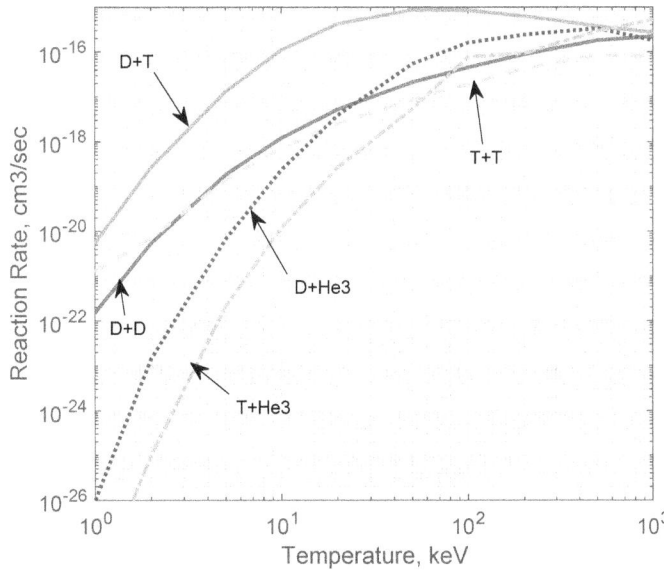

Fig.5. Reactivity is requested for thermonuclear reaction.

The deuterium cannot be used in the laser reactor because one requests in 100 times more ignition criterion then D + T. But, as you see in (23), one may be used in AB reactor (Fig.5).

9. *The power density released in the form of charged particles* is

$$P_{DD} = 3.3 \times 10^{-13} n_D^2 (\overline{\sigma v})_{DD}, \quad \text{W cm}^{-3}$$

$$P_{DT} = 5.6 \times 10^{-13} n_D n_T (\overline{\sigma v})_{DT}, \quad \text{W cm}^{-3} \qquad (8)$$

$$P_{DHe^3} = 2.9 \times 10^{-12} n_D n_{He^3} (\overline{\sigma v})_{DHe^3}, \quad \text{W cm}^{-3}$$

Here in P_{DD} equation it is included D+T reaction.

10. *Cost of the nuclear fuel (2012).*

Deuterium. The sea water contains about $1.55 \cdot 10^{-4}$ %. The World produces about tens thousend tons in year. Cost 1 $/g.

Tritium. The special nuclear reactors can produced it. Now the cost is 30,000 $/g. In future an expected cost will be from 100K÷200K $/g.

Helium-3. Very rare isotop. The Helium-4 contains $1.3 \cdot 10^{-6}/1$ of the Helium-3. Cost is 30K $/g. One project offers to extract it on Moon and delivery to Earth.

Litium 6 -7. Nature mixture cost 150 $/kg.

Uranium-238 contains 0.7% of Uranium-235. It cost 90÷250 $/kg.

Plutonium-239. Cost 5600 $/g.

As you see the thermonucler fuel D+D is the cheapest, but D+T has the lowest temperature for themonucler reaction. All the current experimental thermonuclear instellations are using the D+T.

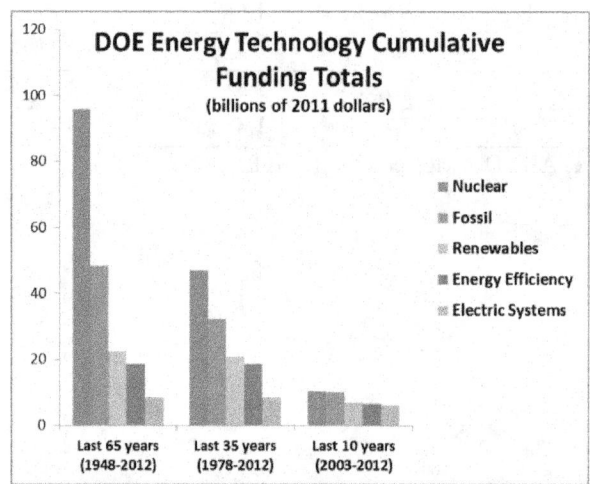

Chapter 2
Inertial Beam Thermonuclear AB Reactor

Abstract

Author offers a new impulse beam hole thermonuclear reactor. Reactor has the following features: one has a power high-current pulse ion accelerator using the thermonuclear fuel (for example, Deuterium) as ions. Accelerator focuses the ion beam into very small focus. The very small fuel capsule covered by shell from heavy and strong elements. The shell has a small hole for fuel beam from accelerator. The capsule contains a solid fuel (example, LiD). The theory and computation give conditions when will be ignition and developing the high intensity thermonuclear reaction. The suggested reactor is small, non-expensive and allows to make small engines.

Key word: Inertial Reactor, Impulse reactor, Beam reactor, Thermonuclear reactor, ICF reactor, AB reactor.

Introduction

The thermonuclear reaction depends from production three magnitudes: density (compression of fuel), temperature and time. At present time the scientists use two main methods for attempts to reach efficiency thermonuclear: Inertial Confinement Fusion (ICF) and Magnetic Confinement Fusion (MCF). In ICF the scientists use the high compression (by laser beam), but low time, in MCF they use low compression, but long time.

Both current main methods (ICF and MCF) are developed more 60 years by the thousands of scientists in all main countries. The governments spent the billions of dollars for their R&D (Research and Development) and are spending hundreds millions dollars every years. But optimist scientists only promise to reach the useful stable nuclear reaction throw 10 – 15 years (after 2018) and build the industrial electric station after the additional 5 – 10 years. The other scientists show: the price of the nuclear energy used tritium fuel (main fuel for current reactors is T+D) will be cost ten times more than the natural fuel using in present electric stations (tritium costs 30,000 \$/gram, trend up 100,000 \$/g).

Description of offered Reactor

The author offers the new reactor (method). Main idea is getting a high temperature by high-current impulse fuel ion focused accelerator and a special fuel capsule. The special fuel capsule has layer from the heavy elements, cover from strong material and the small hole for focused fuel ion beam. Reactor can easy to get the initial very high temperature up 30 keV (about 300 million K), has enough compression (up 600 - 1000 atm) of plasma and confinement (10^{-7} sec). One can work on cheap D+D nuclear fuel (1 gram of deuterium cost only 1\$), is very cheap and has a small installation. The main test (getting the thermonuclear reaction) costs only some ten thousand dollars. If test will be successful, we can immediately design the engines for ships, trains, submarines, electric stations and propulsions for rockets.

Electric impulse ion beam hole method. Early author offered five the new methods (reflex, cumulative, impulse, ultra-cold, electric)[1 ÷ 6], which are cheaper by thousands of times, more efficiency and does not have many disadvantages of the laser and magnetic methods. In given article the author offers a version of the electric impulse ion beam hole reactor. Detailed consideration of advantages the new methods and computation proofs are in next paragraphs.

Outline of the new electric impulse ion hole reactors.
The offered version of the ion beam hole AB thermonuclear reactor is presented in figures 1 – 3.
The new thermonuclear reactor contains (fig.1): spherical strong body (diameter 0.3 – 3 m), high-current ion

focusing accelerator, special fuel capsule and installation for delivery a fuel capsule into reactor body.

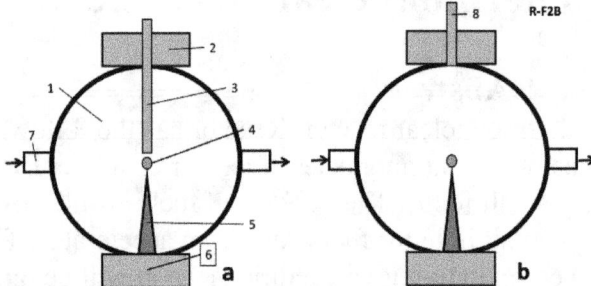

Fig. 1. Thermonuclear ion beam hole AB-reactor. *Notations*: **a** - reactor during the installation of the fuel capsule; **b** - reactor after installation of the fuel capsule. 1 – spherical reactor (diameter 0.3 – 3 m), 2 – installation, 3 - delivery rod in work position (installing of fuel capsule), 4 - fuel capsule, 5-6 – high-current ion focusing accelerator, 7 – ender for compression gas (air), 8 – delivery rod after in installation the fuel capsule.

The fuel capsule (fig.2) contains: the thermonuclear fuel in liquid, solid or a compressed gas form, layer from heavy elements (atom number A ≈ 200, for example, lead), and strong cover important for compressing gas fuel and heavy cover is important for increasing traction time. All have a spherical form. Capsule also has protrusion for mounting and fixing and small hole. Diameter of fuel is about 1 mm, cover in 2 -3 times more. Fuel has a groove for ion beam and can be separated by thin partition if we used a compressed gas fuel. This partition is destroyed by ion beam. The capsule is destroyed by thermonuclear explosion.
Small spherical thermonuclear capsule (other names: fuel cartridge, ampule, granule, beat, pellet) is shown in fig.2.

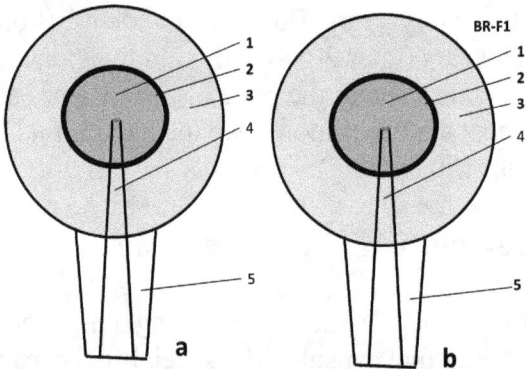

Fig.2. Fuel capsule for the offered ion hole reactor. **a**- forward view; **b** – side view. Notation: 1 – thermonuclear fuel, 2 – layer from heavy elements; 3 – strong cover, 4 – hole for ion beam, 5 - protrusion for mounting and fixing. For fuel mass $M = 10^{-7}$ kg, the internal diameter is about 1 mm, pressure of gas fuel is up $600 \div 1000$ atm.

The thermonuclear reactor is sphere of the diameter 0.3 – 3 m. (Fig.1). Reactor has two Version 1 - 2. In Version 1 the reactor has the additional installations for converting the nuclear energy into an electric, mechanical energy (MHG, turbine), in Version 2 the reactor converts by nozzle the thermonuclear energy in a rocket thrust (fig. 3).
The offered thermonuclear reactor works the next way (one example):
Version 1 for an electric or mechanic energy.
The internal volume of reactor is filled the atmospheric or compression air (enter 7 of Fig.1).
The fuel capsule (Fig.2) is installing by holder 3 (Fig.1) into reactor. Turn on the ion accelerator 6 (up 50 -100 kV, 30 – 60 kJ) (Fig 1). The ion (or neutral beam) from the accelerator are ionized the fuel molecules into capsule. In particular, they positive ionize and dissociate the fuel molecules (for example, D and T are contained into capsule). The positive ionized nucleus of the thermonuclear fuel (having small mass) are quick collectively

accelerated by accelerator up very high temperature (up 30 – 60 keV), focused, neutralized by electron and collide with fuel nucleus into capsule. They ignite the thermonuclear reaction.

The cover from heavy molecules (nuclear mass A ≈ 200) reflect the light (A ≈ 2÷3) fuel nucleus and increase the fusion (reaction) time of the fuel nucleus. In results (as show computation) the fuel nucleus merge and produce a thermonuclear reaction. The thermonuclear reaction (explosive) heats the air into reactor body. For increasing the efficiency, work mass, decreasing explosive temperature and protection from neutrons, the liquid 7 (for example, water, fig.3a) may be injected into reactor.

After thermonuclear explosion the hot gas flow out into the magneto-hydrodynamic generator (MHG) 10 and produces electric energy or runs to the gas, steam turbine and produces a useful work (Fig. 3a). Or the hot compressed gas runs to rocket nozzle and produces the rocket trust (fig. 3b).

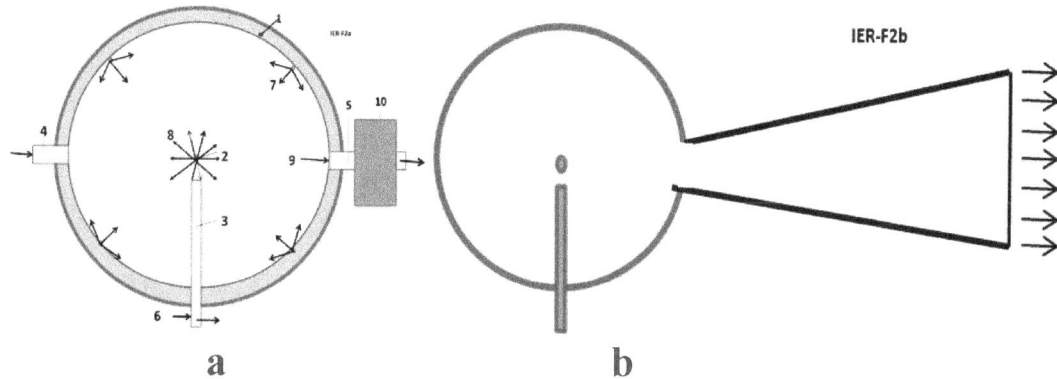

a   b

Fig.3. Final (industrial) work of Impulse Electric ion beam hole AB thermonuclear reactors. *a*) Hot compressed gas from sphere runs to the magneto-hydrodynamic generator (MHG) 10 and produces electric energy or runs to gas turbine and produces a useful work (Fig. 3a). *b*) Hot compressed gas runs to rocket nozzle and produces the rocket trust. *Notation*: 1-strong spherical (shell) of reactor, 2 -fuel capsule, 3 – holder for capsule, 4 – enter of compressed gas (air), 5-exit for hot gas, 6 – electricity, 7 – injection the cooling liquid (for example, water) (option); 8 – thermonuclear explosive of fuel pellet,9 – exit for explosive gas, 10 – MHG or gas (steam) turbine, 11 - exit of hot gas.

The main difference the offered electric reactor from the published cumulative reactors [2, 7] is type of explosive for getting the temperature, pressure and cumulative effect in fuel. On [2 - 4] author used the chemical explosive. The offered reactor uses the strong electric field for acceleration of nucleos, getting high temperature and cumulative effect. The electric method leads to practically unlimited cheap power. In [2, 4] the explosive is located into main spherical body 1 (fig.1) (or gun in [2]). In [4] version 1 (fig.2, [4]) the explosive 3 is small and located in the special fuel cartridge (fig.2, [6]). In current version no special compression explosive. The pressure and high temperature of the fuel are reached the high-current ion focusing accelerator. It is easier and it is more comfortable in using.

In the current cartridge version, the fuel pullet is filling by the compressed gas fuel (up 600÷1000 atmosphere or more). Reactor not has the explosive for an additional compressing of fuel. The fuel is compressed primary and heating only by strong ion focusing accelerator. The computation shows that is possible. We can also use the conventional pellet with frozen (gas) fuel, solid or liquid fuel.

AB Reactors are cooled using well-known methods between explosives or by an injection of water into sphere (fig. 3a).

Advantages of the suggested hole reactor in comparison with ICF Laser method.

The offered reactor and method have the following advantages in comparison with the conventional ICF laser reactor:
1. The high-current ion accelerator allows reaching the needed thermonuclear temperature.
2. Ion hole AB-reactors are cheaper by thousands of time because one does not have the gigantic very

expensive laser or magnetic installations (see [1]-[6]).

3. They more efficiency because the laser installation converts only 1 - 2.5% the electric energy into the light beam. In suggested AB reactors, the all underused energy remains in the spherical reactor and utilized in MHG or turbine.

AB reactors cannot have coefficient Q (used energy) significantly less 1. Moreover, one has heat efficiency more than conventional heat engines because it has very high temperature and compression ratio. One can use as the conventional very high-power engine in civil and military transportation.

4. The offered very important innovation (accelerating of exhaust rocket gas) allows increasing the top speed of the exhaust mass up very high speed. This makes this method available for thermonuclear rockets.
5. Electric AB-reactors give temperature of the fuel much more than the current ICF laser installations.
6. The compression has longer time (up to $10^{-6} - 10^{-7}$ s) than a laser beam pressing ($10^{-9} - 10^{-12}$ s), because molar mass ($\mu \approx 200$) of heavy molecules (cover of capsule in fig.2) is many times ($50 \div 100$) heavier than fuel molar mass ($\mu = 2 \div 3$).

This pressure is supported by shock wave coming from moving gas and pinch effect. This pressure increases the temperature, compressing and probability of thermonuclear reaction.

7. The heavy mass of the cover of capsule (fig.2) (having high nuclear numbers $Z \approx 80$ and $A \approx 200$) not allow the nuclear particles easily to fly apart. That increases the reaction time and reactor efficiency.
8. The suggested AB-thermonuclear reactor is small (diameter about $0.5 \div 3$ m or less up 0.3 m), light (mass is about some tons or less up 100 kg) and may be used in the transport vehicles and aviation.
9. The water may protect the material of the sphere from neutrons.
10. It is possible (see computations) the efficiency of AB reactors will be enough for using as fuel only the deuterium (or others) which is cheaper then tritium in thousands of times (One gram of tritium costs about 30,000 US dollars. One gram of deuterium costs 1$) (see Estimations of a fuel cost).
11. The offered hole AB reactor has high temperature. That allows to use the fuels do not give many neutrons and gamma radiation. These fuels are safety for humanity and installations.
12. Offered reactor may be used for syntheses elements.

Theory, estimation and computation of hole AB-reactor

Let us to estimate the need parameters of ion beam (jet) the ion accelerator. As thermonuclear fuel we take the ^{6}LiD (deuterate of lithium). That is the solid crystals having the melting temperature 692 ºC. We will only consider the thermonuclear reaction D+D because the probability reactions Li+D and Li+Li are small. That reaction is:

$$D+D \rightarrow T(1.01)+p(3.02) \quad 50\% \tag{1}$$
$$D+D \rightarrow {}^3He(0.82)+n(2.4) \quad 50\%$$

Here are: D=^{2}H is deuterium, T=^{3}H is tritium, ^{3}He is helium, p is proton, n is neutron, numbers into brackets are million electron volts. The full average energy is $E_1 = 3.62$ MeV. The energy into capsule (without neutron) is $E_2 = 2.42$ MeV. Using LiD in solid form, we can escape the pressing of fuel in gas form.

Let us take the fuel mass into capsule M = 10^{-7}kg. Number N of nuclear into capsule is

$$N = \frac{M}{\mu m_p} = \frac{10^{-7}}{8 \cdot 1.67 \cdot 10^{-27}} = 7.6 \cdot 10^{18} \tag{2}$$

where N = number of fuel nuclear into capsule; M = mass of fuel, kg; m_p=$1.67 \cdot 10^{-27}$ is mass of nucleon (proton) [kg]. $\mu = 6+2=8$ is molar mass of fuel LiD. Half of N is nuclear D and another half is nuclear of Li. For reaction D+D the other half D and an ignited energy must delivery the ion beam from accelerator. If ignition temperature is $E_o = 10$ keV (, the request heat energy of beam is

$$E = 1.5 \cdot N \cdot E_o = 1.5 \cdot 7.6 \cdot 10^{18} \cdot 10^4 = 1.14 \cdot 10^{23} eV = 1.14 \cdot 10^{23} \cdot 1.6 \cdot 10^{-19} \approx 18 \, kJ \quad . \tag{3}$$

The full thermonuclear energy of fuel D+D is

$$E = N \cdot E_1 = 0.5 \cdot 7.6 \cdot 10^{18} \cdot 3.62 \cdot 10^6 = 1.38 \cdot 10^{25} eV = 1.38 \cdot 10^{25} \cdot 1.6 \cdot 10^{-19} \approx 2.21 \, MJ \,, \quad (4)$$

The full thermonuclear energy of fuel D+D into capsule (without neutron) is

$$E = N \cdot E_2 = 0.5 \cdot 7.6 \cdot 10^{18} \cdot 2.42 \cdot 10^6 = 0.92 \cdot 10^{25} eV = 0.92 \cdot 10^{25} \cdot 1.6 \cdot 10^{-19} \approx 1.47 \, MJ \,, \quad (5)$$

Easy to calculate the size of spherical capsule for fuel $M = 10^{-4}$ g, having specific mass $\gamma = 0.82$ g/cm^3. They are: volume $v = M/\gamma = 1.22 \cdot 10^{-4}$ cm^3, diameter $D = 6.16 \cdot 10^{-2}$ cm; gross section $S = 3 \cdot 10^{-3}$ cm^2.

We take the hole diameter $d = 0.1D = 0.616 \cdot 10^{-2}$ cm.

Let us to estimate the parameters of accelerated D jet from ion accelerator. The need mass of jet is $m = 0.25M = 0.25 \cdot 10^{-7}$ kg. The speed of jet accelerated up $E_o = 18$ kJ is:

$$E_o = \frac{mV^2}{2}; \quad V = \left(\frac{2E_o}{m}\right)^{0.5} = 1.2 \cdot 10^6 \, m/s \quad (6)$$

Electric charge Q and electric current I of D jet for impulse time $t = 10^{-9}$ sec are:

$$Q = Ne = 3.75 \cdot 10^{18} \cdot 1.6 \cdot 10^{-19} = 0.6 \text{ C}, \quad I = Q/t = 0.6/10^{-9} = 600 \text{ MA}. \quad (7)$$

If the impulse time $t - 10^{-8}$ sec., the impulse current will be $I = 60$ MA.

The length of jet for impulse $t = 10^{-9}$ sec is

$$L = V \cdot t = 1.2 \cdot 10^6 \cdot 10^{-9} = 1.2 \cdot 10^{-3} \text{ m} \quad (8)$$

Assume, the focused diameter of ion jet equals the diameter of hole into capsule $d = 2r = 0.1D = 0.616 \cdot 10^{-2}$ cm.

Let us estimate the negative electric pressure and positive magnetic pressure of jet at hole.

The linear electric charge is $\tau = Q/L = 0.6/1.2 \cdot 10^{-3} = 500$ C/m. Electric intensity is

$$E = k\frac{2\tau}{r} = 9 \cdot 10^9 \frac{2 \cdot 500}{3.08 \cdot 10^{-5}} = 2.92 \cdot 10^7 \, N/C \text{ or } V/m \,, \quad (9)$$

The magnetic intencity is

$$H = \frac{I}{2\pi r} = \frac{6 \cdot 10^8}{2 \cdot 3.14 \cdot 3.08 \cdot 10^{-5}} = 3.1 \cdot 10^{12} \text{. A/m} \quad (10)$$

Positive magnetic and negative electric pressure at focused jet are

$$p_m = \frac{\mu_o H^2}{2} = \frac{4\pi 10^{-7}(3.1 \cdot 10^{12})^2}{2} = 3.02 \cdot 10^{18} \, \frac{N}{m^2} \approx 3 \cdot 10^{13} \text{ atm.,}$$

$$p_e = \frac{\varepsilon_o E^2}{2} = \frac{8.85 \cdot (2.92 \cdot 10^7)^2}{2} = 3.31 \cdot 10^{15} \, \frac{N}{m^2} = 3.3 \cdot 10^{10} \text{ atm.} \quad (11)$$

As you see, the positive magnetic pressure is significantly more then negative electric pressure. We can decrease the pressure, if we add in jet after focusing the electrons.

Discussion

The offered beam AB reactor requests a high-current ion accelerator, having the energy $30 \div 100$ kJ, voltage about 50 kV and the impulse time about $2 \div 3$ ns. At present time the ion accelerators widely are used in technology, in coating
, the introduction of elements, the production of chips and electronics. Especially the get big R&D in Program SDI (Strategic Defense Initiative). Charged particle beams diverge rapidly due to mutual repulsion, so neutral particle beams are more commonly proposed. A neutral-particle-beam weapon ionizes atoms by either stripping an electron off of each atom, or by allowing each atom to capture an extra electron. The charged particles are then accelerated, and neutralized again by adding or removing electrons afterwards.

SDI got a beam (ions) up 1 gigajoule of kinetic energy and ion speed closed to the light speed. Our requests are in thousand time less.

RÉFÉRENCES
(READER CAN FIND PART OF THESE ARTICLES IN WEBS:
HTTPS://ARCHIVE.ORG/DETAILS/LIST5OFBOLONKINPUBLICATIONS, HTTP://BOLONKIN.NAROD.RU/P65.HTM,
HTTP://ARXIV.ORG/FIND/ALL/1/AU:+BOLONKIN/0/1/0/ALL/0/1, HTTP://VIXRA.ORG).

[1] Bolonkin A.A., Small, Non-Expensive Electric Impulse Thermonuclear Reactor with colliding jets. 7 11 16, 11 19 16, LULU 2017, 144 ps, http://viXra.org/abs/1611.0276, https://archive.org/download/ArticleThermonuclearReactorOfCollisingJets10416

[2] Bolonkin A.A. , Electric Cumulative Thermonuclear Reactors. 7 17 16. http://vixra.org/abs/1610.0208 , https://archive.org/details/abolonkin_gmail_201610 ,

[3] Bolonkin A.A., "Inexpensive Mini Thermonuclear Reactor". International Journal of Advanced Engineering Applications, Vol.1, Iss.6, pp.62-77 (2012). http://viXra.org/abs/1305.0046, http://archive.org/details/InexpensiveMiniThermonuclearReactor.

[4] Bolonkin A.A. , Cumulative Thermonuclear AB-Reactor. Vixra 7/ 8/2015, http://viXra.org/abs/1507.0053 , https://archive.org/details/ArticleCumulativeReactorFinalAfterCathAndOlga7716 .

[5] Bolonkin A.A., Ultra-Cold Thermonuclear Synthesis: Criterion of Cold Fusion. 7 18 2015. http://viXra.org/abs/1507.0158, GSJornal: http://gsjournal.net/Science-Journals/%7B$cat_name%7D/View/6140 .

[6] Bolonkin A.A., Cumulative and Impulse Mini Thermonuclear Reactors. 3 30 16, http://viXra.org/abs/1605.0309, https://archive.org/download/ImpulseMiniThermonuclearReactors.

[7] Bolonkin, A.A., "Non Rocket Space Launch and Flight". Elsevier, 2005. 488 pgs. ISBN-13: 978-0-08044-731-5, ISBN-10: 0-080-44731-7 . http://vixra.org/abs/1504.0011_v4,

[8] Bolonkin, A.A., "New Concepts, Ideas, Innovations in Aerospace, Technology and the Human Sciences", NOVA, 2006, 510 pgs. ISBN-13: 978-1-60021-787-6. http://viXra.org/abs/1309.0193,

[9] Bolonkin, A.A., Femtotechnologies and Revolutionary Projects. Lambert, USA, 2011. 538 p. 16 Mb. ISBN:978-3-8473-0839-0.http://viXra.org/abs/1309.0191,

[10] Bolonkin, A.A., Innovations and New Technologies (v2). Lulu, 2014. 465 pgs. 10.5 Mb, ISBN: 978-1-312-62280-7. https://archive.org/details/Book5InnovationsAndNewTechnologiesv2102014/

[11] Bolonkin, A.A., Stability and Production Super-Strong AB Matter. International Journal of Advanced Engineering Applications. 3-1-3, February 2014, pp.18-33. http://fragrancejournals.com/wp-content/uploads/2013/03/IJAEA-3-1-3.pdf The General Science Journal, November, 2013, #5244.

[12] Bolonkin, A.A., Converting of Any Matter to Nuclear Energy by AB-Generator. American Journal of Engineering and Applied Science, Vol. 2, #4, 2009, pp.683-693. http://viXra.org/abs/1309.0200,

[13] Kikoin I.K., Tables of Physical Values, Moscow, Atomizdat, 1975 (Russian).

[14] Koshkin N.I., Shirkevich M.G., Handbook of elementary physics, Moscow, Nauka, 1982 (Russian).

[15] AIP, Physics Desk Reference, AIP PRESS. Third Edition.

29 May 2018

Chapter 3
Low Inductive and Resistance Energy Capacitor

Abstract

The paper considers the design of a powerful electric capacitor with a very small induction and resistance. Such capacitors are necessary in many branches of technology, when all the energy of a capacitor must be given out for millionths (10^{-6}) of a second to an object with low electrical resistance.
   The proposed capacitor allows you to reduce the installation of energy supply to a nuclear reactor and its cost a thousand times.
Key words: Low Inductive Capacitor, Low Resistance Capacitor, Energy Capacitor, Capacitor for Fusion Reactor.

Introduction

   In the middle of the last century, there was a need for considerable storage of energy (radar, lasers, nuclear power, etc.) capable of delivering this energy in millionths of a second. Naturally, scientists turned to powerful capacitors. But it turned out to give all the energy to an object with a small electrical resistance in a short time is not so simple. The energy of the simplest triangular pulse is

$$E = 0.5IUt, \qquad\qquad (1)$$

where E is energy, J; I - electric current, A; U - voltage, V; t - time, sec.
   If you try to give a small energy of 50 kJ even at a high voltage of 100 thousand volts per 0.1 micro seconds, then a giant current pulse $I = 2E / Ut$ = 100 / (100x0.0000001) = 10 million amperes will occur, which will create a giant magnetic field. This field will inhibit the transfer of energy to the object, return energy back to capacitor. Begin fluctuations in voltage and current, the loss of energy in the wires. The transmission time will increase many times and only a small fraction of the energy will reach the object. Wires and contacts may burn out. That is why scientists began to invent schemes that would help circumvent this obstacle. One of the latest structures was the Linear Transformer Driver (LTD) - a giant installation with a diameter of 120 meters, a height of three floors and a cost of more than 300 million (Fig. 1). Installation is built in Sandia National Laboratories (USA) and used Z-machine. But Z-machine is not reaching the stable nuclear energy more than it is spending.

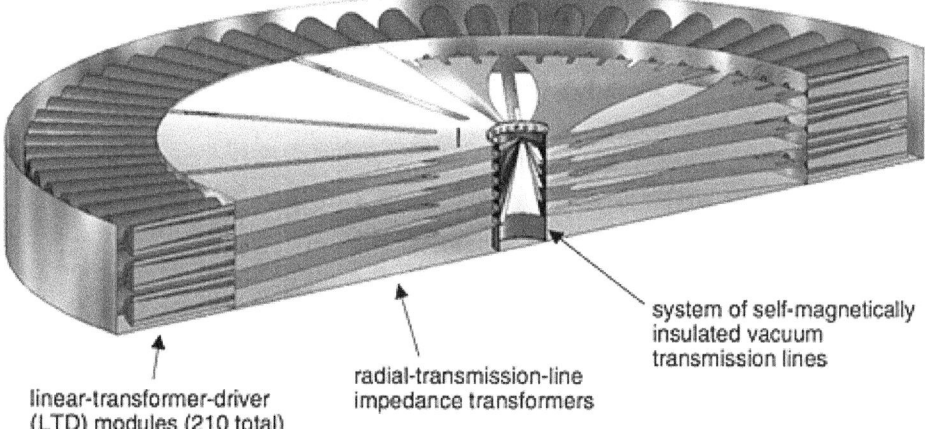

system of self-magnetically insulated vacuum transmission lines

radial-transmission-line impedance transformers

linear-transformer-driver (LTD) modules (210 total)

Fig.1. (LTD) - Linear Transformer Driver. Installation for compressing and transfer energy from capacitors to nuclear capsule. Human being (the black line just left of the center of the LTD) for scale.

The purpose of this article is to propose a new capacitor and show that a new capacitor design can significantly reduce the internal inductance and electrical resistance of a powerful power capacitor, shorten the pulse time, protect the internal wires and capacitor contacts from giant current pulses. And most importantly reduce the size and cost of installation thousands of times.

Description of the installation.

The entire design of the proposed capacitor is subject to the same goal - to reduce the discharge time of the capacitor, bring it to at least $0.1 \div 0.5$ micro seconds and transfer a significant part of this energy to the desired object. Such an object can be, for example, a very small capsule with a nuclear fuel of an inertial thermonuclear reactor, which must be heated to 100 million degrees (10 keV), in order to start a confident thermonuclear reaction. Short transmission time is dictated by the requirement of inertia. With a large heating time the capsule will scatter and the thermonuclear reaction will not start.

The time of the issuance of energy by the capacitor is equal to

$$t = 0.5\pi\sqrt{CL} \qquad\qquad (2)$$

where t is time, sec; C is capacity of capacitor, F; L is an inductance of installation, H.

Since the energy is given, we can only influence the inductance. But the required short discharge time leads to the fact that the current pulse becomes very large - millions of amperes, which can lead to the burning of conductors, contacts and capacitor. In the case of a single pulse, it is not terrible if the cross section of the conductors and contacts is large, because the energy of the capacitor is limited. For example, if the energy of the capacitor is 50 kJ, the transverse conductors of the system are 10 cm^2 and 400 cm long, then when the capacitor is short-circuited, the heating of the copper conductors from a single pulse is about 0.5 degrees, about 7 degrees aluminum.

It is very important that the capacitor energy goes to heat the plasma in the capsule, and not to heat the capacitor, the system conductors (including the wires in the capacitor), the external conductors of the system, the switch, the contacts, and the capsule case.

The ratio r / R has a significant influence on the distribution of energy, where r is the resistance of the capsule and R is the ohmic resistance of the entire system. This ratio shows what proportion of the energy of the capacitor can reach the capsule.

Thus, we see that the only means to quickly discharge a capacitor with a given capsule and the energy of a capacitor is to reduce the inductance and resistance of the system. Otherwise, a large proportion of the energy will go to heating the system and creating a powerful magnetic field. True energy of the internal magnetic field does not disappear. It returns in the form of reverse current and voltage after discharge of the capacitor. But in the inertial reactor it is difficult to use, because the opposite voltage inhibits accelerated fuel cores, i.e. reduces temperature and the likelihood of a thermonuclear reaction.

The inductance of the system can be significantly reduced by arranging the conductors of the forward and reverse currents in the capacitor and the system so that they are as close as possible and have the same opposite current. Then they will create opposite fields and strongly weaken the total magnetic field, i.e. reduce the energy in it. The main magnetic field will be between them. This space should be made minimal. But the conductors are charged in the opposite way and the distance between them is determined by the breakdown ability of the insulator and the configuration of the conductors.

Figure 2a shows two wires (direct and reverse), located side by side to reduce inductance. The minimum distance between them is determined by the penetrating ability of the medium (insulator) in which they are located. Another form of such configurations is a coaxial cable, shown in fig. 2b. It consists of a central wire, insulation and external conductive sheath. Used for protection against radio interference.

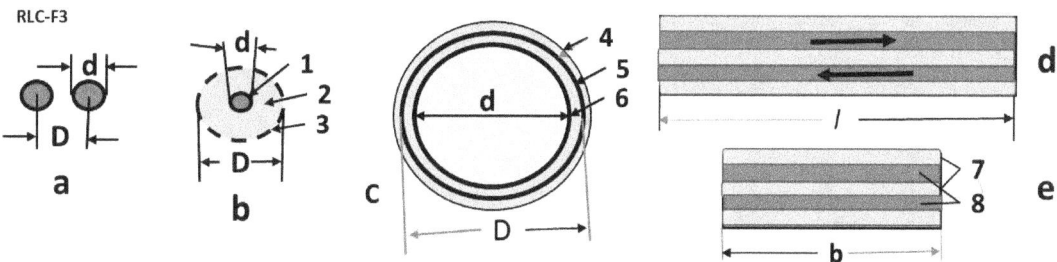

Fig. 2. Cross-sections of conductors for the delivery of electrical energy to the consumer. *Legend*: **a** - two round conductor, located alongside; **b** - coaxial cable: 1 - center wire, 2 – insulator, 3 - surface wire (fly-through); **c** - coaxial cylinders: 4 - insulator, 5 - outer cylinder, 6 - inner cylinder; **e, d** - two lanes located one above the other at the minimum distance: 7 – insulator, 8 - conductive strips, *l* - long strip, *b* - strip width.

The author proposes the other two configurations with low inductance (Fig. 2c). One of them is two isolated conductive cylinders located one in another. The other is two molded conductive strips, one above the other. With the right construction, both solutions (c, d) give better results on inductance than the known solutions (a, b).

Reducing the electrical resistance of a system (without a capsule) can be done quite simply - by increasing the cross sections of the conductors. This requires the creation of new capacitors. Powerful capacitors with low inductance are also not produced even by US factories, since before they were not required by industry. Manufacturers can only offer a parallel connection of capacitors, which reduces the inductance of a capacitor bank (reduces the magnetic field energy by reducing the current in individual capacitors when connected in parallel), but increases the inductance of external connections.

Estimates show that the required power capacitor can measure 1.1 × 1.1 × 1.6 m and weigh about 2.5 tons.

Note that the abandonment of huge, very expensive LTD structures and their replacement with capacitors with the correct design can reduce the cost of installing energy delivery to a nuclear capsule by a factor of about a thousand.

Designing a capsule for such a reactor is also not an easy task. This is not Ohm's task. Plasma resistance is strongly dependent on its temperature. The voltage applied to it causes the collective acceleration of separately positive and negative particles in opposite directions. They freely pass between them and strike the electrodes by heating them, causing X-ray and bremsstrahlung. This task will be considered by the author in other papers.

The theory of the proposed energy transfer.

The theory of radar (discharge of a capacitor) is used for the calculation with the difference that the electrical resistance is variable, since the specific electrical resistance of the plasma is very dependent on its temperature, and the temperature strictly depends on the energy received.

The electrical circuit for discharging a capacitor is shown in Fig. 3. Please note that for our design of the capsule L and $R1$ we consider constant, $R2$ is the resistance of the capsule variable, and the inductance of the capsule is so small that it can be neglected.

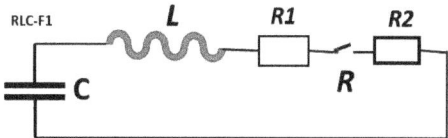

Fig.3. The schematic diagram of our installation. *Notation*: C is capacitor, L is Internal inductance of capacitor

and external wiring; *R1* is Internal resistance of the capacitor and external wiring; *R2* is resistance of the fusion fuel capsule; *R = R1+R2(t)*.

Calculated formulas.

The main well-known calculation formulas and some calculation results for a capacitor having a voltage of *V* = 100 kV, energy *E* = 50 kJ are given below.

The initial voltage capacitor is *U*(0) = 100kV, energy is *E*(0) = 50 kJ:

$$E = 0.5CU^2 \ kJ, \ C = \frac{2E}{U^2} = \frac{2 \times 50}{100^2} = 10^{-5} \ F,$$

(3) where *C* is capacity of condenser, F.

Charge of capacitor is:

$$q = \frac{2E}{U} \ [C].$$

(4)

Capacity of plat capacitor is:

$$C = \frac{\varepsilon_o \varepsilon S}{a},$$

(5)

where $\varepsilon_o = 8.85 \cdot 10^{-11} \ F/m$ – electric constant; ε is dielectric constant of isolator, *S* is capacitor area, sq. m; *a* is distance between sheets, m.

Differential equations of discharging the capacitor is [4] p. 450:

$$ri - U = -L\frac{di}{dt}, \ U = \frac{q}{C}, \ i = -\frac{dq}{dt}.$$

(6)

where *r* is electric resistance , Ohm; *i* is electric current, A; *L* is induction, H; *t* is time, sec.

This system we can re-write as one equation of the second order:

$$L\frac{d^2q}{dt^2} + r\frac{dq}{dt} + \frac{q}{C} = 0 \ \ or \ \ \frac{d^2q}{dt^2} + 2\alpha\frac{dq}{dt} + \omega_o^2 q = 0.$$

(7)

Where

$$\alpha = \frac{r}{2L}, \ \omega_o^2 = \frac{1}{LC}.$$

(8)

Electric resistance of wire is:

$$R1 = \rho\frac{l}{s},$$

(9)

Here ρ is specific resistance, $\Omega.cm$ (for copper $\rho = 1.75 \times 10^{-6}$, $\Omega.cm$); *l* is wire length, cm; *s* is cross section of wire, sq. sm.

Spitzer resistance of plasma is:

$$R2 = \eta\frac{l}{s} \ where \ \eta = \frac{o.1Z}{T^{-3/2}},$$

(10)

Here *Z* is charge of nuclear fuel: for nuclear fuel T+D, D+D *Z* = 1; *T* is plasma temperature in eV.

1 eV = $1.6 \cdot 10^{-19}$ J. 1 eV = 11,604 K.

A typical curve for charging and discharging a capacitor *V = V(t)* is shown in Fig. 4 .

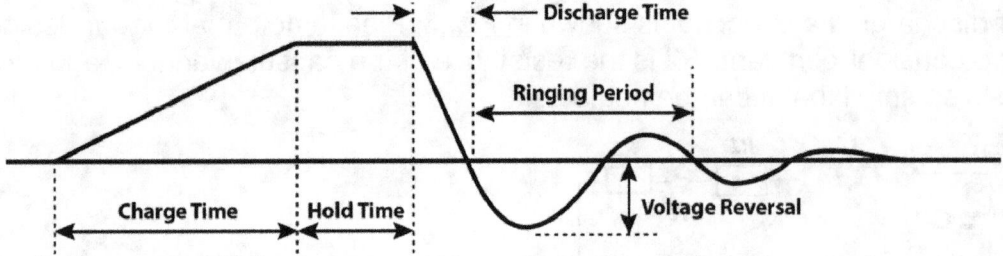

Fig.4. A typical curve for charging and discharging a capacitor *V = V(t)*, where *V* is voltage, V.

A useful parameter is the *damping factor*, ζ, which is defined as the ratio of these two; although, sometimes α is referred to as the damping factor and ζ is not used.

$$\zeta = \frac{\alpha}{\omega_0} . \qquad\qquad (11)$$

In the case of the series RLC circuit, the damping factor is given by

$$\zeta = \frac{R}{2}\sqrt{\frac{C}{L}} . \qquad\qquad (12)$$

The value of the damping factor determines the type of transient that the circuit will exhibit.[6]

Transient response

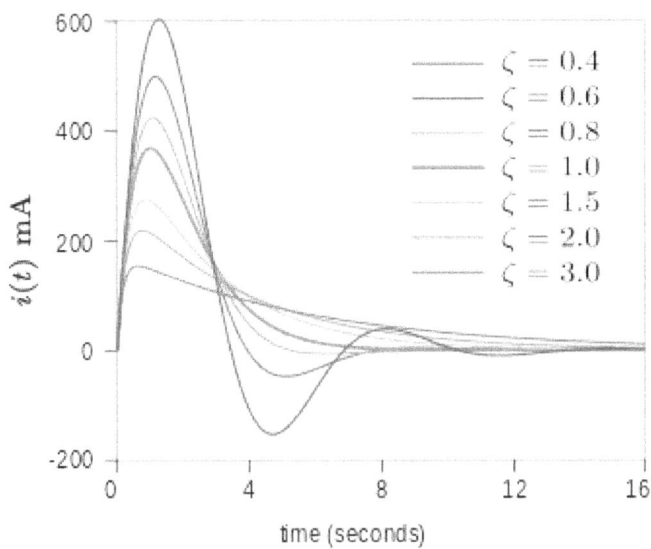

Fig.5. Influence of damping factor in the discharge curve. Plot showing underdamped and overdamped responses of a series RLC circuit. The critical damping plot is the bold red curve. The plots are normalized for L = 1, C = 1 and ω_0 = 1.

The differential equation for the circuit solves in three different ways depending on the value of ζ (Fig.5). These are underdamped ($\zeta < 1$), overdamped ($\zeta > 1$) and critically damped ($\zeta = 1$). The differential equation has the characteristic equation [5].

Estimation of inductivity and electric resistance.

The inductivity of different wires design which are shown in Fig.2 may be estimated by equations:
1. Two round column/wires (Fig. 2a):

$$L = \frac{\mu_o l}{\pi}(\frac{1}{2} + \ln \frac{D}{0.5d}) . \qquad\qquad (13)$$

Here $\mu_o = 4\pi10^{-7}$ - magnetic constant, H/m; l – length of tape, m. In conventional capacitor the insulated long foil tape rolls into a roll. This simplifies production, but greatly increases the inductance.

2. Коаксиальный кабель (Fig.2b).

$$L = \frac{\mu_o l}{\pi} \ln \frac{D}{d}.$$

(14)

This design decreases the inductivity and protect radio lines from interference.

3. Big cylinder into cylinder (Fig.2c). This design offers the author.

$$L = \frac{\mu_o l}{\pi} \ln \frac{D}{d} = \frac{\mu_o l}{\pi} \ln \frac{r + \delta}{r}.$$

(15)

Here r is radius small cylinder, m; δ is thickness of isolator, m.

For $\delta << r$ the equation (13) we can re-write in form:

$$L \approx \frac{\mu_o l \delta}{\pi r}.$$

(16)

For big r the ratio δ/r will be small and L small.

4. Electric energy can deliver by two thin sheets having thin isolator layer between them (Fig.2d,e). If δ_1 is the thickness of isolator, δ_2 is the thickness of sheet and b is width of the sheet, m, for $(\delta_1 + \delta_2) \ll b$, the estimation of inductivity is:

$$L \approx \frac{\mu_o l}{\pi} \frac{\delta_1 + \delta_2}{b}.$$

(17)

This is not big, because the magnetic field will be only between sheets.

Example: For l = 1 m, $\delta_1 = \delta_2 = 0.004\,m$, b = 0.5 m, from (15) we get $L = 6.4 \cdot 10^{-9}$ H. This value we must sum with others.

The Inductivity from thin film/folk and capsule we can neglect because thin folk connected parallel one to other, inductivity of capsule is small.

For average triangle peak of the current may be large.

Example: for energy E = 50 kJ, time $t = 10^{-6}$ sec, $V_{max} = 10^5\,V$ the maximum pick current is

$$I_{max} \approx \frac{4E}{V_{max} t} \approx 2\,MA.$$

(18)

In our capacitor any current peak is not problem, because the heating of the capacitor depends not on the current peak, but only on the energy of the capacitor itself and the mass of conductors inside it and the correctness of the design.

Example: Let us the estimate the heating from single "shot" the capacitor having the size of the central copper conduction Fig.7b. Mass of this conductor is 27 kg. If all energy of capacitor will be spent only for heating this conductor its temperature will increases only in:

$$\Delta T = \frac{E}{C_p M} = \frac{50}{7 \times 27} = 0.26 \; degree \; of \; C.$$

(19)

Here $C_p \approx 7\,kJ/kgK$ is heap capacity of copper. That means that we can test our capacitor in short circuit and we can measure its internal resistance of capacitor as

$$r = \frac{U(0)}{I_{max}},$$

(20)

where r is internal capacitor resistance, Ohm; $U(0)$ is initial voltage of capacitor, V; I_{max} is maximum of current, A. If we measure the I_{max} in short circuit, we calculate the internal capacitor resistance.

The author draws attention to another problem of power capacitors with a very short pulse and a large pulse current. Inertial thermonuclear reactors are needed in very short pulse. This problem does not exist in conventional capacitors for ordinary industrial needs. Therefore, such a problem is not written in textbooks and many manufacturers do not know about it.

The problem is that a strong opposite current in two adjacent wires generates a very significant repulsive force. This force is used in the railgun to accelerate the projectile to hypersonic speeds.

For two round conductors, located side by side, with the opposite direction of the current, this force is equal to:

$$F = \frac{\mu_o i^2 l}{2\pi d},$$ (21)

where $\mu_o = 4\pi 10^{-7} - magnetic\ constant, H/m$; i – current, A; l – wire length, m; d – distance between of wire centers, m.

If we take the average current i = 1 MA, wire length l = 1 m and d = 0.008 m (Fig.8b), the force is F = 2.5×10^7 N/m =2.5x1000 ton/m.

This is gigantic force, which can destroy the contacts. They must have a special design.

Fortunately, the duration of the action is very small and the wire shift is small. If all the energy of the capacitor $E = 10^5$ J will be spent on moving two vertical plates of Fig. 8b with a force F (19), then the displacement s of each plate will be only

$s = E/(2F) = 1$ mm. (22)

Such an offset can compensate for folds at the point of contact and elastic rubber rivets (3) as shown in Fig. 8c(c2).

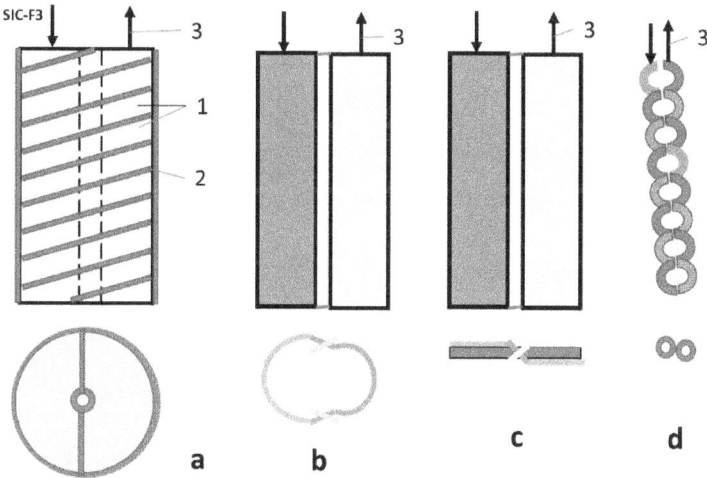

Fig.6. Reduction of inductance (reduction of magnetic fields) by interlacing conductors with equal opposite current. *Legend*: a - two twisted spirals, c - two intersecting cylinders, c - two intersecting planes, two twisted wires. 1 - two spirals, 2 - insulator, 3 - current direction.

Example
Small inductance capacitor having low resistance and short impulse (SIC)

We want to create the capacitor with can deliver the impulse of the energy about 30 kJ in a small object (the length is <1 cm) having the small electric resistance (<0.0001 Ohm) of Installation in a very small time 100 nsec (10^{-7} sec, 0.0000001 sec). That request the very small inductance (<2×10^{-8} Hz), the very small electric resistance (<10^{-4} Ohm) of the installation and very high impulse of current (MA).

The most current capacitors not satisfy these requirements. They have inductance >10^{-6} Hz and the resistance

> 0.1 Ohm. They and their contacts burn in a short circuit of capacitors.

The scientists of thermonuclear engineering try to solve this problem by the gigantic very expensive Max generators (MG) or the **Linear Transformer Driver (LTD)**. **For example, the LTD for Z-machine has diameter 120 m and cost the hundreds of millions of dollars (Fig.1). But the current design of MG or LTD do not alloy to get the stable or good thermonuclear reaction.**

Attention! This material is not a detailed instruction for construction SIC. Only the IDEA of such a condenser is stated here. It is supposed that an experienced creative engineer (or group) will make detailed drawings and computed parameters (*Bolonkin A.A., Low Inductive and Resistance Energy Capacitors*): Initial data are: Voltage U = 100 kV, capacity C = 10^{-5} F, energy E = 50 kJ. Final data: Inductance L<2 × 10^{-8} Hz, resistance <10^{-4} Ohm, discharge time about <4×10^{-7} sec, heating of object is >10 keV.

It is desirable that the developer pre-agreed their drawings and data with the author Alexander Bolonkin (<*abolonkin@gmail.com*>).

For example, any patent lays out only the **idea** of innovation, placing the detailed design and manufacturing on the user of the invention.

Short description of problem.

The schematic diagram of our installation is shown in Fig. 3. **Here:** C is capacitor, L is Internal inductance of capacitor and external wiring; $R1$ is Internal resistance of the capacitor and external wiring; $R2$ is resistance of the fusion fuel capsule; $R = R1 + R2$.

Our goal is to heat 0.0001 grams of fuel to a temperature of 10 keV. To do this, we have to deliver 30 kJ of energy to the capsule fuel from 50 kJ of energy, that is in the capacitor, ASAP (< 4×10^{-7} sec). Otherwise, the capsule will have time to explode, expand, and the ignition of thermonuclear fuel will not occur.

The schematic diagram of our installation is sown in Fig. 3.

A typical curve for charging and discharging a capacitor is shown in Fig.4.

We are satisfied with the data:

Charge time about 10 min (now) and 1 sec (in future).

Hold time about 1 - 5 min.

Discharge Time 4×10^{-7} sec.

Ringing Period – any now.

Voltage Reversal <10 ÷ 20%.

Author offer new innovation design of the capacitor which allow to get the need requirements and alloy to have any single impulse of the electric current.

He reaches these by:

1) The opposed closed currents which create the opposed magnetic fields. These **fields neutralize each other and spend little energy on their creation.** These made in main sheets and all wiles.

2) All wires are made in the form of wide strips with the opposite direction of the currents and with a minimum distance between them.

3) The capacitor is divided into the maximum number of individual parallel films/plates with alternating current directions.

4) Special low-ohm strip contacts of the thin films to the main plates with the main wiring are made.

5) Wiring has a small equal resistance everywhere except for two cuts and the contact area of the plates in the output wire.

6) The mass and thickness of the plates of wires is sufficiently for the permissible heating.
   Conventional capacitors are made of a long-insulated tape rolled into a roll (Fig. 7a).
   The proposed capacitor is made from a set of insulated thin plates/films. These plates have a special arrangement and a separate connection with special central leads/sheets.

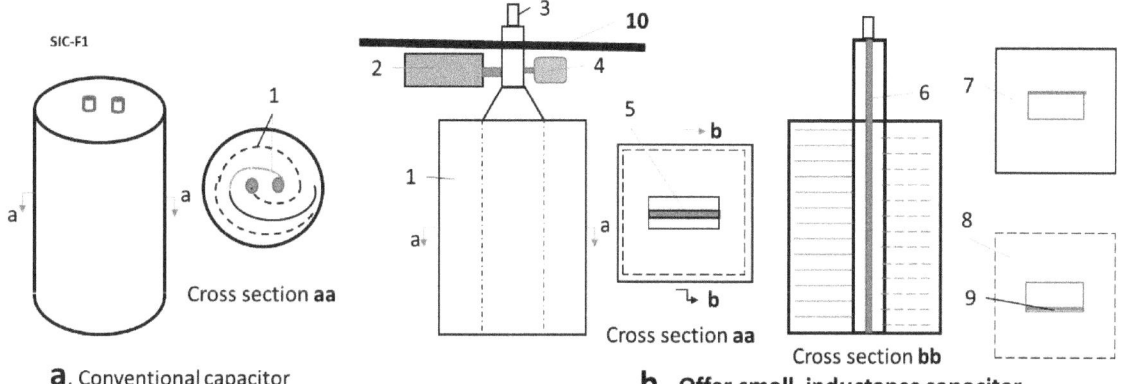

a. Conventional capacitor   b. Offer small inductance capacitor

Fig.7. Conventional and offered capacitor. *Notifications*: **a** – conventional capacitor. 1- a long insulated tape rolled into a roll; **b** - proposed small inductance capacitor. Capacitor is made from a set of insulated thin plates: 1 – capacitor, 2 –high voltage switch, 3 – connection to variable object, 4 – charger, 5 – insulator between main exit/enter plates; 6 – insulator of the main plate; 7 – the first thin film, 8 – the second thin film, connection film, 9 – the second thin film, 10 – steel plate (10 mm), which separate the capacitor, high voltage switch and charger from an explosive area.

Estimation. Example of the proposed Installation

Recommended (computed) sizes, thickness and material (electric engineers and designers can offer the better):
Voltage U = 100 kV, capacity C =10^{-8} F, energy E = 50 kJ, discharge time t <4×10^{-7} sec.

Offered material:
1) Electric **copper** for the thin film and main plate. Data: specific electric resistance ρ =1.7×10^{-6} Ohm.cm; specific mass γ = 8.91 gr/cm^3; Heat capacity Cp =1.99 kJ/kg·K.
2) Isolator PTFE. **Teflon** $(C_2F_4)_n$: dielectric strength (1 MHz) 60 ÷ 173 MV/m; ε = 2.1; specific electric resistance ρ =1×10^{23} ÷ 1×10^{25} Ohm·m; specific mass γ = 2200 kg/m^3; yield strength 23 MPa, melting temperature is 327°C.
   Computed parameters (see theory and computation in given article Bolonkin A.A., *Low Inductive and Resistance Energy Capacitor*): Inductive of installation is <$2 \times 10-8$ Hz, resistance <10^{-4} Ohm, discharge time about <4×10^{-7} sec, heating of object is about 10 keV.
   Requested area of thin film for capacitor is:

$$S = \frac{Cd}{\varepsilon_o \varepsilon} = 215 \text{ sq.m}, \qquad (23)$$

where S – area of capacitor, m^2; C = 10^{-5} – capacity of capacior, F; d = 4 mm – thickness of isolator, m; ε_o= 8.85×10^{-11} F/m – electric constant; ε = 2.1 – dielectric constant of Teflon.

This area requests about 215 thin copper film/foil **105x105** cm (or 430 copper film 57.5x 105 cm) and > 216 the **110x110** cm sheets of Teflon having the thickness 4 mm.

Size (option) of the capacitor/installation box is about 120x120x120 cm. Mass is about 3.5 tons. Average $R2$ $\approx 5 \times 10e - 3$ *Ohm.* *(R2 < 5·10⁻³ Ω).*

Let us estimate the thickness of the coper thin film/foil and main plate.

If we take the cross section area of internal wires 20 sq.cm and width of main sheet is 50 cm, then thickness if the thin film will be about **0.25 mm** and main sheet is about **4 mm**, the $R1 \approx 0.0001$ *Ohm, L* $\approx 2 \times 10e - 8$ Hz (2·10⁻⁸), *time of "shot"* $\Delta t = 10e - 6$ *sec.* (<10⁻⁶). The heating of main sheets is about **0.2°C** after "shot". Connection 3 (Fig.8) must be checkup in tensile stress because the strong impulse current will try disconnect them. That may reach hundreds kg.

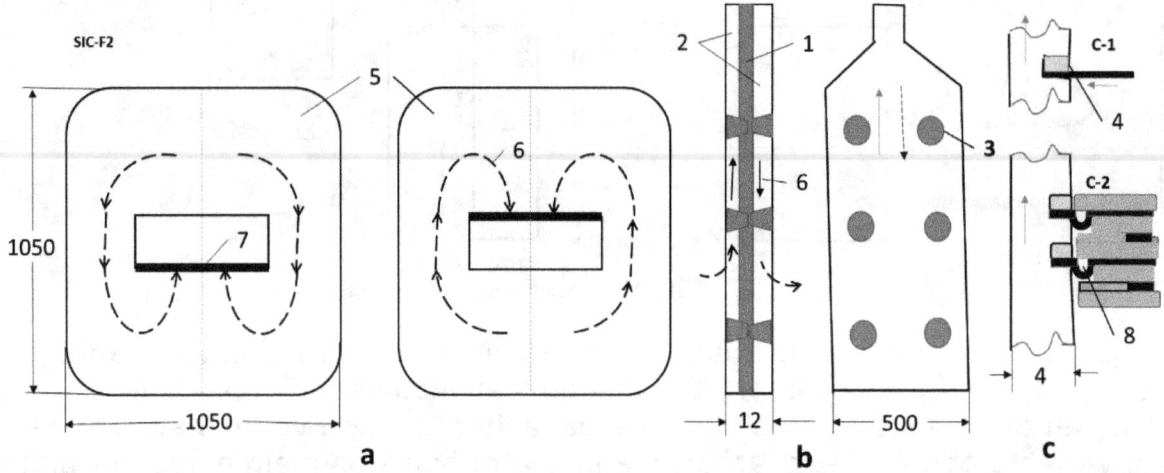

Fig.8. Direction of current in the first and second thin films (**a**) and main plates (**b**). (**c**) – connection the thin films to the main plates. *Notification:* 1 – insulator; 2 - insulator of the main plate, insulator between thin film/foil has form "a"; 3 – connection main plates by insulator; 4 (c1, c2) – connection (by copper) thin film and main plate; 5 – thin films; 6 – direction of current; 7 – connection thin film to main plate; 8 – compensation of the thin film.

Selected initial data are not optimal. Creative electric engineer can offer and recalculate the better version. I think the thin film 145x145 cm ≈ 2 *sq.m* decrease the capacitor height up 80 cm. Decreasing the Teflon thickness from 4 mm to 3 mm decreases the mass in 25%. Decreasing the wire cross section to 10 sq.cm decrease the foil thickness up 0.13 ÷ 0.15 mm. (Voids at the ends are filled with Teflon tape of the same thickness).   Increasing of voltage can improve the main parameters of heating and time of "shot". And so on.

Offered capacitor must be tested in a short circuit. Capacitor inductive and resistance must be measured.

Look also attention in the high voltage (100 kV) switch (and charger). Switch must work very fast (10⁻⁷ sec) and have a small resistance.

Some Results of Computations

Below some results of computation, the heating of thermonuclear fuel into capsule and influence of inductive and electric resistance are presented. Assume, the volume of capsule is constant in heating, the fuel is LiD – sold crystals, mass of fuel is about 0.0001 grams.

The result of integration system differential equations (4)-(8) are below:

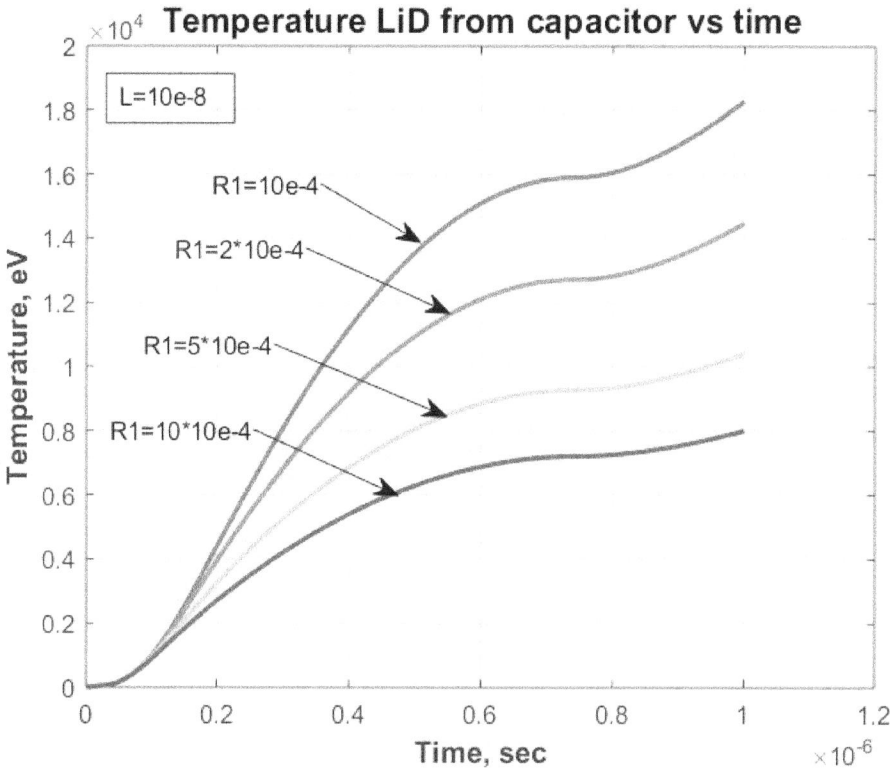

Fig.9. Temperature into capsule vs time from capacitor having C=20 μF, $V(0)$=100 kV for induction 10^{-8}H and different outer resistance $R_1 = 10^{-4} - 10^{-3}$ Ohm.

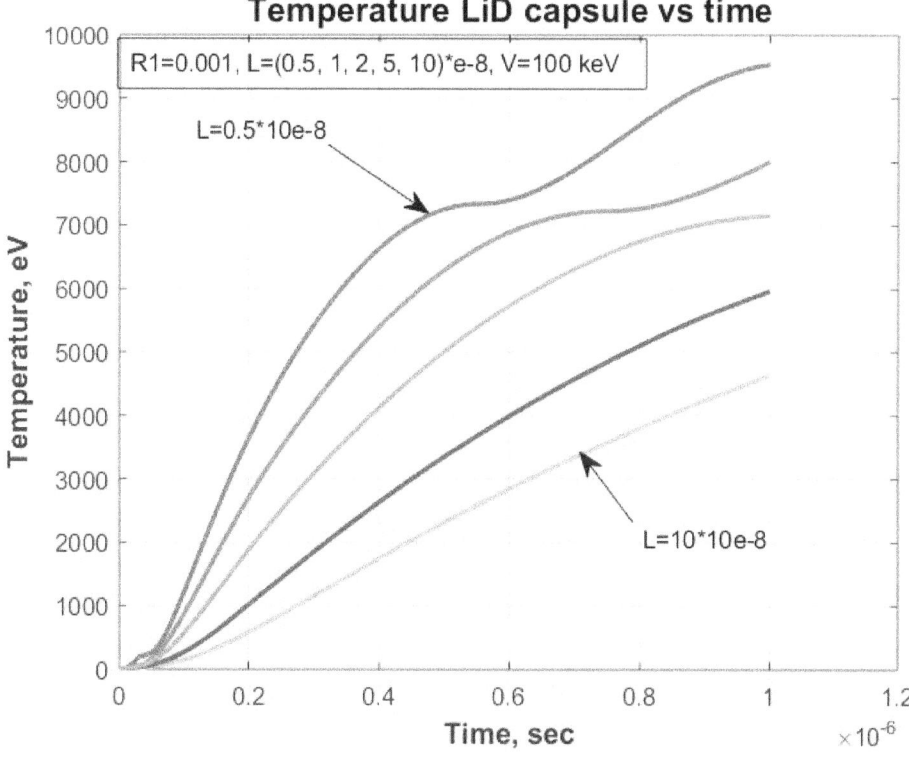

Fig.10. Temperature into capsule vs time from capacitor having 20 μF, V(0)=100 kV for resistance $R_1 = 0.001$ Ohm. and different induction $L = (0.5 \div 10) \cdot 10^{-8}$ H .

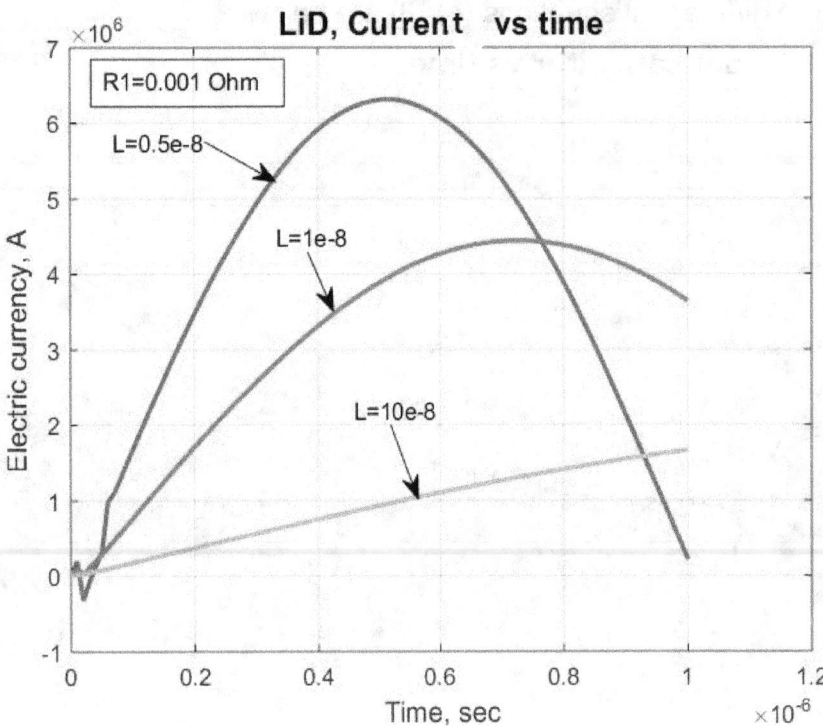

Fig.11. Electric current vs time from capacitor having C = 20 μF, $V(0)$=100 kV for resistance R_1 =0.001 Ohm. and different induction L = (0.5 ÷ 10)·10^{-8} F.

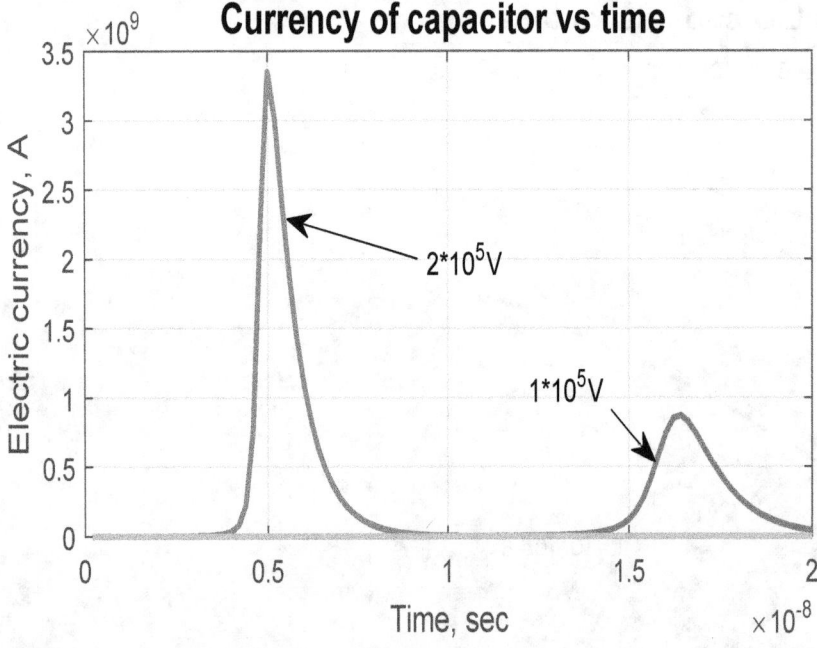

Fig.12. Peak of capacitor current vs time for initial voltage $U(0)$=100 kV and 200 kV. Capacity is C = 20·10^{-6} Farad. $R1$= 0, L= 0, $R2(t)$= 0.1·Z/T(t)$^{-3/2}$.

Fig.12 shows, the heat time may be decreased the maximum up 0.5x10^{-8} sec, if we decrease the inductive up zero.

Description of capacitors and thermo-reactors are in [1]-[5].

References

1. Bolonkin A.A., Small non-Expensive Electric Cumulative Thermonuclear Reactors, USA, Lulu, 2017.
2. Bolonkin A.A., Femtotechnologies and Innovative Projects, USA, Lulu, 2011, p.140.
3. Bolonkin A.A., Computation the Inertial Fusion Reactors and Capacitor Driver, USA, LULU, 2019, 103 ps.
4. Кошкин Н.И., Ширкевич М.Г., Справочник по элементарной физике, Москва, Наука, 1982.
   Koshkin N.I., Shirkevich M.G., Handbook of elementary physics, Moscow, Science, 1982.
5. Калашников С.Г., Электричество. Москва, Наука, 1985.
   Kalashnikov C.G., Electricity, Moscow, Nauka, 1985.
6. Wikipedia, Inertial Thermonuclear Reactor.
   20 November 2018

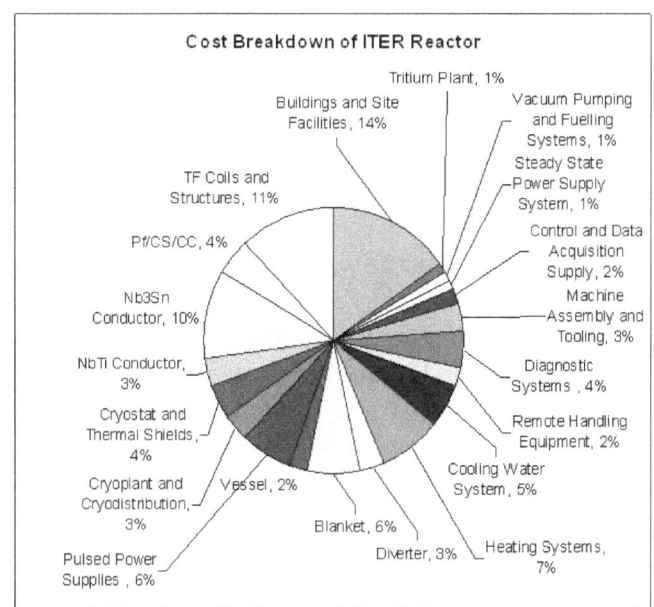

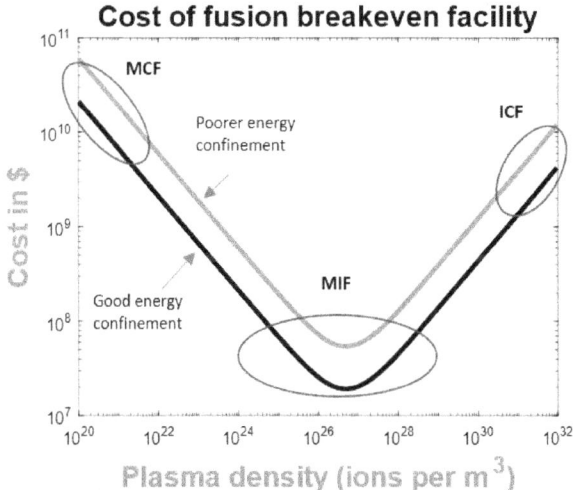

Results based on Lindemuth, I. R. and Siemon, R. E.
"The fundamental parameter space of controlled
thermonuclear fusion," *Am. J. Phys.* **77**, 407, 2009

Chapter 4
Controlled Artificial Nucleosyntheses and Electron Nuclear Reactor

Abstract.

By now, about 3000 nuclides are known, which decay and pass into each other. Many of them can be obtained in nuclear reactors. The author offers a method for searching for the desired nuclides and a simple controlled method for obtaining artificial nuclides, such that in the chain of subsequent decays they contain an alpha decay. With alpha decay, a large amount of nuclear energy is generated that can easily be converted into electricity and reactive traction. This method is simple and safe, does not require large, expensive laser, magnetic installations, million temperatures, it can be used in small and medium engines for cars, airplanes, rockets, space vehicles and for unlimited energy on Earth.

Key words: controlled nucleosyntheses, artificial nucleosyntheses, electron nuclear reactor.

Introduction.

A **radionuclide** (**radioisotope, radioactive nuclide** or **radioactive isotope**) is an atom that has excess nuclear energy, making it unstable. During those processes, the radionuclide is said to undergo radioactive decay. The radioactive decay can produce a stable nuclide or will sometimes produce a new unstable radionuclide which may undergo further decay. Radioactive decay is a random process at the level of single atoms: it is impossible to predict when one particular atom will decay. However, for a collection of atoms of a single element the decay rate, and thus the half-life ($T_{1/2}$) for that collection can be calculated from their measured decay constants. The range of the half-lives of radioactive atoms have no known limits and span a time range of over 55 orders of magnitude.

Radionuclides occur naturally or are artificially produced in nuclear reactors, cyclotrons, particle accelerators or radionuclide generators. (see list of nuclides [19]). More than 2400 radionuclides have half-lives less than 60 minutes. Most of those are only produced artificially, and have very short half-lives. For comparison, there are about 253 stable nuclides. (In theory, only 146 of them are stable, and the other 107 are believed to decay (alpha decay or beta decay or double beta decay or electron capture or double electron capture)).

All chemical elements can exist as radionuclides. Even the lightest element, hydrogen, has a well-known radionuclide, tritium. Elements heavier than lead, and the elements technetium and promethium, exist only as radionuclides. (In theory, elements heavier than dysprosium exist only as radionuclides, but the half-life for some such elements (e.g. gold and platinum) are too long to found).

On Earth, naturally occurring radionuclides fall into three categories: primordial radionuclides, secondary radionuclides, and cosmogenic radionuclides.

Radionuclides are produced in stellar nucleosynthesis and supernova explosions along with stable nuclides. Secondary radionuclides are radiogenic isotopes derived from the decay of primordial radionuclides. Cosmogenic isotopes, such as carbon-14, are present because they are continually being formed in the atmosphere due to cosmic rays.

Radionuclides are produced as an unavoidable result of nuclear fission and thermonuclear explosions. The process of nuclear fission creates a wide range of fission products, most of which are radionuclides. Further radionuclides can be created from irradiation of the nuclear fuel (creating a range of actinides) and of the surrounding structures, yielding activation products. This complex mixture of radionuclides with different chemistries and radioactivity makes handling nuclear waste and dealing with nuclear fallout particularly problematic.

Synthetic radionuclides are deliberately synthesised using nuclear reactors, particle accelerators or radionuclide generators.

Radionuclides are used in two major ways: either for their radiation alone (irradiation, nuclear batteries) or for the combination of chemical properties and their radiation (tracers, biopharmaceuticals).

Theory of radionuclides.
Radioactive decay.

Radioactive decay is the process by which an unstable atomic nucleus loses energy (in terms of mass in its rest frame) by emitting radiation, such as an alpha particle, beta particle with neutrino or only a neutrino in the case of electron capture, or a gamma ray or electron in the case of internal conversion. A material containing such unstable nuclei is considered **radioactive**. Certain highly excited short-lived nuclear states can decay through neutron emission, or more rarely, proton emission.

Radioactive decay is a stochastic (i.e. random) process at the level of single atoms. Except for gamma decay or internal conversion from a nuclear excited state, the decay is a nuclear transmutation resulting in a daughter containing a different number of protons or neutrons (or both). When the number of protons changes, an atom of a different chemical element is created.

Alpha decay occurs when the nucleus ejects an alpha particle (helium nucleus). This is the most common process of emitting nucleons, but highly excited nuclei can eject single nucleons, or in the case of cluster decay, specific light nuclei of other elements. Beta decay occurs in two ways: (i) beta-minus decay, when the nucleus emits an electron and an antineutrino in a process that changes a neutron to a proton, or (ii) beta-plus decay, when the nucleus emits a positron and a neutrino in a process that changes a proton to a neutron.

Types of decay.

Early researchers found that an electric or magnetic field could split radioactive emissions into three types of beams. The rays were given the names alpha, beta, and gamma, in increasing order of their ability to penetrate matter. Alpha decay is observed only in heavier elements of atomic number 52 (tellurium) and greater, with the exception of beryllium-8 which decays to two alpha particles. The other two types of decay are produced by all of the elements. Lead, atomic number 82, is the heaviest element to have any isotopes stable (to the limit of measurement) to radioactive decay. Radioactive decay is seen in all isotopes of all elements of atomic number 83 (bismuth) or greater. Bismuth-209, however, is only very slightly radioactive, with a half-life greater than the age of the universe; radioisotopes with extremely long half-lives are considered effectively stable for practical purposes.

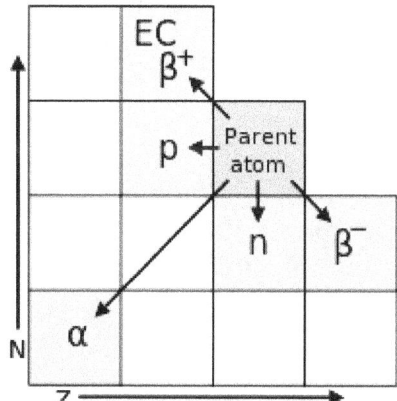

Fig. 1. Transition diagram for decay modes of a radionuclide, with neutron number N and atomic number Z (shown are α, $\beta^{\pm}$, p^{+}, and n^{0} emissions, EC denotes electron capture). That is Rule of Soddy.

Types of decay related to N and Z numbers.

Alpha-Decay.

The Soddy displacement rule for α-decay:

$$_Z^A X \rightarrow {}_{Z-2}^{A-4} Y + {}_2^4 He. \tag{1}$$

Here A is atom number,

Example (alpha decay of uranium-238 into thorium-234):

$$_{92}^{238} U \rightarrow {}_{90}^{234} Th + {}_2^4 He. \tag{2}$$

As a result of α-decay, the atom is displaced by 2 cells to the beginning of the periodic table (that is, the charge of the Z-nucleus decreases by 2), the mass number of the daughter nucleus decreases by 4.

Beta (β⁻)-minus decay

Becquerel proved that β-rays are an electron flow. Beta decay is a manifestation of weak interaction.

Beta decay (more precisely, beta-minus decay, β-decay) is a radioactive decay accompanied by the emission of an electron from the nucleus and electron antineutrinos. Beta decay is an intra-nucleated process. Beta-minus decay occurs due to the transformation of one of the d-quarks in one of the neutrons of the nucleus into a u-quark; In this case, the neutron converts into a proton with the emission of an electron and antineutrinos:

$$_0^1 n \rightarrow {}_1^1 p + {}_{-1}^0 e + \bar{\nu}_e. \tag{3}$$

Free neutrons also undergo beta decay, turning into a proton, an electron, and an antineutrino (see Beta decay of a neutron).

The Soddy displacement rule for β decay: Free neutrons also undergo beta decay, turning into a proton, an electron, and an antineutrino (see Beta decay of a neutron).

The Soddy displacement rule for β decay:

$$_Z^A X \rightarrow {}_{Z+1}^A Y + {}_{-1}^0 e + \bar{\nu}_e. \tag{4}$$

Example: (β⁻-decay of Tritium in Helium-3):

$$_1^3 H \rightarrow {}_2^3 He + {}_{-1}^0 e + \bar{\nu}_e. \tag{5}$$

After β-decay, the element shifts by 1 cell to the end of the periodic table (the charge of the nucleus increases by one), while the mass number of the nucleus does not change.

Beta-plus (β⁺) decay and electron capture.

The Soddy displacement rule for β⁺ decay and electron capture:

$$_Z^A X \rightarrow {}_{Z-1}^A Y + e^+ + \nu_e. \tag{6}$$

Example (ε-electron capture of Beryllium-7 в Lithium-7):

$$_4^7 Be + e^- \rightarrow {}_3^7 Li + \nu_e. \tag{7}$$

After β-decay, the element shifts by 1 cell to the end of the periodic table (the charge of the nucleus increases by one), while the mass number of the nucleus does not change.

Reactions alpha, beta-decay and electron capture produced many energy (up some MeV), but in in case beta-decay and electron capture (from lowest electron orbit of atom) the neutrino carries away this energy.

Electron bombardment of a nucleus usually converts a proton into a neutron. This means that the number of protons decreases by one, and the number of neutrons increases by one. In the Z-N coordinate system, the element is shifted by one to the left and up, and the mass number A remains unchanged.

$$_Z^A X + e^- \rightarrow {}_{Z-1}^A Y + \nu_e. \tag{8}$$

Computing the total disintegration energy given by the equation:

$$E = (m_i - m_f - m_p)c^2 . \tag{9}$$

Where m_i is the initial mass of the nucleus, m_f is the mass of the nucleus after particle emission, and m_p is the mass of the emitted particle.

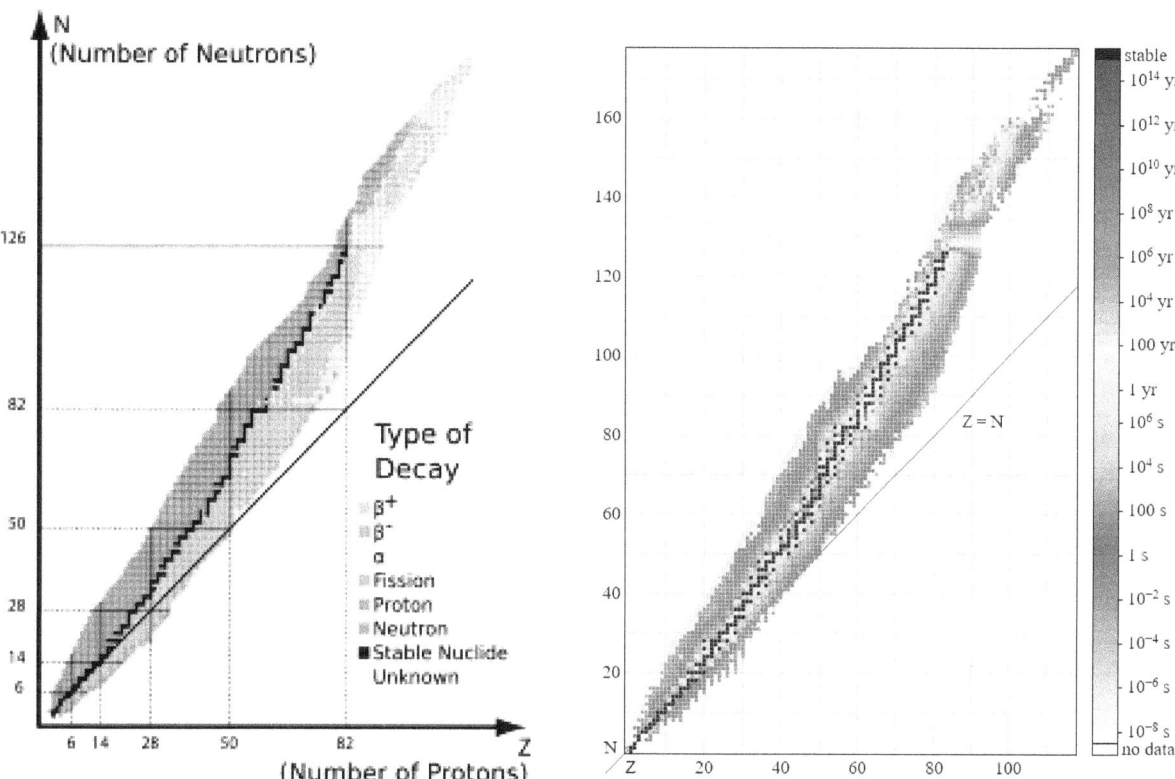

Fig. 2(left). Type of Decay.   https://en.wikipedia.org/wiki/Radioactive_decay#Decay_modes_in_table_form
Fig. 3 (right). Half – live. https://upload.wikimedia.org/wikipedia/commons/8/80/Isotopes_and_half-life.svg

Half-life of radionuclides.

Half-life (symbol $T_{1/2}$ or $t_{1/2}$) is the time required for a quantity to reduce to half its initial value. There are several formulations of the law, for example, in the form of a differential equation:

$$\frac{dN}{dt} = -\lambda N,$$

(10)

which means that the number of decays -dN, which occurred in a short time interval dt, is proportional to the number of N atoms in the sample.
The solution of this differential equation has the form:

$$N(t) = N_0 e^{-\lambda t},$$

(11)

where N_0 – initial number of atoms in initial time $t = 0$.

$$T_{1/2} = \frac{\ln 2}{\lambda} = \tau \ln 2 \approx 0,693\tau.$$

(12)

$$\frac{N(t)}{N_0} \approx p(t) = 2^{-t/T_{1/2}}.$$

(13)

where $\tau = 1/\lambda$ is average (mean) life of particles.

Decay chain

In nuclear science, the **decay chain** refers to a series of radioactive decays of different radioactive decay products as a sequential series of transformations. It is also known as a "radioactive cascade". Most radioisotopes do not decay directly to a stable state, but rather undergo a series of decays until eventually a stable isotope is reached.

Decay stages are referred to by their relationship to previous or subsequent stages. A *parent isotope* is one that undergoes decay to form a *daughter isotope*. One example of this is uranium (atomic number 92) decaying into thorium (atomic number 90). The daughter isotope may be stable or it may decay to form a daughter isotope of its own. The daughter of a daughter isotope is sometimes called a *granddaughter isotope*.

The time it takes for a single parent atom to decay to an atom of its daughter isotope can vary widely, not only between different parent-daughter pairs, but also randomly between identical pairings of parent and daughter isotopes. The decay of each single atom occurs spontaneously, and the decay of an initial population of identical atoms over time t, follows a decaying exponential distribution, $e^{-\lambda t}$, where λ is called a decay constant. One of the properties of an isotope is its half-life, the time by which half of an initial number of identical parent radioisotopes have decayed to their daughters, which is inversely related to λ. Half-lives have been determined in laboratories for many radioisotopes (or radionuclides). These can range from nearly instantaneous to as much as 10^{19} years or more.

Example of natural Decay change (Thorium series):

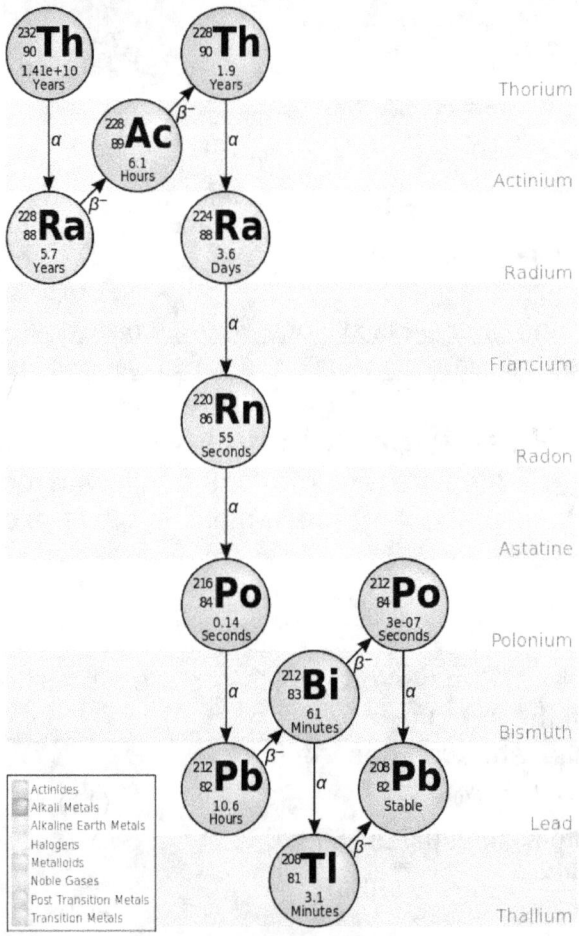

Fig. 4. Thorium series.

The 4n chain of Th-232 is commonly called the "thorium series" or "thorium cascade". Beginning with naturally occurring thorium-232, this series includes the following elements: actinium, bismuth, lead, polonium, radium, radon and thallium. All are present, at least transiently, in any natural thorium-containing sample, whether metal, compound, or mineral. The series terminates with lead-208.

The total energy released from thorium-232 to lead-208, including the energy lost to neutrinos, is 42.6 MeV.

Table #1. Typical isotope-decay.

nuclide	historic name (short)	historic name (long)	decay mode	half-life (a=year)	energy released, MeV	product of decay
^{252}Cf			α	2.645 a	6.1181	^{248}Cm
^{248}Cm			α	3.4×10^5 a	5.162	^{244}Pu
^{244}Pu			α	8×10^7 a	4.589	^{240}U
^{240}U			β⁻	14.1 h	.39	^{240}Np
^{240}Np			β⁻	1.032 h	2.2	^{240}Pu
^{240}Pu			α	6561 a	5.1683	^{236}U
^{236}U		Thoruranium	α	2.3×10^7 a	4.494	^{232}Th
^{232}Th	Th	Thorium	α	1.405×10^{10} a	4.081	^{228}Ra
^{228}Ra	MsTh$_1$	Mesothorium 1	β⁻	5.75 a	0.046	^{228}Ac
^{228}Ac	MsTh$_2$	Mesothorium 2	β⁻	6.25 h	2.124	^{228}Th
^{228}Th	RdTh	Radiothorium	α	1.9116 a	5.520	^{224}Ra
^{224}Ra	ThX	Thorium X	α	3.6319 d	5.789	^{220}Rn
^{220}Rn	Tn	Thoron, Thorium Emanation	α	55.6 s	6.404	^{216}Po
^{216}Po	ThA	Thorium A	α	0.145 s	6.906	^{212}Pb
^{212}Pb	ThB	Thorium B	β⁻	10.64 h	0.570	^{212}Bi
^{212}Bi	ThC	Thorium C	β⁻ 64.06% α 35.94%	60.55 min	2.252 6.208	^{212}Po ^{208}Tl
^{212}Po	ThC′	Thorium C′	α	299 ns	8.955	^{208}Pb
^{208}Tl	ThC″	Thorium C″	β⁻	3.053 min	4.999	^{208}Pb
^{208}Pb	ThD	Thorium D	stable			

Base idea. Control Artificial Nucleosyntheses.

The modern theory of the universe says: after the Big Bang and the Great Synthesis, thousands of unstable isotopes with different periods of existence arose. Over the billions of years of the existence of the universe, their time of life has come to an end and almost two hundred isotopes, whose lifetime is longer than the time of existence of the universe (or a given planet), have practically remained. Some of the short-lived isotopes are formed in the hot subsurface of the stars or by the action of cosmic rays.

Using nuclear reactors, people have learned to create artificial radioactive isotopes, which are known to date about 3000. They can turn into each other and give off giant energy. Unfortunately, the lifetime (decay) of these isotopes is strictly prescribed and it is practically impossible to regulate it with modern technologies. This greatly limits their practical application. In addition, many of the artificial isotopes are expensive and there is a very short time.

The author's task is to circumvent these obstacles. But not in order, as in ancient alchemy, to turn a certain matter into gold, but in order to solve the main problem of our time - to have an abundance of cheap energy.

Currently, the government and scientists are trying to solve the problem of energy using nuclear power by the Fission or Fusion methods. The Fission method has been implemented, dozens of powerful uranium nuclear power plants have been constructed, but their energy is not cheaper than fuel cells on natural fuel. In addition, they were very dangerous in the event of an accident. Over the Fusion method, scientists have been working for more than 60 years in all industrialized countries. Governments have spent hundreds of billions of US dollars, but there are no results and will not be recognized by scientists in the next 15-20 years.

   The essence of the author's idea: in the Fusion method, we connect two negatively charged nuclei, naming T + D (tritium and deuterium) to obtain energy. This task is very complicated. Both atoms are positively charged and repelled from each other. We heat the mixture to tens of millions of degrees and compress it to hundreds of thousands of atmospheres to make the reaction happen.

   In the proposed method, negatively charged electrons bombard a positively charged nucleus and attract each other, acquiring an additional velocity (kinetic energy). In the event of an electron falling into the nucleus, we get a new isotope shifted in the isotope table to the cell to the left and up. In many cases, nuclear energy is released. But for us it is useless, for it immediately takes away from it. The neutrino has almost zero mass and therefore takes almost all energy. Neutrinos have a giant penetrating power and its energy cannot be used by modern technology. That's why scientists did not pay attention to the study of such a reaction.

   But a new isotope can have alpha-decay or become the beginning of a chain that has alpha decay in its composition (see Fig. 4). We can use alpha decay for direct conversion to electrical energy and reactive traction with a very high specific impulse.

   The ability (probability) to get into the nucleus (to cause the desired reaction) is characterized by the so-called cross section of the reaction. The author was unable to find experimental data on the cross section of electron-nuclides. The scientists of NIST USA could not help him either. They only indicated that this cross section cannot be less than the target area of the target nucleus. The fact that the reaction of the electron-nucleus of the nuclide exists no doubt, for the natural reactions of the electron-nucleus exist. Moreover, this is the only reaction of natural decay in nature, the probability of which depends on the temperature and pressure. This influence is small (about 1%), but the electron energy in the nearest orbit is small (about 10 eV). We can easily disperse the attacking electrons to tens and hundreds of thousands of eV. By regulating the number and energy of electrons, we can regulate the number of new nuclides that are obtained, and hence the amount of energy received.

   But there are some technical limitations: we must start with an inexpensive nuclide that has half-time at least a month or more. Otherwise, the initial fuel will quickly deteriorate. Isotopes in the chain and obtained before the first desired isotope (alpha-decay or neutron) should have a small half-time (less 1 sec.). Otherwise, the engine (rocket) will react slowly to control, which in many cases is unacceptable. Note that we adjust not the probability of decay of the isotope λ, but the amount of the initial isotope participating in the reaction, i.e. coefficient N_0 in the formula (10).

The method of searching for the desired nuclides.

   The number of known and studied (in the sense of decay) isotopes to date is about 3,000. Many of them can decay simultaneously through many channels. Each isotope channel has its own probability percentage, its decay time and its decay energy. Therefore, the search for the desired decay chain is a laborious task (graph theory). The solution to this problem is possible on a computer (super-computer). Branches leading to a dead end (large half-time) should be cut off. The search ends when we get the right amount of energy (alpha process) or the desired element (or isotope with half-time satisfying us).

   The selected nuclear fuel chain must satisfy three conditions:
1) Fuel should be inexpensive.

2) The initial element of the chain must have sufficient disintegration time (for vehicles - a storage period of at least some days).

3) In the final element of the chain should quickly alpha decay (less than one second for vehicles).

To find the right fuel, author suggests two methods:

1) Start from a sufficiently stable element and bombard it with electrons to move down the A-constant chain in the hope that it will contain an element with a short alpha half-life.

2) Start from an element with a short alpha decay period and bombard it with electrons to move up the A-constant chain in the hope that it will have a fairly stable element.

If you have a source of slow (thermal) neutrons, then you can explore the side branches of this chain.

You examine the series:

$$^AY_Z \leftarrow {}^AY_{Z+1} + e \leftarrow {}^AY_{Z+2} + e \leftarrow \dots \qquad (14)$$

and select in it the starting element (initial nuclear fuel, that has need half-time which is the satisfaction of you). Here e means the bombing this element by electron.

The reader can find the initial data for selection, in particular, in [18]. Using the filter [18] you can get the list of nuclides ^AY_Z having α-decay mode and the half-time less given value. You examine the series (14) and select in it the starting element (initial nuclear fuel, that has need half-time which is the satisfaction of you). Here e means the bombing this element by electron.

If you are selecting fuel for a transport engine, where the response time is important, then half-time of the isotopes between the initial and final isotope should in sum have a time of less than 1 second.

In the first method: you start with a stable or long-term isotope and by bombarding electrons, neutrons, it is possible to obtain an alpha-isotope with the right time.

Advantages of the proposed method.

The proposed method for obtaining nuclear energy (alpha decay) has huge advantages over the Fission and Fusion methods currently used:

1. It uses scanty amounts of nuclear fuel (a fraction of a milligram) and can be used on small and medium-sized engines (cars, airplanes, missiles) [1-3]. The engine can be switched on and off at any time.
2. It is safe. Destruction of the engine will not lead to contamination of the terrain.
3. It does not need huge and expensive reactors. The technology of production of the focused electron beam is well worked out (we recall the old volume TVs).
4. Compared to the Fusion method, the proposed method does not require giant laser or magnetic installations, in quasi-zero and millionth temperatures.
5. In principle, it allows you to immediately receive electrical energy and, in most cases, does not produce neutrons that make the materials radioactive (the beta radiation is delayed even by a sheet of paper and air).
5. In principle, it allows to obtain artificial elements.

Discussing

Mankind has long been dreaming of mastering unlimited energy. But such an opportunity arose when scientists discovered huge reserves of nuclear energy. Two methods of obtaining it are possible: the method of decay of large complex nuclei into simpler ones and the method of merging simple nuclei into more complex ones.

Scientists quickly developed and mastered the technology of decay (Fission) methods. But it turned out that it is complex, expensive, and most importantly very dangerous. Technology requires a large mass of radioactive matter, which in the event of an accident infects large areas.

The second method - the fusion of light nuclei, is being developed for more than 60 years. Governments spent tens of billions of dollars on it, but there is no result and it is not expected soon. In addition, estimates show that the cost of energy received by the current Fusion method will be much greater than the current energy prices.

The author proposes a new method for obtaining nuclear energy. For a long time (~ 100 years) it is known that isotopes decay themselves with the release of energy. But we cannot in any way influence this natural disintegration. In addition, only recently with the advent of nuclear reactors, it became possible to obtain isotopes.

The author proposes to obtain isotopes by bombarding nuclei with *electrons*. The method of bombarding nuclei with nuclei or neutrons for the purpose of obtaining energy is known and developed long ago. But electron bombardment leads in most cases to beta decay, which does not give practical energy, and apparently nobody studied the bombardment of the nucleus by electrons. Apparently the choice of the initial nucleus can be obtained immediately by the nucleus subject to alpha decay or a chain of decays, which in the end contains a rapid alpha decay.

Essentially, another proposal is the management of disintegration. The author does not encroach on the law of independence of the rate of decay from the temperature and pressure of the environment (although in the case of beta decay such a dependence exists). Changing the speed and intensity of the electron beam, it changes the number (concentration) of the initial and daughter nuclei in the isotope, and thus the amount of energy released.

Note that the implementation of the author's proposals is possible only in a simple reactor proposed by the author in [1-8], using the proposed isotope fuels.

Since the proposed method does not use high nuclear fusion temperatures, the proposed method can be attributed to the so-called "cold nuclear reactor"[8] (room temperature).

RÉFÉRENCES

[1] Bolonkin A.A., Inertial Impulse Beam Hole Thermonuclear AB Reactor. www.IntellectualArchive.com, #1972, 8 7 18. http://GSJournal.net. http://viXra.org/abs/1808.0089 .

[2] Bolonkin A.A., Provisional (patent) application # 62729486. Inertial Impulse Electric Thermonuclear Reactor and Method of It. https://archive.org/details/ ProvisionalPatentInertialElectricReactor81218 , www.IntellectualArchive.com, #1977.

[3] Bolonkin A.A., Small, Non-Expensive Electric Impulse Thermonuclear Reactor with colliding jets. 7 11 16, 11 19 16, LULU 2017, 144 ps, http://viXra.org/abs/1611.0276, https: //archive.org/download/ArticleThermo nuclearReactorOfCollisingJets10416,

[4] Bolonkin A.A. , Electric Cumulative Thermonuclear Reactors. 7 17 16. http://vixra.org/abs/1610.0208 , https://archive.org/details/abolonkin_gmail_201610 ,

[5] Bolonkin A.A., "Inexpensive Mini Thermonuclear Reactor". International Journal of Advanced Engineering Applications, Vol.1, Iss.6, pp.62-77 (2012). http://viXra.org/abs/1305.0046, http://archive.org/details/InexpensiveMiniThermonuclearReactor.

[6] Bolonkin A.A. , Cumulative Thermonuclear AB-Reactor. Vixra 7/ 8/2015, https://archive.org/details/ArticleCumulativeReactorFinalAfterCathAndOlga7716 . http://viXra.org/abs/1507.0053 .

[7] Bolonkin A.A., Ultra-Cold Thermonuclear Synthesis: Criterion of Cold Fusion. 7 18 2015. http://viXra.org/abs/1507.0158, GSJornal: http://gsjournal.net/Science-Journals/%7B$cat_name%7D/View/6140 .

[8] Bolonkin A.A., Cumulative and Impulse Mini Thermonuclear Reactors. 3 30 16,
https://archive.org/download/ImpulseMiniThermonuclearReactors .
http://viXra.org/abs/1605.0309,

[9] Bolonkin, A.A., "Non Rocket Space Launch and Flight". Elsevier, 2005. 488 pgs. ISBN-13:
978-0-08044-731-5, ISBN-10: 0-080-44731-7 . http://vixra.org/abs/1504.0011 v4,

[10] Bolonkin, A.A., "New Concepts, Ideas, Innovations in Aerospace, Technology and the Human
Sciences", NOVA, 2006, 510 pgs. ISBN-13: 978-1-60021-787-6. http://viXra.org/abs/1309.0193.

[11] Bolonkin, A.A., Femtotechnologies and Revolutionary Projects. Lambert, USA, 2011. 538 p.
16 Mb. ISBN:978-3-8473-0839-0.http://viXra.org/abs/1309.0191,

[12] Bolonkin, A.A., Innovations and New Technologies (v2). Lulu, 2014. 465 pgs. 10.5 Mb, ISBN:
978-1-312-62280-7.
https://archive.org/details/Book5InnovationsAndNewTechnologiesv2102014/

[13] Bolonkin, A.A., Stability and Production Super-Strong AB Matter.
International Journal of Advanced Engineering Applications. 3-1-3, February 2014, pp.18-33.
http://fragrancejournals.com/wp-content/uploads/2013/03/IJAEA-3-1-3.pdf
The General Science Journal, November, 2013, #5244.

[14] Bolonkin, A.A., Converting of Any Matter to Nuclear Energy by AB-Generator.
American Journal of Engineering and Applied Science, Vol. 2, #4, 2009, pp.683-693.
http://viXra.org/abs/1309.0200.

[15] Kikoin I.K., Tables of. Physical Values, Moscow, Atomizdat, 1975 (Russian).

[16] Koshkin N.I., Shirkevich M.G., Handbook of elementary physics, Moscow, Nauka, 1982 (Russian).

[17] AIP, Physics Desk Reference, AIP PRESS. Third Edition.

[18] Data of nuclides: https://www-nds.iaea.org/relnsd/vcharthtml/VChartHTML.html .

[19] Wikipedia: https://en.wikipedia.org/wiki/List_of_nuclides .

24 September 2018

Some common radioisotopes that decay by electron capture include (d-day, y-year) :

Radioisotope	Half-life
^{7}Be	53.28 d
^{37}Ar	35.0 d
^{41}Ca	1.03×10^{5} y
^{44}Ti	60 y
^{49}V	337 d
^{51}Cr	27.7 d

Radioisotope	Half-life
^{53}Mn	3.7×10^{6} y
^{55}Fe	2.6 y
^{57}Co	271.8 d
^{59}Ni	7.5×10^{4} y
^{67}Ga	3.260 d
^{68}Ge	270.8 d
^{72}Se	8.5 d

For a full list, see the table of nuclides.   For a full list, see the table of nuclides.

Chapter 5
Problems of Energy Utilization the Inertial Fusion Reactor

Abstract

The paper proposes and considers the utilization of an inertial electro-kinetic thermonuclear reactor. Such a rector turns the kinetic energy of charged particles directly into electricity or thrust. This greatly simplifies, facilitates, increases efficiency and reduces the cost of construction of the reactor and its entire electrical installation, since electrical energy can be easily transmitted over long distances and converted to any other type of energy. Author also considers the inertial thermo-nuclear reactor as propulsion system in space.

Key words: Inertial thermonuclear reactor, Electro-kinetic generator, electric generator of the thermonuclear reactor, propulsion system of inertial thermo-nuclear reactor.

Introduction

Any nuclear reactor produces nuclear particles with high kinetic energy. In a conventional fission nuclear reactor, charged particles appear in a solid or liquid medium and, colliding with the almost stationary surrounding matter, quickly lose their kinetic energy by heating the environment. It turns out the circuit: kinetic energy → thermal energy → electricity. The link "thermal energy" (for example, steam engine) requires complex expensive devices and mechanisms (for example, heaters, boilers, pumps, turbines) and has a low efficiency. In addition, the fission reactor is based on neutrons with high penetrating power and, therefore, must be bulky and heavy.

Thermonuclear fusion reactor has one important feature - the reacting fuel is small (micrograms) and is located in a vacuum or highly rarefied plasma, which has virtually no braking effect (resistance) to the movement of the charged particle.

This circumstance makes it possible to create an Electro-Kinetic Generator (EKG or KEG - Kinetic Electric Generator) that allows to separate negatively and positively charged particles (electrons and nuclei) and to create a difference in electrical potentials on the electrodes. The electric current, although it turns out to be pulsed, but it has one direction (from center). The work of KEG is described in more detail below.

An ordinary electric generator turns mechanical work into electricity. It is extensively researched and widely applied in industry. But in our case, it is not applicable, because we have the kinetic energy of the nuclei, not the mechanical energy of devices created by man.

The kinetic energy of charged particles can be concentrated in a given direction and used to create thrust.

Description of KEG and his work. Innovation and advantages.

This is a typical spherical chamber of an inertial fusion reactor described in [1]. It added electron collectors 2 and positively charged nuclei 3, a current and voltage Converter 8 and a layer of insulation 4 from the neutron absorber 5.

KEG works as follows. The fuel chamber 1 explodes and forms a plasma clot (electrons + positive particles). But their average speeds are different. Electrons have the energy (temperature) of the ignition reaction (about 10 Kcv). The particles have energy (temperature) resulting from a nuclear reaction, i.e. millions of eV. This means that the speed of electrons is much less than the speed of particles, even if their speed during the explosion have time to be compared, the speed of electrons cannot exceed the speed of particles, because the electron mass is 1836 times less than the proton – the minimum mass of charged particles. This means that the energy (temperature) of the electron will be at least $(1836)^{0.5} = 43$ times less than any charged particle and electrons from the plasma flow flowing from the epicenter of the explosion (capsule) can be detained by a thin conductive film 2 (for example, aluminum). Massive charged particles having the same speed will easily break through the thin film and will be detained by a thicker conductive wall 3. As a result, the film 2 will be charged negatively, and the wall 3 is positive. Between them there will be a current used by the payload 7. Since the current and voltage of the generator pulsate, an 8 Converter can be included in the electrical circuit, which smoothies these pulsations.

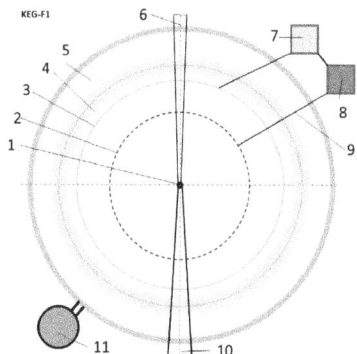

Fig.1. Kinetic electric generator for thermonuclear reactor. *Notations*: 1 – fuel capsule of inertial fusion reactor; 2 – the conductive thin film for collection electron; 3 – the conductive wall (film) for collection positive nucleus; 4 – insulator; 5 – collector of neutrons (heater); 6 - deliver of fuel capsule; 7 – electrical useful load; 8 - **current and voltage converter**; 9 – wire connection; 10 – holder of fuel capsule; 11 – customer of heat energy.

Other types of KEGS are shown in Fig. 2A, b. It is an electric generator that uses solar wind energy and is designed for spacecraft. The idea of this generator is similar to the idea of an engine that uses solar wind energy to create traction. The theory of this engine is developed in [2], Chapter 12, pp. 245-270. It is known that the Sun emits a rarefied plasma flow (protons and electrons) at an average speed of 500 km/s, called the solar wind. On the path of the wind (Fig. 2A) perpendicular to the flow put two conductive films 3, 4. The first film retards the electrons and negatively charged, the second film keeps the protons and positively charged. Between the films is included a payload of 5. For stability, the films are rotated and connected by a Central spacer.

Unfortunately, to obtain significant energy, the film should be large. The author therefore offers the option of two (Fig. 2B), in which space the generator includes two powerful charge 8-9 forming Dipole. These charges attract to the films 3 – 4 electrons and protons from huge areas.

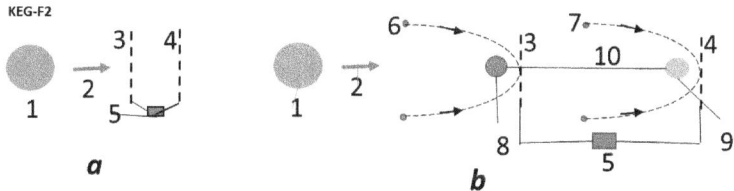

Fig.2. Kinetic electric generator for solar wind. *Notations: a* – simple case; *b* – complex case; 1 – Sun, 2 – solar wind, 3 – film-collector of electrons, 4 – film-collector of protons, 5 – electrical load, 6 – electrons, 7 – protons, 8 – positive charge, 9 – negative charge, 10 – connection negative and positive charges.

There are other unusual applications of the proposed generator. For example, obtaining electric energy from the wind on Earth (or another planet) without a propeller and a Dynamo [3]. In the wind, electrons are injected, which are picked up, carried away by the wind and settle, for example, to the ground. As a result, there is a potential difference between the injector (electron source) and the Earth.

KEGs, when they produce energy, they always slow down the flow. But it can be reversed. When it consumes energy, it accelerates the flow. This is the basis of the air-jet engine proposed by the author [4]. In the air flow in a pipe having an electric field, charged particles are injected, which are accelerated by the field and transmit their kinetic energy to the air molecules (accelerate the air flow), i.e. create traction. This method is suitable for hypersonic engines, where the traction and economic characteristics are limited by the temperature of the incoming air. For hypersonic engines, this temperature is high and modern combustion chamber materials do not allow it to significantly increase due to fuel. As a result, the efficiency of a conventional jet engine is reduced.

Theory and calculations

   As already mentioned, the idea of direct conversion of the kinetic energy of plasma into electrical energy (voltage and current) is based on different permeability of thin conductive films by charged plasma particles. It was proposed by the author about 35 years ago. This method is fundamentally different from the electromagnetic generator (EMG) method, since it does not require a powerful magnetic field and all related problems (for example, heating heavy magnets with neutrons and infecting them with radioactive isotopes). The high speed and pulse power of charged particles is also big problem for EMG.

   Let us consider the evaluation of this method by the example of the most popular reaction: tritium + deuterium with laser compressing and heating.

$$T + D \rightarrow 4He(3.5 \text{ MeV}) + n(14.1 \text{ MeV}), \qquad (1)$$

The result of this reaction appears positively charged helium with energy of 3.5 of the MeV and a neutron with energy 14.1 the MeV. Although the methods of neutron energy utilization are complex, they are technologically well developed at existing nuclear fission reactors, and we will not consider them. In this method, neutron energy is utilized through the heat cycle. The kinetic energy of charged particles can also be utilized through the heat cycle, which is complex and has a low efficiency (about 30%). But there are nuclear reactions that do not produce neutrons (for example: $^{11}B + p \rightarrow 3^4He$ (8.7 MeV)). And it makes sense to look for methods that allow you to directly convert nuclear energy into electricity.

   First of all, pay attention to the fact that the plasma particles can have a different temperature (energy). If the particles are dispersed (accelerated) by an electric field, the energy (temperature) of electrons and nuclei will be the same, because they pass the same potential difference. But if there is a reaction with the release of nuclear energy, the plasma nuclei receive huge kinetic energy (speed), and the electrons get nothing. However, at the first collision with the nuclei electrons get the speed of the nucleus. But their mass in 1836 is less than the mass of a proton. This means that the velocity of the electron is equal to the velocity of the nucleus, and the kinetic energy in

$$(m_p/m_e)^{0.5} = (1836)^{0.5} \approx 43 \text{ times} \qquad (2)$$

less energy core (at the same speed). Since the explosion is almost point, and the plasma soon becomes very sparse, this speed becomes a group, i.e. the same for all nuclei and electrons moving along the radius, and constant speed.

   The breakdown ability of the aluminum film by electrons and protons is shown in Fig. 44.3, p. 953, Fig. 44.17, p. 857 [5] . It can easily be converted to other mass by the formula

$$R_x (E) = (m_x/m_p) \cdot R_p ((m_p/m_x)E), \quad \text{gr / cm}^2 \qquad (3)$$

and other films material

$$r = R / \gamma, \quad \text{cm}, \qquad (4)$$

where γ is specific weight of film, gr/cm^3; r is thickness of film, cm. For aluminum $\gamma = 2.7$ g / cm^3.

The penetration capacity of the aluminum film by protons is shown in Table 1.
The range of protons in atmospheric air is given in Table 2.
The breakdown capacity of the aluminum film electrons is shown in Table 3.
A mileage of α-particles in air under normal conditions is shown in Table 4.
The tables are made on the basis of the figures 44.3, 44.17 and 44.8 (Kikoin [5], p. 953 - 957).
Penetration of electrons can also be estimated on the basis of the formula of Flammersfeld [5], p.957:
$$E = 1.92(r^2 + 0.22r)^{0.5}, \qquad (5)$$
here E – energy of electron, MeV; r - the extrapolated mileage, cm.

Table 1. Proton mileage in aluminum

Energy [MeV]	10^{-1}	1	10	10^2	10^3
Mileage R g/cm^2	$2 \cdot 10^{-4}$	$3 \cdot 10^{-2}$	$2 \cdot 10^{-1}$	10	$3 \cdot 10^3$

Table 2. Proton mileage in atmospheric air

Energy [MeV]	10^{-1}	1	10^1	10^2	10^3
Mileage r, cm	10^{-1}	2	10	$8 \cdot 10^3$	$3 \cdot 10^5$

Table 3. Extrapolated mileage of electron in aluminum

Mileage R mg/cm^2	0.1	1	10	10^2	10^3	10^4
Energy [MeV]	$8 \cdot 10^{-4}$	$2 \cdot 10^{-2}$	$8 \cdot 10^{-2}$	0.4	2.5	15

Table 4. Mileage of α-particles (helium) in atmospheric air

Energy [MeV]	0.1	0.5	1	5	10	20
Mileage [cm]	0.1	0.2	0.5	5	10	20

Estimates made for typical fusion reactions show that the permeability of electrons with energy (temperature) equal to the energy of nuclei is about 10 times higher than that of nuclei. But electron, having the same speed with proton, has energy in 43 time less that proton. It means, that the first film should delay the electrons and pass nucleus. But in TR the nucleus carry a huge energy flux density is large, and they will immediately destroy (burn) the film. However, in the expanding plasma flow after nuclear synthesis, the group velocity of electrons cannot exceed the velocity of nuclei. This according to (2) 43 times reduces their energy and allows to make the first film as a collector of electrons. But nevertheless the first film will continue to slow down 15-30% of high-energy nuclei and their heat will immediately melt and destroy the film.

Therefore, with the dream of using this direct method of extracting electrical energy from the plasma flow of a thermonuclear reaction, it is necessary to temporarily part.

Remains the magneto-dynamic method. But it is so complex and difficult that it is unlikely to be feasible in the near future.

It is known that the practical use of the thermonuclear reaction (1), even if it can be carried out in full, is unprofitable because of the huge cost (10 times higher than the energy from fossil fuels (see [1], Chapter 8).

But attempts to investigate in detail the possibility of using inertial thermonuclear energy for transport have revealed a new problem, which will be a serious obstacle to such use regardless of nuclear fuel.

Future major obstacle to the use of inertial thermonuclear energy for transport.

The obstacle is due to the inertial nature of the explosive method of obtaining energy. Suppose an aircraft or a ship needs an engine with a capacity of 10,000 kW. This means that in 1 second it should produce E = 10 MJ of energy. It is easy to calculate that $M = 10^{-7}$ kg of nuclear fuel T + D reacting according to the formula (1) once per second produce energy

$$E_1 = M \cdot 17.6 \cdot 10^6 / (5 \cdot 1.67 \cdot 10^{-27}) = 2.11 \cdot 10^{26} \text{ eV} = 2.11 \cdot 10^{26} \cdot 1.6 \cdot 10^{-19} \text{ J} = 3.38 \cdot 10^7 \text{ J.} \qquad (6)$$

Taking into account the efficiency and all sorts of losses is enough. If you need more power, we can increase the frequency of the cycles.

Now determine the average current at efficiency equal to 1, the frequency of 1 Hz. and the typical reaction time is t = 10^{-7} sec. in the complete transformation of all energy into electricity and the voltage is 14.1 MB (see (1)):

$$I = E_1/Vt = 3.38 \cdot 107 / (14.1 \cdot 106 \cdot 10^{-8}) = 2.4 \cdot 10^8 \text{ A} \qquad (7)$$

Here V is the voltage, Volts. If we take into account the transformation of the capsule shell with a mass of 1 gram into plasma, the voltage will drop 10^4 times (up to 1410 volts), but the current pulse will increase 10^4 times to $2.4 \cdot 10^{12}$ A, and the pulse time will remain negligible 10^{-7} seconds.

The resulting energy E we must save and use until the next cycle. But the problem is that modern technology does not have the means to instantly take on the storage of such a huge amount of energy and its rapid issuance to the user. You need a huge capacitor Bank or a large superconducting magnetic toroid, which is unacceptable, for example, for an aircraft. In addition, the superconducting magnetic toroid itself requires a long development.

The author considered the case of transformation of all energy instantly into water vapor under high pressure and further transformation into an acceptable form of energy with the help of powerful steam turbines. This is acceptable for large fixed installations and large ships, although it requires large amounts of distilled water.

Inertial thermonuclear reactor as a driving system in space.

One of the promising areas of use of the inertial thermonuclear reactor is its use as a highly economical pulse controlled engine for space flights with low fuel consumption in near solar space. Like the nuclear engine of collapse it is unlikely to be used for interstellar
flights with the man. But because of the simplicity can deliver spacecraft over long distances and, importantly, can use any matter that will occur on the way (meteorites, asteroids, planets, etc.) to increase their capabilities.

The scheme of the inertial thermonuclear rocket engine is shown in Fig. 3. It consists of a microcapsule with a thermonuclear fuel capable of exploding, releasing nuclei with huge kinetic energy; a parabolic electrostatic mirror consisting of a double conductive grid with an adjustable electrical voltage between the grids, 100V ÷ 4 MV.

The mirror reflects the cores in a given direction. The thrust is regulated by the composition and frequency of explosions of microcapsules.

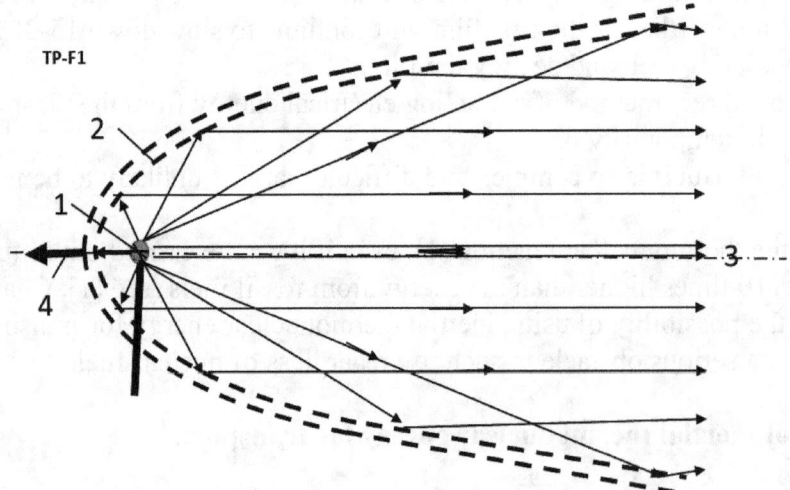

Fig.3. Scheme of inertial thermonuclear rocket engine on charged nuclei. *Designations:* 1- fuel microcapsule, 2 - parabolic electrostatic mirror, 3 - the flow of nuclei with captured electrons, 4 -engine thrust.

To estimate the momentum acquired by the aircraft, the amount of thermonuclear fuel in the capsule is taken $m_1 = 0.0001$ grams, the capsule shell has a weight $m_2 = 1$ gram, and the mass of the aircraft is taken $M = 1000$ kg. Reaction (1):

$$T + D \rightarrow {}^4He\ (3.5\ MeV) + n\ (14.1\ MeV).$$

Only ^{4}He having energy 3.5 MeV helium is emitted in the specified direction.
The speed of ejection of the fuel, mass flow and get the Space Apparatus momentum (velocity increment ΔV) was calculated according to the formulas:
1. Initial speed of nuclear

$$V = (2E/m)^{0.5}. \tag{8}$$

Here E is energy of fuel nuclear in J; m is ejected mass, kg.
2. Numbers of nuclear ^{4}He in 10^{-7} kg fuel is

$$n = m_1/(\mu m_p) = 10^{-7}/(5 \cdot 1.67 \cdot 10^{-27}) = 1.2 \cdot 10^{19}. \tag{9}$$

Here $m_p = 1.67 \cdot 10^{-27}$ kg is mass of 1 nucleon.
3. Massa of nuclear ^{4}He in 10^{-7} kg fuel is

$$m = 4 \cdot m_p \cdot n = 8 \cdot 10^{-8} \text{ kg}. \tag{10}$$

4. Energy of nuclear ^{4}He in 10^{-7} kg fuel is

$$E = E_1 n = 3.5 \cdot 10^6 \cdot 1.2 \cdot 10^{19} \text{ eV} = 3.5 \cdot 10^6 \cdot 1.2 \cdot 10^{19} \cdot 1.6 \cdot 10^{-19} = 6.72 \text{ MJ}. \tag{11}$$

Here 1 eV $= 1.6 \cdot 10^{-19}$ J.
5. Speed Increment of space apparatus, having mass $M = 1000$ kg is

$$\Delta V = (2Em)^{0.5}/M. \tag{12}$$

where m is an ejected mass, kg.
Let us to consider 3 cases:
1. Fuel capsule (FC) without cover.
2. Fuel capsule having the 1-gram cover
3. Cover of fuel capsule having any cover mass 1 kg.
  Cover of fuel capsule increases the burning time, efficiency of nuclear reaction, but increases the mass consumption.
  Result of computation is presented in Table 5.

Table 5. Result of computation the jet speed, jet mass, increment of apparatus speed, apparatus is having mass $M = 1000$ kg, initial mass of jet needed for increasing apparatus speed in 1 m/s for reaction T+D, fuel mass 0.0001 gram and energy of charge particle fuel $E = 8 \cdot 10^8$ J, frequency cycle 1 Henry.

Case/Value	Case 1, FC without cover	Case 2. FC having 1 gram cover	Case 3. Cover of FC includes any mass 1 kg
Jet speed, V km/sec	130,000	116	3,67
Jet mass, m kg	$8 \cdot 10^{-8}$	10^{-3}	1
Increment speed of apparatus 1000 kg, ΔV m/s	10^{-3}	0.116	3.67
Mass needs for increasing apparatus speed 1 m/s, kg	$7.7 \cdot 10^{-5}$	$8.62 \cdot 10^{-3}$	0.272

  Case 1 may be used (for special Thermo-reactor) for flight to end of Solar system, Case 2 is suitable for flight inside Solar system, case 3 may be used for staring from planets, asteroids.
For starting from Earth, the additional any mass must be in 4 – 10 times more than mass of apparatus. In start time from Earth the frequency cycle must be increased in 5 – 10 times.

Discussion

A detailed study of thermonuclear energy began about 60 years ago. It generated great enthusiasm and great hopes for rapid development and breakthrough in the field of energy. Most of the leading technology States have allocated billions of dollars to research thermonuclear installations. Space scientists began to ask for grants for the development of nuclear spacecraft in the hope of rapid advances in thermonuclear technology.

But 60 years of hard work did not allow to obtain even a stable thermonuclear energy exceeding the energy expended. Attempts to heat the plasma in a strong grip – the laser beam is started the problem (as it came to be recognized more and more scientists) to a standstill, requiring further huge amounts of money, time with an uncertain outcome. Scientists thought that it is necessary to overcome the difficulties encountered on the way, and they will get the desired result. There was no in-depth study of possible difficulties. The author tried to draw attention to them [1], Chapter 8. But the editor of the leading magazine on nuclear energy, said that this topic is not interested in them.

The author believes that sooner or later, thermonuclear energy will be widely developed. In this article, he draws attention to another obstacle that will inevitably arise on the way to the realization of the energy of the transport inertial thermonuclear reactor and calls to start looking for ways to overcome this obstacle. This obstacle is the explosive nature of inertial thermonuclear fusion and the instantaneous release of a huge amount of energy, for the storage of which at least a fraction of a second, the technology of the present time has no means.

References.

1. Bolonkin A.A., Small Non-Expensive Electric Cumulative Thermonuclear Reactors, LULU, USA, 2017, 143 ps.
2. Bolonkin A.A., Non-Rocket Space launch and Flight. Ch. 13, Electrostatic Solar Wind Propulsion, Elsevier, 2002, pp.245-270.
3. Bolonkin A.A., Innovations and New Technology (v.2). Ch. 7. Non-Turbo Electric Wind Generator, LULU, USA, 2005, pp.102-112.
4. Bolonkin A.A., ., Innovations and New Technology (v.2). Ch. 8. Electron Air Hypersonic Propulsion, LULU, USA, 2005, pp.113-120.
5. Kikoin I.K., (red.) Tables of Physical Values, Moscow, Atomizdat, 1976. Russian, 1007 ps. (Кикоин И.К.,(редактор). Таблицы Физических Величин, Справочник. Москва, Атомиздат, 1976, 1007 стр.).

9 February 2019

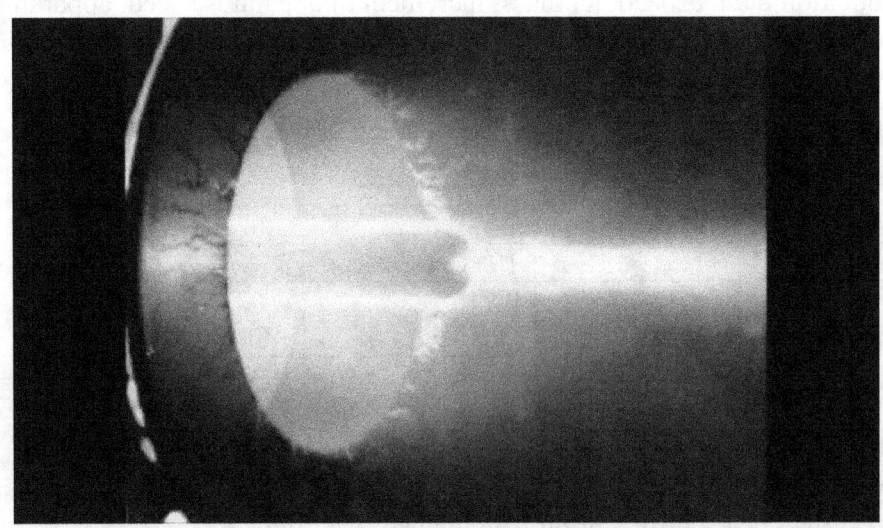

Chapter 6.
Electric Impulse Thermonuclear Reactor with the Opposed Jets.

Abstract

Chief in electric thermonuclear reactor is using electric voltage $50 \div 1000$ kV (an electric condenser discharge), which allows to heat the compressed fuel in special pellet by electric field (impulse) up hundreds millions degrees of temperature.

In electric impulse and cumulative version of AB thermonuclear reactors the fuel nucleus are heated by high electric voltage ($50 \div 1000$ kV) and magnetic or cumulative compressed into center of the special cylindrical or spherical fuel cartridge.

The additional compressing and combustion time the fuel nucleus may have from electric pinch-effect and heavy nucleus of the fuel cartridge cover. The main advantages of the offered method are very small electric fuel cartridge (11-18 mm) and small of the full reactor installation (reactor has the spherical diameter 0.3 - 3 m), using the many thermonuclear fuels at room temperature and possibility of using the offered thermonuclear reactor for transportation (ships, trains, aircrafts, rockets, etc.). Author gives theory and estimations of the suggested reactors.

Author also is discussing the problems of converting the received thermonuclear energy into mechanical (electrical) energy and into rocket thrust.

Offered small micro-reactors may be used as heaves (ignition, fusee) for small artillery nuclear projectiles and bombs.

Description and Innovations of Electric Impulse AB reactors
Description.
Laser method. Disadvantages.

Thermonuclear reactors and, in particular, Laser methods are have been under development for about 60 years. Governments have already spent tens billions of US dollars, but it is not yet seen as an industrial application of thermonuclear energy for the coming 10-15 years. The laser has very low efficiency (2- 3%), high-pressure acts every shot time (10^{-9}– 10^{-10} s), enough energy not delivered to the center of the spherical fuel pellet (low temperature), there are many future problems the radioactivity and converting the thermonuclear energy into useful energy.

Electric impulse method. Early author offered five the new methods (reflex, cumulative, impulse, ultra-cold, electric)[1 ÷ 5], which is cheaper by thousands of times, more efficiency and does not have many disadvantages of the laser and magnetic methods. In given chapter the author offers a version of the electric impulse improved reactor. Detailed consideration of advantages the new methods and computation proofs are in next paragraphs.

Outline of the new electric impulse reactors and method.

The improved version of the electric impulse AB thermonuclear reactor is presented in figures 1 – 3.

The new thermonuclear reactor contains:

1) Small cylindrical thermonuclear cartridge (cylindrical fuel ampule (granule, beat, pellet) having two cameres, (Fig.1). For fuel mass $M = 10^{-7}$ kg, the internel diameter is about 1 mm, the camera length is about 1 mm; pressure of gas fuel is up $200 \div 1000$ atm.

2) The thermonuclear reactor (sphere diameter is 0.3 – 3 m. (Fig.2). Reactor has two Version 1 - 2. In Version 1 the reactor has the additional installations for converting the nuclear energy into an electric, mechanical energy, in Version 2 the reactor converts the thermonuclear energy in a rocket thrust (fig.   3).

The fuel cartridge has (fig.1): the strong cylindrical shell from non-conductive heavy matter (A ≈ 200); left fuel camera 1, right fuel camera 2, thermonuclear fuel in left camera 3, thermonuclear fuel in right camera 4 (one may be different from left camera), power electric condenser 5 (50÷150 kJ, voltage 100 ÷1000 kV), negative electric contact 6, two positive electric contacts 7, thin electric conductivity partition 8 delete the left and right fuel cameras. Partition is burned by electric impulse.

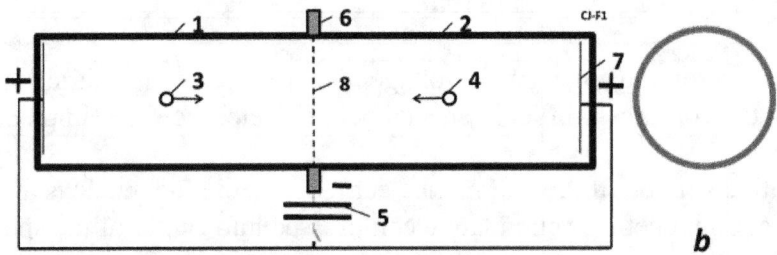

Fig.1. Principal schema of the fuel cartridge. *Notation: **a** –* side view; ***b** –* forward view. 1- left fuel camera, 2 - right fuel camera, 3 – thermonuclear fuel in left camera, 4 – thermonuclear fuel in right camera (one may be different from left camera), 5 – power electric condenser (50÷150 kJ, voltage 100 ÷1000 kV), 6 – negative electric contact, 7 –two positive electric contacts, 8 – thin elastic electric conductivity partition.

Body of nuclear reactor is shown in Fig.2. One contains: strong spherical body (shell) of reactor 1 (diameter about 0.3 – 3 m); the fuel cartridge 2 (It is described in Fig.1); holder (electric conductor) 3 of fuel cartridge; enter of compressed air (gas) 4; exit of a hot compressed air (gas) after thermonuclear heating 5; electric voltage from condenser 6.

The offered thermonuclear reactor works the next way (Figs. 1 – 3):
1) Version 1 for an electric or mechanic energy.

The internal volume of reactor body is filled the atmospheric or compression air (enter 4 of Fig.2).

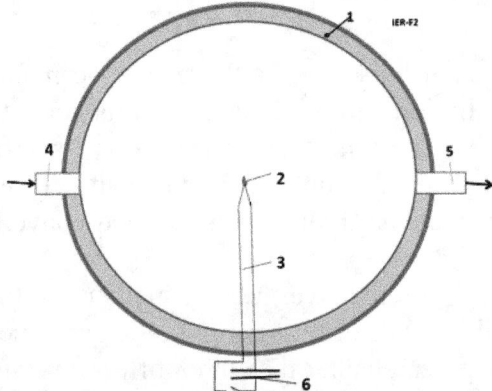

Fig. 2. AB thermonuclear electric impulse reactor. *Notations*: 1 - strong spherical body (shell) of reactor (diameter about 0.3 – 3 m); 2 - the fuel cartridge (it is described in Fig.1); 3 - holder of fuel cartridge; 4 – enter of compressed air (gas); 5 – exit of a hot compressed air (gas) after thermonuclear heating; 6 – electric conductor.

The fuel cartridge (Fig.1) lifts by holder 3 (Fig.2) into reactor body. Turn on the charged (up 50 -500 kV) electric condenser 6 (Fig 2). The electrons from the contacts 6-7 (Fig.1.) are ionised the fuel molecules (fig.1) into the left and right cameras. In particle, they positive ionize and dissociate the fuel molecules (for example, D and T are contained into cameras 3 – 4 of cartridge). The positive ionized nucleus of the thermonuclear fuel (having small mass) are quick collectively accelerated up very high temperature (up 50 – 500 keV) and collide the collectively moving nucleus of opposed camera. Partition 8 is burned. The high electric currency produces the strong pinch effect, delete the charged particles from the pellet walls and compress the thermonuclear fuel. The cover from heavy molecules (mass A ≈ 200) reflect the light (A ≈ 2÷3) fuel nucleus and increase the fusion (reaction) time of the fuel nucleus. In results (as show computation) the fuel nucleus merge and produce a thermonuclear reaction. The thermonuclear reaction (explosive) heats the air into reactor body. For increasing the efficiency, work mass, decreasing explosive temperature and protection from neutrons, the liquid 7 (for example, water, fig.3a) may be injected into reactor.

After thermonuclear explosion the hot gas flow out into the magneto-hydrodynamic generator (MHG) 10 and produces electric energy or runs to the gas, steam turbine and produces an useful work (Fig. 3a). Or the hot compressed gas runs to rocket nozzle and produces the rocket trust (fig. 3b).

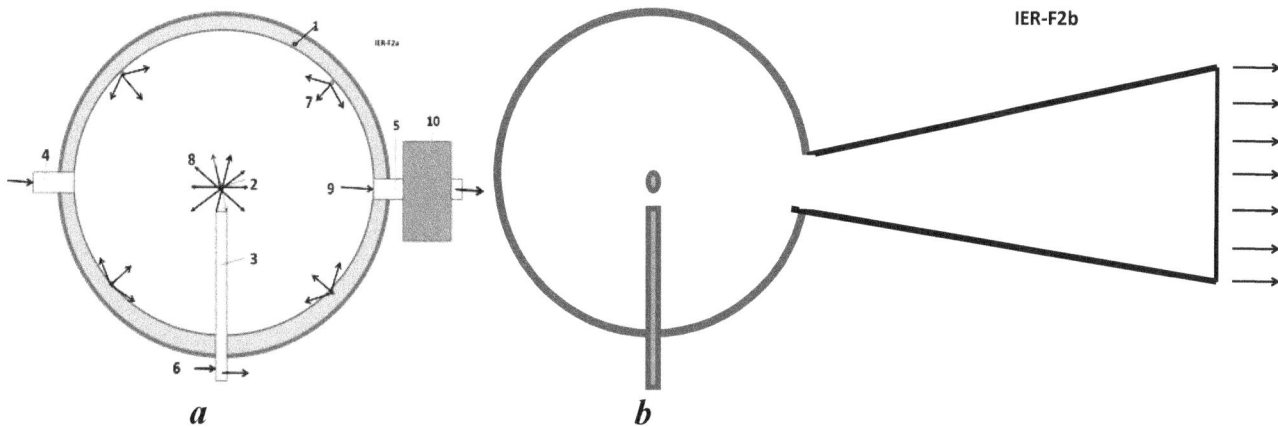

a   *b*

Fig.3. Final (industrial) work of Impulse Electric AB thermonuclear reactors. *a*) Hot compressed gas from sphere runs to the magneto-hydrodynamic generator (MHG) 10 and produces electric energy or runs to gas turbine and produces an useful work (Fig. 3a). *b*) Hot compressed gas runs to rocket nozzle and produces the rocket trust. *Notation*: 1 – 6 are same Fig.2; 7 – injection the cooling liquid (for example, water)(option); 8 – thermonuclear explosive of fuel pellet; 10 – MHG or gas (steam) turbine; 11 - exit of hot gas.

The main difference the offered electric reactor from the published cumulative reactors [2, 7] is type of explosive for getting the temperature, pressure and cumulative effect in fuel. On [2 - 4] author used the chemical explosive. The offered reactor uses the strong electric field for acceleration, getting high temperature and cumulative effect. The electric method leads to practically unlimited cheap power. In [2, 4] the explosive is located into main spherical body 1 (fig.1) (or gun in [2]). In [4] version 1 (fig.2, [4]) the explosive 3 is small and located in the special fuel cartridge (fig.2, [6]). In current version no special compression explosive. The pressure and high temperature of the fuel are reached the high voltage condenser. It is easier and it is more comfortable in using.

In the current cartridge version) the fuel pullet is filling by the compressed gas fuel (up 200÷600 atmosphere or more). Reactor not has the explosive for an additional compressing of fuel. The fuel is compressed primery and heating only by strong electric charge of a condenser. The computation shows that is possible. We can also use the conventional pellet with frozen fuel.

AB Reactors are cooled using well-known methods between explosives or by an injection of water into sphere (fig. 3a).

Advantages of the suggested reactors in comparison with ICF Laser method.

The offered reactor and methods have the following advantages in comparison with the conventional ICF laser reactor:
1. The high voltage electric condenser allows reaching the needed thermonuclear temperature.
2. Cumulative, Impulse Electric AB-reactors are cheaper by thousands of time because they do not have the gigantic very expensive laser or magnetic installations (see [1]-[6]).
3. They more efficiency because the laser installation converts only 1 - 2.5% the electric energy into the light beam. In suggested AB reactors, the all underused energy remains in the spherical reactor

and utilized in MHG or turbine. AB reactors cannot have coefficient Q (used energy) significantly less 1. Moreover, one has heat efficiency more than conventional heat engines because it has very high temperature and compression ratio. One can use as the conventional very high power engine in civil and military transportation.

4. The offered very important innovation (accelerating of exhaust rocket gas) allows increasing the top speed of the exhaust mass up very high speed. This makes this method available for thermonuclear rockets.

5. Electric AB-reactors give temperature of the fuel much more than the current ICF laser installations.

6. The compression has longer time (up to $10^{-3} - 10^{-6}$ s) than a laser beam pressing ($10^{-9} - 10^{-12}$ s), because molar mass ($\mu \approx 200$) of heavy molecules (cover of cartridge in fig.1) is many times (50 ÷ 100) heavier than fuel molar mass ($\mu = 2 \div 3$).This pressure is supported by shock wave coming from moving gas and pinch effect. This pressure increases the temperature, compressing and probability of thermonuclear reaction.

7. The heavy mass of the cover of cartridge (fig.1) (having high nuclear numbers $Z \approx 80$ and $A \approx 200$) not allow the nuclear particles easily to fly apart. That increases the reaction time and reactor efficiency.

8. The suggested AB-thermonuclear reactor is small (diameter about 0.5÷3 m or less up 0.3 m) light (mass is about some ton or less up 100 kg) and may be used in the transport vehicles and aviation.

9. The water may protect the material of the sphere from neutrons.

10. It is possible (see computations) the efficiency of AB reactors will be enough for using as fuel only the deuterium (or others) which is cheaper then tritium in thousands times (One gram of tritium costs about 30,000 US dollars. One gram of deuterium costs 1$) (see Estimations of a fuel cost).

11. The offered AB reactors have high temperature. That allows to use the fuels do not give the heutrons and gamma-radiation. These fuels are safety for humanity and installations.

12. Offered reactor may be used for syntheses elements.

Theory, computation and estimation of Electric, Cumulative and Impulse AB-reactors and comparison them with current laser ICF.

Estimation of Laser method (ICF).

For comparison the laser and offer Electric, Cumulative and Impulse AB methods, we estimate the current ICF laser method.

Typical laser installation for ICF has the power 5 MJ and deliver to pellet about 20÷50 kJ energy. The pullet has the 1 – 10 mg liquid (frozen) fuel D+T (density 200 kg/m³), diameter of the spherical fuel pullet about1- 2 mm, diameter of an evaporative coating 4 – 10 mm.

Let us take the delivered energy E = 50 kJ, volume of the coating v = 5 mm³, specific weight of coating γ = 400 kg/m³ (molar weight μ = 10).

For these data and instant delivery of laser energy the maximum pressure in cover is

$$p = \frac{E}{v} = \frac{5 \times 10^4}{5 \cdot 10^{-9}} = 10^{13} \frac{N}{m^2} = 10^8 atm \qquad (1)$$

But we don't know what part this pressure transfer to the fuel pellet.

Number of nuclear in 1 m³ of covering is

$$n = \frac{\gamma}{\mu m_p} = \frac{0.4 \cdot 10^3}{10 \cdot 1.67 \cdot 10^{-27}} = 2.4 \cdot 10^{28} \quad [m^{-3}] \qquad (2)$$

Here $m_p = 1.67 \cdot 10^{-27}$ is mass of nucleon (proton) [kg].

Temperature of evaporating cover is

$$T = \frac{p}{nk} = \frac{10^{13}}{2.4 \cdot 10^{28} 1.38 \cdot 10^{-23}} = 3 \cdot 10^7 \quad [K]$$

(3)

Here $k = 1.38 \times 10^{-23}$ Boltzmann constant, J/K.

Speed of evaporated covering is

$$V = \left(\frac{8kT}{\pi \mu m_p}\right)^{0,5} = \left(\frac{8 \cdot 1.38 \cdot 10^{-23} 3 \cdot 10^7}{3.14 \cdot 10 \cdot 1.67 \cdot 10^{-27}}\right)^{0.5} = 2.51 \cdot 10^5 \ m/s = 251 \ km/s$$

(4)

Time of evaporating for thickness of covering $l = 2 \cdot 10^{-3}$ m is

$$t = \frac{l}{V} = \frac{2 \cdot 10^{-3}}{2.51 \cdot 10^5} = 8 \cdot 10^{-9} \quad s$$

(5)

Let us to consider now the process into pellet.

The density of fuel particles is

$$n_f = \frac{\gamma}{\mu m_p} = \frac{200}{2.5 \cdot 1.67 \cdot 10^{-27}} = 4.8 \cdot 10^{28} \quad \frac{1}{m^3}$$

(6)

where $\mu = 2.5$ is average molar mass of fuel D+T.

The frozen (liquid) fuel, after converting in gas, has a temperature of about $T = 4$ K.

The pressure average speed V_n of particles after conversion of the fuel into gas (plasma) and sound speed V_f to fuel gas at temperature 4K are:

$$p_f = n_f kT = 4.8 \cdot 10^{28} \times 1.38 \cdot 10^{-23} \times 4 = 2.65 \cdot 10^6 \quad N/m^2 = 26.5 \quad atm,$$

$$V_n = \left(\frac{8kT}{\pi \mu m_p}\right)^{1/2} = \left(\frac{8 \cdot 1.38 \cdot 10^{-23} \cdot 4}{3.14 \cdot 2.5 \cdot 1.67 \cdot 10^{-27}}\right)^{1/2} = 183 \quad \frac{m}{s},$$

(7)

$$V_f = \left(\frac{p_f}{\rho_f}\right)^{1/2} == \left(\frac{2.65 \cdot 10^6}{200}\right)^{1/2} = 115 \quad m/s.$$

Additional fuel pressure in *center* of pellet from two opposing sound wave bump-up is

$$p_s = \rho_f (2V_f)^2/2 = 200 \cdot (2 \cdot 115)^2/2 = 5.3 \cdot 10^6 \quad N/m^2 = 53 \quad atm$$

(8)

Fuel temperature in *center* of small mass pellet where two opposing sound (shock) wave bump-up happens is

$$T = \frac{\pi \mu m_p (V_n + V_f)^2}{8k} = \frac{3.14 \cdot 2.5 \cdot 1.67 \cdot 10^{-27} (183 + 115)^2}{8 \cdot 1.38 \cdot 10^{-23}} = 10.5 \ K$$

(9)

In reality, the full pressure and temperature in center of capsule is much more. We compute ONLY the sound wave. Any shock wave becomes fast at short distance the sound wave. However, in our case this computation is very complex.

Current inertial reactors have the maximal rate of fuel compressing in center of pellet about

$$\xi \approx 600$$

(10)

Criterion of ignition (for radius of pullet $R_0 = 0.02$ sm and solid or liquid fuel $\rho_0 = 0.2$ g/cm³) is

$$\rho R = \rho_o R_o \xi^{2/3} = 0.2 \cdot 0.02 \cdot (600)^{2/3} = 0.28 < 1$$

(11)

where ρ in g/cm³, R in cm. That value is not enough ($0.28 < 1$).

You can imagine – with just a small effort and we will fulfill the criterion of ignition! Look your attention in very low temperature of fuel (9). For this temperature, the criterion may be wrong, or area of the ignition located into center of pullet may be very small, that energy is very few for ignition of all fuel?

Estimation of some parameters the proposed Electric Impulse AB reactor.

The proposed Electric Impulse AB Reactor accelerates the fuel 3-4 (fig.1) by a strong electric field (100÷1000 keV) and heats the fuel up 100÷1000 keV. The counter-flows and electric pinch-effect compresses and additional heats the fuel up to very high values, producing a nuclear reaction. Inlike [1 – 4] the cumulative explosionis produced not chemical explosive but a strong electric impulse.

Below is not mega-project. Instead, below, is the estimations of the typical parameters of electric impulse AB reactors.

1. Suitable thermonuclear reactions.
  The corresponding reactions are:
  D + T →^{4}He (3.5MeV) + n (14.1MeV);
  D + D →T (1.01MeV) + p (3.02MeV) 50%
  D + D →^{3}He(0.82MeV) + n (2.45MeV) 50%
The deuterium cannot be used in the laser reactor because one requests in 100 times more ignition criterion then D + T. But D+D may be used in AB reactors with an additional heating by electric field.
  The ^{3}He is received in deuterium reaction may be used in next reactions:
        D + ^{3}He →^{4}He (3.6MeV) + p (14.7MeV);
        ^{3}He + ^{3}He →^{4}He +2p (12.9MeV).
They produce only high-energy protons which can be directly converted in electric energy. Last reactions do not produce radio isotopic matters (no neutrons).
  Reaction D + D has the other distinct advantages:
1. One produces the protons which energy can be converted directly to electric energy.
2. One produces the tritium which is expensive and may be used for thermonuclear reaction.
3. One produces less and low energy neutrons which create radioactive matters.
  The other important advantage is using the pellets with compression gas fuel. Let us take a micro-balloon (pellet) having fuel gas with p_o>200 atm., radius 0.05 cm., temperature 300K. The mass fuel will be less 1 mg.
  Compressed micro-balloon (pellet) is more comfortable for working because it is unnecessary to store the fuel at lower (frozen)) temperature.

2. Cumulative nucleus speed, temperature and pressure in the fuel cartridge after electric impulse.
  When we turn on the high voltage electric impulse, the power electron flows into pellet vaporize, ionize and dissociate the fuel.
  The average ion (nuclear) temperature. The average voltage U = 15 kV, 100 kV of condenser is accelerated fuel ion in vacuum . The ion temperatures
$$T = 15 \cdot 10^3 \cdot 1.18 \cdot 10^4 = 177 \cdot 10^6 \ K, \quad T = 10^5 \cdot 1.18 \cdot 10^4 = 1.18 \cdot 10^9 \ K. \quad (12)$$
  This temperature will have the fuel gas is filled the cartrige. The energy of ionization and dissociation is small (3 ÷ 15 eV) in comparison with energy from acceleration (15 ÷ 100) keV. We can neglect it. The full ionized ions are moving as one whole.That means no gas resistance for fuel ion acceleration (electron mass is only 1/1836 of mass proton) . Any atom in internal space of cartridge will be ionized and accelerated in two counter-flow directions.
  The average speed of ionsand ion and electrons for U = 15 kV is:
$$V_i = \sqrt{\frac{2eU}{\mu m_p}} = \sqrt{\frac{2 \cdot 1.6 \cdot 10^{-19} 15 \cdot 10^3}{2.5 \cdot 1,67 \cdot 10^{-27}}} \approx 10^6 \ \frac{m}{s}, \quad V_e = \sqrt{\frac{2eU}{\mu m_e}} = \sqrt{\frac{2 \cdot 1.6 \cdot 10^{-19} 15 \cdot 10^3}{2.5 \cdot 9.1 \cdot 10^{-31}}} \approx 4.6 \cdot 10^7 \ \frac{m}{s} ,(13)$$

Here $e = 1.6 \cdot 10^{-19}$ is charge of ion, C; U is condenser voltage, V; μ is relative mass of molar fuel D+T; $m_p = 1.67 \cdot 10^{-27}$ is proton weight, kg, $m_e = 9.1 \cdot 10^{-31}$ is electron mass, kg.

The average time of ion and electron moving in distance $L = 1$ mm and speed $V = 10^6$ m/s are:

$$T_i = \frac{L}{V_i} = \frac{1 \cdot 10^{-3}}{10^6} = 1 \cdot 10^{-9} \ s, \quad T_e = \frac{L}{V_e} = \frac{1 \cdot 10^{-3}}{4.6 \cdot 10^7} = 2.17 \cdot 10^{-11} \ s \tag{14}$$

3. Maximal deviation of fuel ion from cartridge center is

$$r_a = V_0 T_i = 1750 \cdot 1 \cdot 10^{-9} = 1.75 \cdot 10^{-6} \ m, \tag{15}$$

Where V_0 is speed of nucleon for temperature at 300 K.

4. Ion free path in the center of the cartridge

$$l = \frac{1}{\sqrt{2} \pi n d^2} = \frac{1}{\sqrt{2} \cdot 3.14 \cdot 2.4 \cdot 10^{25} (5 \cdot 10^{-12})^2} = 3.76 \cdot 10^{-4} \ cm \tag{16}$$

Here $n = N/v = 2.4 \cdot 10^{19}/10^{-6} \text{cm}^{-3}$ is density of fuel ion in center of cartridge (v is volume 1 cm^3), d is ion diameter, cm. Than means, the ions will collisions many times at center of the cartridge.

5. *Thermonuclear energy*. One/tenth mg (10^{-7} kg) of thermonuclear fuel D+T has energy:
   Number of nucleus:

$$n_1 = \frac{M}{\mu m_p} = \frac{10^{-7}}{2.5 \cdot 1.67 \cdot 10^{-27}} = 2.4 \cdot 10^{19} \tag{17}$$

One pair of nuclear D+T produces energy $E_1 = 17.6$ MeV. The n_1 nuclear particles contain the energy

$$E = 0.5 \, n_1 E_1 = 0.5 \cdot 2.4 \cdot 10^{19} 17.6 \cdot 10^6 = 21.1 \cdot 10^{25} \ eV = 21.1 \cdot 10^{25} 1.6 \cdot 10^{-19} \ J = 3.38 \cdot 10^7 \ J \tag{18}$$

One pair of nuclear D+D produces energy $E_1 = 3.64$ MeV. The $n_{1} = 3 \cdot 10^{19}$ nuclear particles contain the energy

$$E = 0.5 \, n_1 E_1 = 0.5 \cdot 3 \cdot 10^{19} 3.64 \cdot 10^6 = 5.46 \cdot 10^{25} \ eV = 5.46 \cdot 10^{25} 1.6 \cdot 10^{-19} \approx 8.74 \cdot 10^6 \ J \tag{19}$$

If coefficient efficiency of the Electric Cumulative AB Reactor is $\eta = 0.3$, 0.1 mg of fuel T+D produces the energy 10 million joules, D+D produced 2.62 million joules. If we make one explosion per sec, installation has the power of 10 million watts (T+D). The part of this energy will be produced inside fuel microcapsule-fuel pellet (3.5 MeV from ^{4}He, $E = 6.76 \cdot 10^6$ J) the most of energy (14.1 MeV from neutrons, $E = 2.7 \cdot 10^7$ J) will be produced into the big containment sphere .

Conventional coefficient of nuclear reactor efficiency is about $0.3 \div 0.5$, the steam (gas) turbine is about 0.9.

6. *Estimation of pressure and temperature after nuclear explosion in reactor (more precisely, inside reactor sphere).*

Let us to find the pressure and temperature after thermonuclear explosive the 0.1 mg fuel D+T into reactor having sphere 1 m^3 filled by air.

Number of nuclear particles in sphere 1 m^3 is

$$n_n = \frac{M}{\mu m_p} = \frac{10^{-7}}{2.5 \cdot 1.67 \cdot 10^{-27}} = 2.4 \cdot 10^{19} \ \frac{1}{m^3} \tag{20}$$

Full thermonuclear energy ($\eta = 1$):

$$E_n = 0.5 n_n E_1 = 0.5 \cdot 2.4 \cdot 10^{19} 17.6 \cdot 10^6 = 21.1 \cdot 10^{25} \ eV = 3.38 \cdot 10^7 \ J \tag{21}$$

Number of air particles with air density $\rho = 1.225$ kg/m^3 in pressure $p = 1$ atm. $\mu \approx 28$ is

$$n_o = \frac{M}{\mu m_p} \approx \frac{1.225}{28 \cdot 1.67 \cdot 10^{-27}} = 2.6 \cdot 10^{25} \ \frac{1}{m^3}. \tag{22}$$

If coefficient efficiency of thermonuclear reaction is $\eta = 0.3$ in volume 1 m^3:

$$p = \frac{\eta E_n}{v} = \frac{0.3 \cdot 3.38 \cdot 10^7}{1} = 1 \cdot 10^7 \approx 10^7 \frac{N}{m^2} = 100 \text{ atm} \qquad (23)$$

Total pressure– nuclear explosive is $p \approx 100$ atm.
Temperature of gas mixture of explosive plus nuclear fuel is

$$T = \frac{p}{(n_0 + n_n)k} = \frac{10^7}{(2.6 \cdot 10^{25} + 2.4 \cdot 10^{19}) \cdot 1.38 \cdot 10^{-23}} = 27.9 \cdot 10^3 \text{ K} \qquad (24)$$

If we increase the initial pressure into reactor body up 10 atm, that the temperature decreases to 2790K. The same temperature is in a combustion chamber of conventional engine of the internal combustion. We can use the conventional cooling system.

The same method may be used for estimation of injection water into installation body or any garbage material in a space ship.

7. Possibility charging the condenser .

If we use the fuel D+D in our reactor can directly produce protons having $E_1 = 3.03$ MeV. For fuel 10^{-7} kg ($N = 3 \cdot 10^{19}$) in one explosion (cycle) and efficiency $\eta = 0.5$ that gives the electric energy

$$E = 0.5\eta E_1 eN = 0.5 \cdot 0.5 \cdot 3.03 \cdot 10^6 \cdot 1.6 \cdot 10^{-19} 3 \cdot 10^{19} = 3.6 \cdot 10^6 \; J$$

This energy in 50 times more than energy $72 \cdot 10^3$ J, which is requests for heating fuel for thermonuclear reaction.

8. Thickness of sphere cover. Assume the spherical cover of reactor is made from conventional steel having safety tensile stress $\sigma = 50$ kg/mm^2 = $5 \cdot 10^8$ N/m^2, pressure is 100 atm. The full tensile force is $F = \pi r^2 p = 3.14 \cdot 0.5^2 \cdot 10^7 = 0.785 \cdot 10^7$ N. Requested area of steel is $S_r = F/\sigma = 0.785 \cdot 10^7/5 \cdot 10^8 = 0.0157$ m^2. The thickness of sphere wall is $\delta = S_r/2\pi r = 0.0157/2 \cdot 3.14 \cdot 0.5 = 0.005$ m. Mass of sphere is $M_c \approx \gamma S_s \delta = 7800 \cdot 3.14 \cdot 0.005 = 122.5$ kg. Here $S_s = 4\pi r^2 = 4 \cdot 3.14 \cdot 0.5^2 = 3.14$ m^2 is average surface of sphere.

If we use the more strong material for sphere wall, for example: 1μm iron whisker having safety tensile stress $\sigma \approx 400$ kg/mm^2 = $4 \cdot 10^9$ N/m^2, we decrease the sphere's mass by 4 – 8 times. We can also make the sphere wall from composite materials (example: an artificial fiber carbon or glass having safety stress $\sigma \approx 100 \div 150$ kg/mm^2 and density $\gamma = 1500 \div 2700$ kg/m^3). There are many other reqests to sphere cover.

9. Cooling the sphere by water. If explosions are very frequent, we then can decrease the wall or/and gas temperature by injection of the chilled or room temperature water. The water also protects our installation from high-energy neutrons in other words, it behaves as a shielding materials.

Let us estimate the amount of water which decreases the temperature and pressure of gas (at most steam H_2O) into sphere for magnitudes acceptable for current steam turbines: $T = 400^oC = 672$ K. The critical point of water (triple point) is $T = 273^oC$, $p = 22$ MPa.

Heating 1 kg water from 20°C to 100°C requests energy $E = C_p \Delta T = 4.19 \cdot 80 = 333$ kJ, evaporation – $r = 2260$ kJ, heating of steam up 400°C - $E = C_p \Delta T = 1.05 \cdot 300 = 315$ kJ. Total amount of water heat energy is $E_w = 333 + 2260 + 315 = 2908$ kJ/kg. Total mass of water for nuclear efficiency $\eta = 1$ equals $M_w = E/E_w = 3.4 \cdot 10^7/2.9 \cdot 10^6 = 11.7$ kg. For $\eta = 0.3$ $M_w = 3.5$ kg. The 2 – 3 cm of water thickness protects the installation from high energy of neutrons produced by reaction D+T.

Unfortunately, the injection of water before decompressing strongly decreases the efficiency of installation.

10. Run protons and heavy nuclear particles.

The physic directory by Kikoin, Moscow, 1975, p. 953[12] gives the following equation for running the protons and charged heavy particles inside gas at pressure 1 atm

$$R_x(E) = \frac{m_x}{m_p} R_p \left(\frac{m_p}{m_x} E \right) \quad , \qquad (25)$$

Where R_x is run of the investigated particles, m_x is mass of investigated particles, m_p is mass of proton, R_p is run of known particles in a known environment, E is energy of particles in MeV. The run of proton in H_2 at pressure 1 atmosphere is in Table 6:

Table 6. Run (range) of proton in gas H_2 at pressure 1 atmosphere

Energy E [MeV]	1	10	100
Run R [cm]	10	$5 \cdot 10^2$	$2 \cdot 10^4$

For particles ^{4}He (3.5 MeV) in reaction D+T under the pressure $p = 10^7$ atmosphere the run is

$$R_x(E) = \frac{m_x}{m_p} R_p \left(\frac{m_p}{m_x} E \right) / p \approx \frac{4}{1} R_p \left(\frac{1}{4} 1 \cdot 3.5 \right) / 10^8 \approx 4 \cdot 10 / 10^7 = 4 \cdot 10^{-6} \ cm \approx 4 \cdot 10^{-5} \ mm \qquad (26)$$

That means the all energy of the charges particles after nuclear reaction is used for heating other "cold" particles. If probability of an initial reaction is more than 10 keV/3500 keV = 1/ 350, the chain reaction and ignition will occur .

In the Electric Impulse AB Reactor these conditions are in *whole* fuel capsule, in laser reactor of many times lower conditions may be *only* in center of fuel capsule (collision of the imposed shock waves). If reacted particles run out the center of capsule, its energy will wasted.

The run way of neutrons is large and is very complex function of energy and conditions of Environment.

11.*Converting the nuclear energy of Electric Impulse AB reactor to electric, mechanical energy or rocket thrust.*

The best means for converting the nuclear energy of the offered Reactor is magneto hydrodynamic electric generator (MHD-generator) which converts with high efficiency the high temperature and high pressure plasma directly in electric energy. Together with capacitors one can produces continuous electric currency. Impulse work of reactor allows to cool the reactor by injection the cooler (or conventional cooling) and protect the Electric Cumulative AB Reactor installation from very high temperature.

The second way for converting an Electric Cumulative AB Reactor nuclear energy is conventional heat exchanger and gas turbine. As cooler may be used the FLiBe – melted mix of fluoride salts of lithium and beryllium.

The third way is injection of water inside sphere and steam turbine as description over.

12. *Using the Electric Impulse reactor as an impulse space rocket engine.*

There are good prospects (possibility) to use the suggested Electric Impulse AB Reactor as an impulse rocket engine.

Assume the fuel energy is 10^8 J and mass of cartridge is 5 gram. If plasma will flow from reactor to space the average speed V of jet is

$$\text{From} \quad E = \frac{mV^2}{2} \quad \text{we get} \quad V = \left(\frac{2E}{m} \right)^{1/2} = \left(\frac{2 \cdot 10^8}{5 \cdot 10^{-3}} \right)^{1/2} = 2 \cdot 10^5 \ \frac{m}{s} = 200 \ km/s \ . \qquad (27)$$

Here $E = 10^8$ is nuclear energy in one impulse one mg nuclear fuel, J; $m = 5$ g is the mass injected to outer space (fuel cartridge), kg.

Received speed $V = 200$ km/s is in many times more than a current exhaust chemical speed 3 km/s. If of space apparatus has mass $m_2 = 1$ ton, the ship speed changes in $V_2 = (m/m_2)V = 1$ m/s in one impulse. If we spend 10 kg of fuel cartridges, the apparatus get speed 2 km/s.

More importantly, the next possibility is of the rocket powered by the Electric Impulse AB Reactor. Any matter from any planets, asteroids, space body may be used as fuel (or addition to emission) for increasing the derivation of impulses. For example, assume the captured solid object moving through space is composed of some water, and we filled rocket tanks using that mined planet, comet or asteroid water. From (27) and Law of equal impulse we have from every impulse

$$V_2 \approx (2Em_1)^{1/2} / m_2 = (2 \cdot 10^8 \cdot 16)^{1/2} / 10^3 = 56.6 \quad m/s \,. \tag{28}$$

Here V_2 is additional speed of space ship; m_1 mass jet, kg, $m_1 = 16$ kg of water; m_2 is mass of space apparatus.

13. *Estimation of the high speed neutron penetration*

$$l = 1/n\sigma, \tag{29}$$

Where l is path of penetration, cm; n is density of material, $1/cm^3$; $\sigma = 10^{-24}$ cm^2 is cross section of the neutron. For steel $l = 12$ cm, for compressed air up 200 atm the $l = 205$ cm.

14. *Requested thickness of the cylindrical "$_c$" and spherical "$_s$" shell is*

$$\left(\frac{D}{d}\right)_c = \left(1 + \frac{p}{\sigma}\right), \quad \left(\frac{D}{d}\right)_s = \left(1 + \frac{p}{\sigma}\right)^{0.5}. \tag{30}$$

Where D is outer diameter of shell, d is inner diameter of shell, p is pressure, atm; σ is safety tensile stress kg/cm^2. Example, if $p = 10$ kg/mm^2, $\sigma = 50$ kg/mm^2, then $(D/d)_c \approx 1.2$.

Detail Estimation of Electric Impulse Reactors for transportation engine

1. **Estimation of nuclear energy (power).** Let us make more detail estimation the Electric Impulse Reactors for engine of transport vehicle having the fuel pellet 0.1 mg ($M_f = 10^{-7}$ kg) with fuel D+T and D+D.

Estimation of energy the D+T fuel for the coefficient efficiency $\eta = 0.5$ is;

The couple nuclei T+D produces nuclear energy $E_1 = 17.6$ MeV.

Number N of fuel nuclei's is:

$$N = \frac{M_f}{\mu m_p} = \frac{10^{-7}}{2.5 \cdot 1.67 \cdot 10^{-27}} = 2.4 \cdot 10^{19} \,.$$

Here μ is average relative (molar) mass of D+T; m_p is mass of proton, kg.

The nuclear energy of 0.1 mg D+T fuel in 1 Hz for efficiency $\eta = 0.5$ is

$$E = 0.5 E_1 e N \eta = 0.5 \cdot 17.6 \cdot 10^6 \cdot 1.6 \cdot 10^{-19} 2.4 \cdot 10^{19} \cdot 0.5 = 16.9 \cdot 10^6 \approx 17 \ MJ / Hz \,.$$

Here $e = 1.6 \cdot 10^{-19}$ is charge of electron, C.

That is power energy of the $2 \div 5$ power aviation turbo-engines. If one cycle in second (1Hz) is not enough, we can decrease the frequency. The piston engine has up 50-70 revolution per second, the high speed aviation gun up 30 shots in second.

If we use the D+D fuel having single energy $E_1 = 3.15$ MeV , $\mu = 2$, $N = 3 \cdot 10^{19}$ the nucler energy

$$E = 0.5 E_1 e N \eta = 0.5 \cdot 3.15 \cdot 10^6 \cdot 1.6 \cdot 10^{-19} 3 \cdot 10^{19} \cdot 0.5 = 3.8 \cdot 10^6 \approx 3.8 \ MJ / Hz$$

is aproximetely in 4.5 times less because E_1 is less.

2. **Size of cartridge and pellet.** Let us estimate the size of the electric impulse cartrige having an internal diameter $d = 1$ mm. The thickness δ of the cartrige wall is 0.5 mm.

Let us estimate the compressed pellet having gas mass $M = 10^{-7}$ kg, pressure $p = 1000$ atm $= 10^8$ N/m^2 and $T = 300$ K. Specific density the gas D+D, D+T in compression $p = 1$atm is $\rho_o = 0.1$ kg/m^3.

The relative outer diameter of pellet for pressure $p = 10$ kg/mm^2 and the safety tensile stress of the pellet cover $\sigma = 50$ kg/mm^2 with according (30-2) is

$$\frac{D}{d} = \left(\frac{p}{\sigma} + 1\right) = \left(\frac{10}{50} + 1\right) = 1.2$$

Outer diameter of cartridge for safety tensile stress 50 kg/mm^2 is $D \approx 1.2$ mm.

3. Estimation of requested an acceleration time, electric currency, energy for fuel heating.

Let us take the fuel D+D, mass $M = 10^{-7}$ kg, the full fuel camera has internel diameter $d = 1$ mm, length $l = 2$ mm separated in two subcameres 1x1 mm having volume $v \approx 1$ mm^3 (fig.1) each.

The number nucleus in subcamera is $N = 1.5 \cdot 10^{19}$, density in 1 cm^3 (m^3) is

$$n = \frac{N}{v} = \frac{1.5 \cdot 10^{19}}{10^{-3}} = 1.5 \cdot 10^{22} \frac{1}{cm^3} = 1.5 \cdot 10^{28} \frac{1}{m^3}$$

The pressure at temperature $T = 300$ K is

$$p = nkT = 1.5 \cdot 10^{28} \cdot 1.38 \cdot 10^{-23} \cdot 300 = 6.21 \cdot 10^7 \ N/m^2 = 621 \ atm$$

Electric charge located in every subcamera

$$Q = Ne = 1.5 \cdot 10^{19} \cdot 1.6 \cdot 10^{-19} = 2.4 \ C$$

Let us take average voltage of condenser $U = 50$ kV and the length of subcamera $l = 1$ mm $= 10^{-3}$ m. Then averege intensity of electric field is

$$E_e = U/l = 5 \cdot 10^4 / 10^{-3} = 0.5 \cdot 10^8 \ V/m.$$

The force acting on the charge is

$$F = QE = 2.4 \cdot 0.5 \cdot 10^8 \ N = 1.2 \cdot 10^8 \ N.$$

Acceleration of charge is

$$a = \frac{F}{M} = \frac{1.2 \cdot 10^8}{0.5 \cdot 10^{-7}} = 2.4 \cdot 10^{15} \frac{m}{s^2}.$$

Acceleration time is

$$t = \left(\frac{2l}{a}\right)^{0.5} = \left(\frac{2 \cdot 0.5 \cdot 10^{-3}}{2.4 \cdot 10^{15}}\right)^{0.5} = 0.42 \cdot 10^{-9} \ s.$$

Here $l = 0.5$ mm is average distanse of the charge acceleration.
Average charge speed in end of acceleration is

$$V = at = 2.4 \cdot 10^{15} \cdot 0.42 \cdot 0^{-8} = 10^6 \ m/s$$

Kinetic energy of fuel is

$$E = \frac{MV^2}{2} = \frac{10^{-7}(10^6)^2}{2} = 0.5 \cdot 10^5 \ J.$$

Electric currency is

$$I = \frac{Q}{t} = \frac{2.4}{0.42 \cdot 10^{-9}} = 5.7 \cdot 10^9 \ A.$$

Energy of electric currency is

$$E = IUt = 5.7 \cdot 10^9 \cdot 5 \cdot 10^4 \cdot 0.42 \cdot 10^{-9} = 1.2 \cdot 10^5 \ J$$

Here $U = 5 \cdot 10^4$ V is everage voltage of the condenser.
Estimation of evarage magnetic pressure (pinch effect).
Intensity H of magnetic field is

$$H = \frac{I}{2\pi r} = \frac{5.7 \cdot 10^9}{2 \cdot 3.14 \cdot 0.5 \cdot 10^{-3}} = 1.8 \cdot 10^{12} \frac{A}{m}.$$

Here $r = 0.5$ mm is the internel radius of fuel pellet.
Pressure p of magnetic field is

$$p = \mu_o \frac{H^2}{2} = \frac{4\pi 10^{-7}\left(1.8\cdot 10^{12}\right)^2}{2} = 20\cdot 10^{17} \ \frac{N}{m^2} = 2\cdot 10^{13} \ atm.$$

Here $\mu_o = 4\pi\cdot 10^{-7}$ is the magnetic constant.

This pressure increases the fuel density in n_{mag} times and separated nucleus from the capsule walls $N_{mag} = p/p_o = 2\cdot 10^{13}/621 = 3.22\cdot 10^{10}$, $n_{mag} = nN_{mag} - 1.5\cdot 10^{22}3.22\cdot 10^{10} = 4.83\cdot 10^{32} \ 1/cm^3$.

The probability P of the nucleus collision for temperature (voltage) $=10^5$ eV (cross section $\sigma = 4\cdot 10^{-26}$ fig.4), nucleus density $n = n_{mag}$ and the length of the opposed jet $\delta = 0.05$ cm) is

$$P = \sigma n\delta = 4\cdot 10^{-26}\cdot 4.83\cdot 10^{32}\cdot 0.5\cdot 10^{-1} = 9.7\cdot 10^5 \geq 1.$$

In reality the probability is less 1, because the number of reactive nucleus decrease as result of reactions

$$N(\delta) = N_0(1 - \exp(-\sigma n\delta)).$$

After collision the left and right collections of charges, the shock wave will move in a cylindrical capsule reflected from its ends. That oscillation may be support by electric impulses and increase the reaction time.

Evaluation time retaining a high temperature and pressure.

Rate of collising the plasma iones after average electric heating $T = 10^5$ eV is

$$\nu = 4.8\cdot 10^{-8} Z^4 \mu^{-1/2} n\cdot \ln\Lambda\cdot T^{-3/2} \ [1/s] = 4.8\cdot 10^{-8}\cdot 1^4\cdot (2)^{-1/2}1.5\cdot 10^{22}\cdot 10\cdot 10^{-7.5} = 1.6\cdot 10^8 \ [1/s].$$

Here $Z = 1$ is charge state, $\mu = m_i/m_p = 2$ is molar mass expressed in units of the proton mass, $n = 1.5\cdot 10^{22}$ is ion density, $1/cm^3$, T is temperature expressed in eV.

If the cover of pellet has $\mu \approx 200$ and ion has $\mu \approx 2$, the fuel ion can pass its energy to the cover ion after $\approx 200/2 = 100$ collisions (impacts). That means the time of the ion transfer its energy encreses in 100 times and will be less 10^{-6} sec. In reality for need mass of fuel cover the additional reactivity time tie may be $10^{-3} \div 10^{-5}$ seconds.

Estimation of received energy and the energy contained in the fuel.

Let us to estimate the received energy from D+D fuel having mass $M = 10^{-7}$ kg (without energy of the first hit) and energy containing in fuel for selected $T = 10^5$ eV and the capcule pressure 620 atm = fuel density $n = 1.5\cdot 10^{22} \ 1/cm^3$.

The thermonuclear power is

$$P_{DD} = 3.3\cdot 10^{-13} n^2\cdot (\overline{\sigma v})_{DD} = 3.3\cdot 10^{-13}\cdot (1.5\cdot 10^{22})^2\cdot 4.5\cdot 10^{-17} = 3.34\cdot 10^{15} \ W/cm^3.$$

Here $(\overline{\sigma v})_{DD} = 4.5\cdot 10^{-17}$ is taken from Table 4.

The energy getting in time $t = 10^{-6}$ sec is

$$E = P_{DD}\cdot v\cdot t = 3.34\cdot 10^{15}\cdot 2\cdot 10^{-3}\cdot 10^{-6} = 6.68\cdot 10^6 \ J.$$

Here $v = 2\cdot 10^{-3} \ cm^3$ is volume of fuel pellet.

The fusion energy of couple D+D nucleus is

$$E_1 = 3.65 \ MeV = 3.65\cdot 10^6\cdot 1.6\cdot 10^{-19} = 5.84\cdot 10^{-13} \ J$$

Full energy $M = 10^{-7}$ kg of fuel D+D is

$$E_f = 0.5\cdot N\cdot E_1 = 0.5\cdot 1.5\cdot 10^{19}\cdot 5.84\cdot 10^{-13} = 4.4\cdot 10^6 \ J.$$

We get the full energy is less than the received energy E_1. That means the time of themonuclear fusion is less them 10^{-6} sec. In fact the fusion time will be significantly smaller 10^{-6} sec because a pinch magnetic effect is strong compressed the gas fuel.

Loss the heat throw a thermal conductivity.

The loss the heat throw pallet wall is

$$Q = \alpha(p)F \cdot \Delta T \cdot \Delta t = 100 \cdot 621 \cdot 7.8 \cdot 10^{-6} \cdot 1.16 \cdot 10^{9} \cdot 10^{-6} = 0.562 \quad kJ/Hz$$

Here $\alpha = 100$ W/(m^2·K·s); $p = 621$ atm; F = internel surface of pallet; m^2; $T = 10^5$ eV $= 1.16 \cdot 10^{19}$ K is avarage temperature, K; t is time, sec.

That is small part from the spent electric energy ≈ 100 kJ.

Bremsstrahlung radiation P_{Br} of a hot plasma having temterature T$= 10^5$ eV.

$$P_{Br} = \frac{16\alpha^3 h^2}{\sqrt{3}m_e^{3/2}} n_e^2 T_e^{1/2} Z_{eff} = 1.69 \times 10^{-32} n_e^2 T_e^{1/2} Z_{eff} = 1.69 \cdot 10^{-32} \cdot (1.5 \cdot 10^{22})^2 (10^5)^{1/2} \cdot 1 = 1.2 \cdot 10^{15} \left[\frac{W}{sm^3} \right]$$ Here α

is the fine structure constant, h is Planck's constant, m_e is mass of electron, n_e is electron density, 1/cm^3; T_e is electron temperature, eV, Z_{eff} is "effective" ion charge.

The loss of energy by volume $v = 0.002$ cm^3 in time $t = 10^{-7}$ sec is

$$E_{br} = P_{Br} \cdot v \Delta t = 1.2 \cdot 10^{15} 2 \cdot 10^{-3} \cdot 10^{-7} = 0.24 \cdot 10^6 \quad J.$$

That is 5.4% from full energy.

4. Estimation of electric condenser. For heating of fuel we use the short strong electric impulse. For impulse the electric condenser may be used. Let us to estimate the condenser parameters for getting the fuel temperature $T = 15$ keV (plasma temperature is 30 keV for opposet jets).

If fuel mass is $M = 0.1$ mg $= 10^{-7}$kg, the number of nucleus for D+T is

$$N = \frac{M}{\mu m_p} = \frac{10^{-7}}{2.5 \cdot 1.67 \cdot 10^{-27}} = 2.4 \cdot 10^{19}.$$

For D+D the $N = 3 \cdot 10^{19}$.

The energy W is needed for heating the fuel D+T up temperature $T = 15$ keV is

$$W = NT \cdot e = 2.4 \cdot 10^{19} 15 \cdot 10^3 1.6 \cdot 10^{-19} \approx 60 \, kJ.$$

For heating D+D fuel is W = 72 kJ.

The minimal specific energy of conventional conductor according [10] p. 368 is $\gamma = 2$ kJ/kg. Consequently, the requested mass of condenser is about 30 – 36 kg. But if we can use the advanced supercapacitor ($\gamma = 10$ kJ/kg) or ultra-capacitor ($\gamma = 20$ kJ/kg) or capacitor EE Stor, having claimed capacity $\gamma = 1000$ kJ/kg, we can decreased the capacitor mass. In any case, the capacitor mass is small part of thermonuclear engine.

5. Estimation of capacitor discharge. The electric schema of connection the condenser to fuel pellet is present in fig.5. That contains: the condenser 1, the source of high voltage 2 (high voltage electric generator or battery), fuel cartridge (pellet) R_1, and connection wires having the electric resistance R_2.

The source charges the condenser has voltage up 50 ÷ 200 kV, energy 70 ÷ 200 kJ. The condenser connects to fuel cartridge.

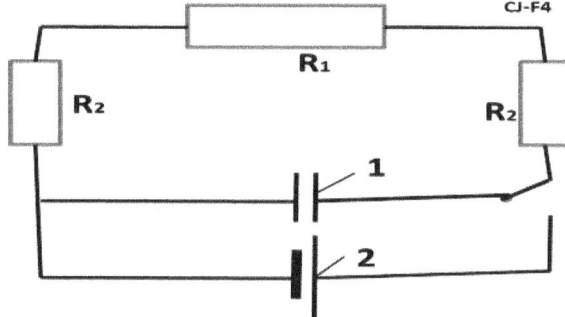

Fig.5. Electric schema of connection the condenser to the fuel pellet. *Notation*: R_1 is the resistance of fuel pellet, R_2 is

resistance of the connection wires from condenser to the fuel pellet, 1 is condenser , 2 is high voltage electric generator (battery).

Electric resistance of a copper wires connected condenser to cartidgeis (fuel pellet) is

$$R_2 = \rho \frac{l_0}{s} = 1.75 \cdot 10^{-6} \frac{200}{5} = 7 \cdot 10^{-5} \ \Omega .$$

Where $l_0 = 200$ cm is length of one wire, cm; $s = 5$ cm^2 is cross-section area of wire, sm^2; $\rho = 1.75.10^{-6}$ is a coefficient electric resistance of cupper.

Estimation the plasma special resistance into cartridge. The plasma resistance coefficients are:

$$\eta_\perp = 1.03 \cdot 10^{-2} Z \cdot \ln \Lambda \cdot T^{-3/2} .$$

For $T = 10$ eV $\eta_\perp \approx 1.03 \cdot 10^{-2} \cdot 1 \cdot 10 \cdot 10^{-3/2} = 3.16 \cdot 10^{-3} \ \Omega \cdot$ cm.

For $T = 10^5$ eV $\eta_\perp \approx 1.03 \cdot 10^{-2} \cdot 1 \cdot 10 \cdot 10^{-7.5} = 3.2 \cdot 10^{-9} \ \Omega \cdot$ cm.

Here: $Z = 1$ is ion charge of fuel D, T; $\ln \Lambda \sim 10$ is Columbu's logarithm; T is temperature, eV.
The plasma electric resistance for fuel cameras $l_0 = 1$ mm, area $s = 1$ mm^2 are:

For $T = 10 \ eV$ $R_1 = \eta_\perp \frac{l_0}{s} = 3.16 \cdot 10^{-4} \frac{0.1}{2 \cdot 0.01} \approx 1.6 \cdot 10^{-2} \ \Omega ,$

For $T = 10^5 \ eV$ $R_1 = \eta_\perp \frac{l_0}{s} = 3.2 \cdot 10^{-9} \frac{0.1}{2 \cdot 0.01} \approx 1.6 \cdot 10^{-8} \ \Omega .$

Let us take the average electric resistance $R = R_1 + 2R_2 = 10^{-3} \ \Omega$. The condenser energy $E = 60$ kJ.
Time of the condenser discharge for initial voltage $U = 30$ kV

$$t = \frac{RE}{U^2} = \frac{10^{-3} 60 \cdot 10^3}{(30 \cdot 10^3)^2} = 6.67 \cdot 10^{-8} \ s$$

The time of discharge is about time of the full thermonuclear reaction (10^{-9} s).
The average electric currency in cartridge

$$I = \frac{E}{Ut} = \frac{6 \cdot 10^4}{3 \cdot 10^4 6.67 \cdot 10^{-8}} = 0.3 \cdot 10^8 \ A$$

Capacity of condenser

$$C = \frac{t}{R} = \frac{6.67 \cdot 10^{-8}}{10^{-3}} = 6.67 \cdot 10^{-5} \ F .$$

The specific energy weight γ_c [J/kg] of the condenser may be estimated by formulas

$$\gamma_c = \frac{\varepsilon_0 \varepsilon E_q^2}{\gamma} .$$

Where $\varepsilon_o = 8.85 \cdot 10^{-12}$ F/m is electric constant; $\varepsilon \approx 3$ dielectric constant; $E_q \approx 160 \div 640$ MV/m is safety electric stress of isolator; $\gamma \approx 1000 \div 3000$ kg/m^3 is specific weight of isolator. The $\gamma_c \approx 0.3$ kJ/kg.
Electric switch must be very fast ($t < 10^{-9}$ s)(electronic or explosive) and pass very high currency.

6. Electric breakdown. For starting fusion reaction the spark gap into the fuel pellet must be less the definite value. According [14] p.123, fig.51for plates in compressed hydrogen the voltage 10^5 volts has the spark gap $pd = 10^4$ kPa·mm (here p is pressure, kPa; d is distance between plates, mm). That means for 10^5 volts and a length fuel chamber $d = 1$mm the maximum hydrogen pressure can be less $p = 10^7/(5 \cdot 10^4 1) = 200$ atm. That is not problem because the plates we can change the spearheads which decrease the need distance in 5 times or put in fuel a conductive yarns .

Look your attention, the offered method allow to get very high thermonuclear temperature. We take $U = 15 \div 50$ kV, but no limit take $U = 100, 200, 500$ kV. The 200 kV produce the temperature $T = 200 \cdot 10^3 \cdot 1.18 \cdot 10^4 = 2.36 \cdot 10^9$ K (two billions!). As you see in fig. 5 Ch.1 and estimations over, that significantly increase the probability of thermonuclear reaction and produce a fuel for the other reactor.We can use the cheap fuel produced few neutrons, many protons, expensive elements, which can be a fuel for thermonuclear reactors.

Cheap and quick testing of the proposed idea and method.

The offered idea and method of the thermonuclear fusion has the huge advantage over other known methods. One does not need in billions dollars and tens of years for cheking up idea and getting a stable thermonuclear reaction. The method may be checked by many enthusiasts and
ameteurs of new technique.

Testing the offered method is very cheap (ten thousand dollars), and need 3-6 months. Experimenter makes or orders only the pullets having different fuel (D+D, T+D, D+^{3}He, etc. , mg) and condenser having a high voltage 50 – 100 kV and an energy 60 – 100 kJ. No complex measure devices. If no thermonuclear reaction, the experimenter gets the electric spark. If there is the thermonuclear reaction, the experimenter gets an explosion quills the energy of 2 – 8 kg of TNT (for mass of fuel 0.1 mg). One successful test allows beginning a design of the engines for transports.

Discussion

About sixty years ago, scientists conducted Research and Development of a thermonuclear reactor that promised then a true revolution in the energy industry and, especially, in humankind's aerospace activities. Using such reactor, aircraft could undertake flights of very long distance and for extended periods and that, of course, decreases a significant cost of aerial transportation, allowing the saving of ever-more expensive imported oil-based fuels. (As of mid-2006, the USA DoD has a program to make aircraft fuel from domestic natural gas sources). The pressure, time and temperature required for any particular fuel to fuse is known as the Lawson criterion L. Lawson criterion relates to plasma production temperature, plasma density and time. The thermonuclear reaction is realized when L is more certain magnitude. There are two main methods of nuclear fusion: inertial confinement fusion (ICF) and magnetic confinement fusion (MCF).

Existing thermonuclear reactors are very complex, expensive, large, and heavy. They cost many billions of US dollars and require many years for their design, construction and prototype testing. They cannot stably achieve the nuclear ignition and the Lawson criterion. In future, they will have a lot of difficulties with acceptable cost of nuclear energy, with converting the nuclear energy to conventional energy, with small thermonuclear installation suitable for transportation or space exploration. Scientists promise an industrial application of thermonuclear energy after 10 – 15 years additional researches and new billions of US dollars in the future. But old methods do not allow us to reach an industrial or transport engine in nearest future.

In inertial confinement many scientists thought that short pressure ($10^{-9} – 10^{-12}$ s), which they can reach by laser beam, compress the fuel capsule, but this short pressure only create the shock wave which produced the not large pressure and temperature in a limited range area in center of fuel capsule. The scientists try to reach it by increasing NIF, but plasma from initial vaporization the cover of fuel capsule does not allow to delivery big energy. After laser beam, the fuel capsule is "naked" capsule. Capsule cannot to keep the high-energy particles of the nuclear ignition and loss them. Producing the power laser beam is very expensive and has very low efficiency (2 - 5%).

The offer method EIF (Electric Impulse Fusion) does not have these disadvantages. One uses the primary high pressed gas fuel ampoules and directly heats them to need high temperature by special electric impulse in special cartridge. The shell of capsule protects the fuel by the heavy elements ($\mu = 200$) having high number of nucleons A and charges Z. They reflect the light protons, D, T, repels high-energy reacted particles (D, T, ^{3}He, ^{4}He, p) back to fuel and significantly increasing the pressure and conformation time.

The laser ICF, MCF ideas cannot be used for thermonuclear reaction in its classical form. Produced temperature and pressure by laser ICF and magnetic MCF are not enough for thermonuclear reaction. The main author innovation is using the electric field [1-2] for reaching the need temperature (up 1 MeV) and using the primary compressing the gas fuel (up 1000 atm) in special ampoules. That increases the intensity of nuclear reaction (and temperature) in hundreds times.

Author noted that the mass of fuel is very small and allows reaching the high speed by opposed high intensity electric fields.

The impotent innovations are the compressed the fuel gas into fuel cartridge at room temperature and an electric impulse for heating of fuel up the thermonuclear temperatures. The current ICF uses the frozen fuel about absolute zero. That is not acceptable for practice. Author also suggested the transport nuclear engine and nuclear rocket.

The method possible allows to use reaction D+D (instead D+T) with cheap nuclear fuel D (Tritium is very expensive – about 30,000 USD per 1 g, deuterium costs 1 \$/g). One also allows using the compressed fuel-gas at room temperature. We can use the nuclear reactions which do not produce the neutrons and gamma radiation. They are dangerous for people.

Testing the offered method is very cheap (ten thousand dollars), need 3-6 months and may be made by many amateur of technique. They make or order only the pullets and condenser.

Conclusion

The author offers a new small very cheap Electric Impulse thermonuclear Reactors (EIR), which increases the temperature of a primary compressed nuclear fuel in thousands times, reaches the ignition and full thermonuclear reaction. Electric Impulse AB Reactor, herein offered by its originator, contains several innovations and inventions.

Main of them is using a electric field, which allows to accelerate the thermonuclear fuel to very high speed which (as it is shown by computations) heating up the hundreds million degrees of temperature. Important innovation is compressed gas fuel at room temperature, instellation for electric and mechanical energy and thermonuclear rocket.

The offered reactor is small, very cheap, may be used for not-expensive electricity, as engine for Earth transportation (train, truck, sea-going ships, aircraft), for space apparatus and for producing small and cheap and powerful weapons. Closed ideas are in [1]-[12].

RÉFÉRENCES FOR CHAPTER 2.
(READER CAN FIND PART OF THESE ARTICLES IN WEBS:
HTTPS://ARCHIVE.ORG/DETAILS/LIST5OFBOLONKINPUBLICATIONS, HTTP://BOLONKIN.NAROD.RU/P65.HTM,
HTTP://ARXIV.ORG/FIND/ALL/1/AU:+BOLONKIN/0/1/0/ALL/0/1, HTTP://VIXRA.ORG).

[1] Bolonkin A.A., Small, Non-Expensive Electric Impulse Thermonuclear Reactor with colliding jets. 7 11 16, 11 19 16, http://viXra.org/abs/1611.0276 ,
https://archive.org/download/ArticleThermonuclearReactorOfCollisingJets10416 .
[2] Bolonkin A.A. , Electric Cumulative Thermonuclear Reactors. 7 17 16.
http://vixra.org/abs/1610.0208 , https://archive.org/details/abolonkin_gmail_201610 ,
[3] Bolonkin A.A., "Inexpensive Mini Thermonuclear Reactor". International Journal of Advanced Engineering Applications, Vol.1, Iss.6, pp.62-77 (2012). http://viXra.org/abs/1305.0046
http://archive.org/details/InexpensiveMiniThermonuclearReactor,
[4] Bolonkin A.A. , Cumulative Thermonuclear AB-Reactor.. Vixra 7/ 8/2015,

http://viXra.org/abs/1507.0053 ,
https://archive.org/details/ArticleCumulativeReactorFinalAfterCathAndOlga7716

[5] Bolonkin A.A., Ultra-Cold Thermonuclear Synthesis: Criterion of Cold Fusion. 7 18 2015.
http://viXra.org/abs/1507.0158, **GSJornal**: http://gsjournal.net/Science-
Journals/%7B$cat_name%7D/View/6140 .

[6] Bolonkin A.A., Cumulative and Impulse Mini Thermonuclear Reactors. 3 30 16,
http://viXra.org/abs/1605.0309, https://archive.org/download/ImpulseMiniThermonuclearReactors ,

[7] Bolonkin, A.A., "Non Rocket Space Launch and Flight". Elsevier, 2005. 488 pgs. ISBN-13:
978-0-08044-731-5, ISBN-10: 0-080-44731-7 . http://vixra.org/abs/1504.0011 v4,

[8] Bolonkin, A.A., "New Concepts, Ideas, Innovations in Aerospace, Technology and the Human
Sciences", NOVA, 2006, 510 pgs. ISBN-13: 978-1-60021-787-6. http://viXra.org/abs/1309.0193,

[9] Bolonkin, A.A., Femtotechnologies and Revolutionary Projects. Lambert, USA, 2011. 538 p. 16 Mb.
ISBN:978-3-8473-0839-0.http://viXra.org/abs/1309.0191,

[10] Bolonkin, A.A., Innovations and New Technologies (v2). Lulu, 2014. 465 pgs. 10.5 Mb, ISBN:
978-1-312-62280-7. https://archive.org/details/Book5InnovationsAndNewTechnologiesv2102014/

[11] Bolonkin, A.A., Stability and Production Super-Strong AB Matter.
International Journal of Advanced Engineering Applications. 3-1-3, February 2014, pp.18-33.
http://fragrancejournals.com/wp-content/uploads/2013/03/IJAEA-3-1-3.pdf
The General Science Journal, November, 2013, #5244.

[12] Bolonkin, A.A., Converting of Any Matter to Nuclear Energy by AB-Generator.
American Journal of Engineering and Applied Science, Vol. 2, #4, 2009, pp.683-693.
http://viXra.org/abs/1309.0200,

[13] Kikoin I.K., Tables of Physical Values, Moscow, Atomizdat, 1975 (Russian).

[14] Koshkin N.I., Shirkevich M.G., Handbook of elementary physics, Moscow, Nauka, 1982 (Russian).

[15] AIP, Physics Desk Reference, 3-rd Edition. Springer, AIP PRESS. Third Edition.

17 July 2016

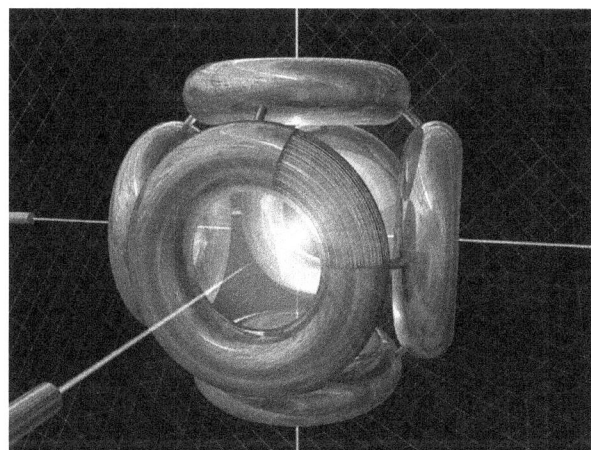

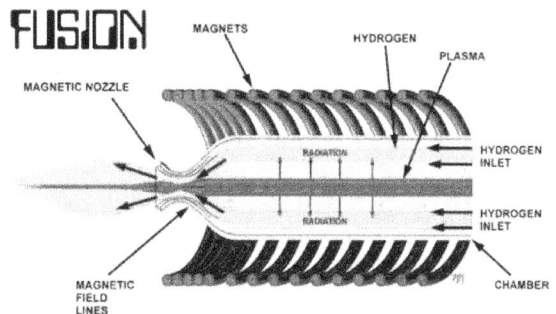

Chapter 7
Electric Cumulative Reactors

Abstract.

The author offers the new, small and cheap electric cumulative thermonuclear reactors, which increases the temperature and pressure of its nuclear fuel by millions of times, reaches the required ignition stage and, ultimately, a constant well-contained thermonuclear reaction. Electric Cumulative AB Reactors contain several innovations to achieve its power output product. Chief among them in electric thermonuclear reactors are using moving cumulative explosives and an electric discharge, which allows to accelerate the fuel and special nucleus to very high speed which (as shown by integral computations) compresses the fuel thousands times and heats the fuel by electric impulse to hundreds of millions degrees of temperature.

In electric cumulative version of AB thermonuclear reactors, the fuel nucleus are accelerated by high electric voltage (15 ÷ 60 kV) up the hundreds millions degree and *cumulative compressed* into center of the *spherical fuel cartridge*. The *additional compressing* and combustion time the fuel nucleus may have from heavy nucleus of the fuel cartridge.

The main advantages of the offered method are very small fuel cartridge (11-18 mm) of the full reactor installation (reactor having spherical diameter (0.3 - 3 m), using the thermonuclear fuel at room-temperature and achieves the possibility of using the offered thermonuclear reactor for transportation (ships, trains, aircrafts, rockets, etc.). Author gives theory and estimations of the suggested reactors. Author also is discussing the problems of converting the received thermonuclear energy into mechanical (electrical) energy and into rocket thrust. Offered small micro-reactors may be used as heaves (propellant after ignition, fuse) for small artillery nuclear projectiles and bombs.

Main difference proposed fuel cartridge from version in Chapter 2 is the spherical form. Spheris form allows to use cumulative effect and increase the fuel pressure. But we loss the pinch-effect and may the problems with thin central electrode.

Keywords: *Micro-thermonuclear reactor, Cumulative electric thermonuclear reactor, Impulse thermonuclear reactor, transportation thermonuclear reactor, aerospace thermonuclear engine, nuclei fuse, thermonuclear rocket*

Description and Innovations of Electric Cumulative AB reactors

Description.
Outline of the new electric cumulative reactors and method.
The improved version 1 of the electric AB thermonuclear reactors is presented in figures 1 – 3.

The new thermonuclear reactor contains: small (*spherical* diameter is 11 – 18 mm) thermonuclear cartridge (special fuel ampoule, granule, beat, pellet, Fig.1), thermonuclear reactor (sphere diameter is 0.3 – 3 m. fig.2, Ch.2). Reactor has two Version 1 - 2. In Version 1 the reactor has the additional installations for converting the nuclear energy into an electric, mechanical energy; in Version 2 the reactor converts the thermonuclear energy in a rocket thrust (fig. 3, Ch.2).

The fuel cartridge has (fig.1): the spherical shell 1; conductive layer 2 connected to condenser 7; layer 3 of heavy material (molar mass about 200), (option); layer 4 contains the thermonuclear fuel in a solid compound (option): internal volume of cartridge 5; edge 6 of a conductive needle connected with charged condenser 7, or very small conductive compressed fuel pellet (option); electric capacitor 7 connected to the layer 2, the edge needle 6 and source of energy.

The cartridge has three versions:

1) The Version 1 contains the layers 3-4 and vacuum 5.
2) The Version 2 not contains the layers 3-4 (or has only 3), but contains the gas thermonuclear fuel in volume 5 (in room temperature); 6 – edge of a conductive needle connected with charged condenser 7.
3) Version 3 is version 1 – 2, but one has the additional very small pellet in center of cartridge. Pellet has a thin conductive layer over thin cover and compressed fuel into pellet located on needle edge (Fig.1).

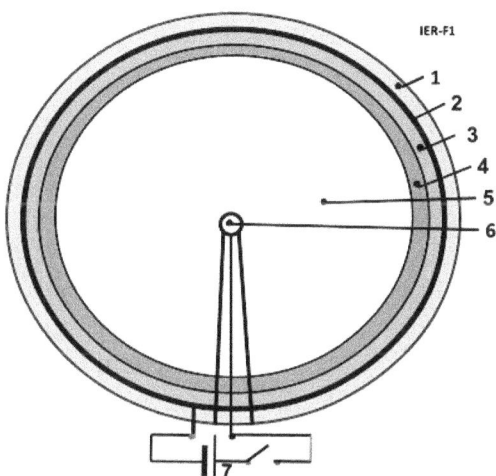

Fig.1. Cartridge (special fuel ampule, pellet) of the thermonuclear fuel, sphere of diameter 1÷2 cm. *Notations:* 1 – spherical shell of fuel cartridge; 2 – conductive layer connected to condenser 7; 3 - layer of heavy material (molar mass about 200), (option); 4 – layer contains the thermonuclear fuel in a solid compound (option);

 Cartridge has three versions:
1) The Version 1 contains the layers 3-4 and vacuum 5.
2) The Version 2 not contains the layers 3-4, but contains the gas thermonuclear fuel in volume 5 (in room temperature); 6 – edge of a conductive needle connected with charged condenser 7.
3) The version 3 has very small compressed fuel pellet having the thin conductive layer and thin cover (option).

  Body of nuclear reactor is shown in Fig.2, Ch.2. One contains: strong spherical body (shell) of reactor 1 (diameter about 0.3 – 3 m); the fuel cartridge 2 (It is described in Fig.1); holder (electric conductor) 3 of fuel cartridge; 4 – enter of compressed air (gas); exit of a hot compressed air (gas) after thermonuclear heating 5; electric voltage from condenser 6.
The offered thermonuclear reactor works the next way (Figs. 1 and 2 – 3,Ch.2):
 1) Cartridge Version 1 for an electric or mechanic energy.
The internal volume of reactor body is filled the atmospheric or compression air (enter 4 of Fig.2, Ch.2).
  The fuel cartridge (Fig.1) lifts by holder 3 (Fig.2, Ch.2) into reactor body. Turn on the charged (up 20 -100 kV) electric condenser 7 (Fig 2). The electrons from the needle edge 6 (Fig.1) are accelerated to the conductive layer 2 (fig.1). They positive ionize and dissociate the fuel molecules (for example, D and T are contained into matter of layers 3-4 (in cartridge Version 1) or into volume 5 (in cartridge Version 2)). The positive ionized nucleus of the thermonuclear fuel (having small mass) are quick accelerated up very high temperature (up 15 – 50 keV) and collide at the needle edge. The heavy and slow molecules from cartridge layer 3 (fig.1) research the region at the needle edge later and compress the thermonuclear fuel and increase the fusion (reaction) time of the fuel nucleus at the needle edge. In results (as show computation) the fuel nucleus merges and produce a thermonuclear reaction. The thermonuclear reaction (explosive) heats the air into reactor body. For increasing the efficiency, work mass, decreasing explosive temperature and protection from neutrons, the liquid 7 (for example, water, fig. 3, Ch.2) may be injected into reactor.

2) In cartridge Version 3, the needle edge contains the very small pellet 6 having the thin conductive layer, thin cover and compressed fuel. The high energy outer fuel flow additional compress the pellet 6, penetrate through the pellet cover and heat the internal part of fuel.

After thermonuclear explosion the hot gas flow out into the magneto-hydrodynamic generator (MHG) 10 and produces electric energy or runs to the gas, steam turbine and produces an useful work (Fig. 3a, Ch.2). Or the hot compressed gas runs to rocket nozzle and produces the rocket trust (fig. 3b, Ch.2).

The main difference the offered electric reactor from the cumulative reactors (versions 2, 4) [2, 4] is type of explosive for getting the temperature, pressure and cumulative effect in fuel. On [1 -2] author used the chemical explosive. The offered reactor uses the strong electric field for acceleration, getting high temperature and cumulative effect. The electric method leads to practically unlimited cheap power. In [2, 4] the explosive is located into main spherical body 1 (fig.1) (or gun in [2]). In [4] version 1 (fig.2, [4]) the explosive 3 is small and located in the special fuel cartridge (fig.2, [4]). In current version no special compression explosive. The pressure and high temperature of the fuel are reached the high voltage   condenser. It easier and it is more comfortable in using.

The version 3 of current cartridge the fuel pullet is filling by the compressed gas fuel (up 100 atmosphere or more) and version 1 - 3 not has the explosive for an additional compressing of fuel. The fuel is compressed by cumulative fuel flow (in Version 1 by additional flow of heavy elements), heating only by strong electric charge of a condenser. The computation shows that is possible. In version 3, we can use the conventional pellet with frozen fuel.

AB Reactors are cooled using well-known methods between explosives or by an injection of water into sphere (fig. 3a, Ch.2).

Detail Estimation of Electric Cumulative reactors for transportation engine

1. **Estimation of nuclear energy (power).** Let us make more detail estimation the Electric Cumulative reactors for engine of transport vehicle having the fuel pellet 0.1 mg ($M_f = 10^{-7}$ kg) with fuel D+T and D+D.

Estimation of energy the D+T fuel for the coefficient efficiency η =0.5 is;
The couple nuclei T+D produces nuclear energy E_1 = 17.6 MeV.
Number N of fuel nuclei's is:

$$N = \frac{M_f}{\mu m_p} = \frac{10^{-7}}{2.5 \cdot 1.67 \cdot 10^{-27}} = 2.4 \cdot 10^{19}$$

Here μ is average relative mass of D+T; m_p is mass of proton, kg.
The nuclear energy of 0.1 mg D+T fuel in 1 Hz is

$$E = 0.5 E_1 e N \eta = 0.5 \cdot 17.6 \cdot 10^6 \cdot 1.6 \cdot 10^{-19} 2.4 \cdot 10^{19} \cdot 0.5 = 16.9 \cdot 10^6 \approx 17 \; MJ / Hz$$

Here e = 1.6·10^{-19} C is charge of electron, C.

That is power energy of the 2 ÷ 5 power aviation turbo-engines. If one cycle in second (1Hz) is not enough, we can decrease the frequency. The piston engine has up 50-70 revolution per second, the high speed aviation gun up 30 shots in second.

If we use the D+D fuel having single energy E_1 = 3.65 MeV , μ = 2, the nucler energy is aproximetely in 5 times less because E_1 is less.

2. **Size of cartridge and pellet**. Let us estimate the size of the electric cumulative cartridge having an internal diameter d = 10 mm. The thickness δ of wall is 0.5 mm. Outer diameter of cartridge for safety tensile stress 100 kg/mm^2 is $D \approx 11$ mm.

Let us estimate the compressed **pellet** having gas mass M = 0.5·10^{-7} kg (Version 3, the other accelerated fuel is located in cartridge), pressure p = 300 atm = 3·10^7 N/m^2 and T = 300 K. Specific density the gas D+D, D+T in compression p = 1atm is ρ_o= 0.1 kg/m^3,atm. The internal radius of gas pellet is:

$$r = \left(\frac{3M}{4\pi p \rho_0}\right)^{1/3} = \left(\frac{3 \cdot 0.5 \cdot 10^{-7}}{4 \cdot 3.14 \cdot 3 \cdot 10^2 0.1}\right)^{1/3} \approx 0.74 \cdot 10^{-3} m \approx 0.74\ mm.$$

The relative outer diameter of pellet for pressure $p = 3$ kg/mm^2 and the safety tensile stress of the pellet cover $\sigma = 50$ kg/mm^2 is

$$\frac{D}{d} = \left(\frac{p}{\sigma}+1\right)^{0.5} = \left(\frac{3}{50}+1\right)^{1/2} = 1.03.$$

3. Nuclear processes in to pellet. After electric cumulative explosive into cartridge the mass of 10^{-7} kg fuel D+T into pellet having diameter $d = 1$ mm ($v \approx 0.8 \cdot 10^{-6}$ cm^3) after electric cumulative compressing is

$$n = \frac{m}{\mu m_p v} = \frac{10^{-7}}{2.5 \cdot 1.67 \cdot 10^{-27}(0.8 \cdot 10^{-6})} = 3 \cdot 10^{25} cm^{-3}.$$

Where p is pressure after cumulative compressing, N/m^2; v is a pellet volume, cm^3; m_p is mass of proton, kg. Density of D+D fuel is $n = 3.75 \cdot 10^{25}$ cm^{-3}.

Time of fuel combustion for $T = 15$ keV is

For D + T $\quad t = \dfrac{0.5\eta E_1}{5.6 \cdot 10^{-13} n <\sigma v>} = \dfrac{0.5 \cdot 0.5 \cdot 2.82 \cdot 10^{-12}}{5.6 \cdot 10^{-13} 3 \cdot 10^{25} 2.65 \cdot 10^{-16}} = 1.58 \cdot 10^{-10}$ s,

For D + D $\quad t = \dfrac{0.5\eta E_1}{3.3 \cdot 10^{-13} n <\sigma v>} = \dfrac{0.5 \cdot 0.5 \cdot 0.58 \cdot 10^{-12}}{5.6 \cdot 10^{-13} 3.75 \cdot 10^{25} 3.2 \cdot 10^{-18}} = 3.7 \cdot 10^{-9}$ s,

Where η is coefficient efficiency, E_1 is energy couple nucleis (for D+T $E_1 = 17.6$MeV$\cdot 1.6 \cdot 10^{-19} = 2.82 \cdot 10^{-12}$ J; for D+D $E_1 = 3.65$MeV$ = 0.58 \cdot 10^{-12}$ J). Here we used the Table 5.

4. Estimation of electric condenser. For heating of fuel we use the short strong electric impulse. For impulse the electric condenser may be used. Let us to estimate the condenser parameters for getting the fuel temperature $T = 15$ keV.

If fuel mass is $M = 0.1$ mg $= 10^{-7}$ kg, the number of nucleus for D+T is

$$N = \frac{M}{\mu m_p} = \frac{10^{-7}}{2.5 \cdot 1.67 \cdot 10^{-27}} = 2.4 \cdot 10^{19}.$$

For D+D the $N = 3 \cdot 10^{19}$.

The energy W is needed for heating the fuel D+T up temperature $T = 15$ keV is

$$W = NT \cdot e = 2.4 \cdot 10^{19} \cdot 15 \cdot 10^3 \cdot 1.6 \cdot 10^{-19} \approx 60\ kJ.$$

For heating D+D fuel is W = 72 kJ.

The minimal specific weight of conventional conductor according [8] p. 368 is $\gamma = 2$kJ/kg. Consequently, the requested mass of condenser is about 30 – 36 kg. But if we can use the advanced supercapacitor ($\gamma = 10$ kJ/kg) or ultra-capacitor ($\gamma = 20$ kJ/kg) or capacitor EEStor, having claimed capacity $\gamma = 1000$ kJ/kg, we can decreased the capacitor mass. In any case, the capacitor mass is small part of thermonuclear engine.

5. Estimation of capacitor discharge.

Electric resistance of a copper wires connected condenser to cartidge is

$$R = \rho \frac{l_0}{s} = 1.75 \cdot 10^{-6} \frac{400}{1} = 7 \cdot 10^{-4} \approx 10^{-3}\ \Omega.$$

Where $l_0 = 400$ cm is length of wire, cm; $s = 1$ cm^2 is cross-section area of wire, sm^2.

The plasma resistance into cartridge we can neglect.

Time of the condenser discharge for initial voltage $U = 30$ kV

$$t = \frac{RW}{U^2} = \frac{10^{-3}60 \cdot 10^3}{(30 \cdot 10^3)^2} = 6.75 \cdot 10^{-8} \ s$$

The time of discharge must be more than time of the full thermonuclear reaction (10^{-9} s).

The average electric currency in cartridge

$$I = \frac{W}{Ut} = \frac{6 \cdot 10^4}{3 \cdot 10^4 6.75 \cdot 10^{-8}} = 0.3 \cdot 10^8 \ A$$

Capacity of condenser

$$C = \frac{t}{R} = \frac{6.75 \cdot 10^{-8}}{10^{-3}} = 6.75 \cdot 10^{-5} \ F$$

The specific energy weight γ_c [J/kg] of the condenser may be estimated by formulas

$$\gamma_c = \frac{\varepsilon_0 \varepsilon E_q^2}{\gamma}$$

Where $\varepsilon_o = 8.85 \cdot 10^{-12}$ F/m is electric constant; $\varepsilon \approx 3$ dielectric constant; $E_q \approx 160 \div 640$ MV/m is safety electric stress of isolator; $\gamma \approx 1000 \div 3000$ kg/m^3 is specific weight of isolator. The $\gamma_c \approx 3$ kJ/kg.

6. Magnetic pressure from electric currency.

a) Pellet having electric cumulative compressing has initial currency $I = 3 \cdot 10^7$ A, radius of pellet (edge of needle) before compressing $r = 0.5 \cdot 10^{-3}$ m has magnetic intensity H and magnetic pressure p:

$$H = \frac{I}{2\pi r} = \frac{3 \cdot 10^7}{2 \cdot 3.14 \cdot 0.5 \cdot 10^{-3}} \approx 10^{10} \ \frac{A}{m},$$

$$p = \frac{\mu_0 H^2}{2} = \frac{4\pi \cdot 10^{-7} (10^{10})^2}{2} = 6.28 \cdot 10^{13} \ \frac{N}{m^2} = 6.28 \cdot 10^8 \ atm$$

That is closed to electric cumulative pressure for diameter capsule (center area) $r = 0.5$ mm ($p = 6 \cdot 10^8$ atm.).

The electric intensity E and electric pressure near center $r = 0.5$ mm is

$$E = \frac{U}{r} = \frac{30 \cdot 10^3}{0.5 \cdot 10^{-3}} = 6 \cdot 10^7 \ \frac{V}{m}, \quad p = \varepsilon_o \frac{E^2}{2} = 8.85 \cdot 10^{-12} \frac{(6 \cdot 10^7)^2}{2} = 1.6 \cdot 10^4 \ \frac{N}{m^2} = 0.16 \ atm.$$

As you see the electrostatic pressure we can neglect.

7. The heating problem of a needle edge.

Let us estimate the cooling of needle edge. Assume the needle edge is made from copper ($\rho = 1.75 \cdot 10^{-6}$ Ohm·cm) and has the length $l = 0.5$ cm and cross-section area $s = 0.04$ cm^2. The electric resistance of edge is

$$R = \rho \frac{l}{s} = 1.75 \cdot 10^{-6} \frac{0.5}{0.04} = 2.2 \cdot 10^{-5} \ Ohm.$$

Let us take the average condenser discharge time $t = 6,75 \cdot 10^{-8}$ s and electric currency $I = 3 \cdot 10^7$ A (see early estimation).

The energy loss in needle is

$$E = I^2 Rt = (3 \cdot 10^7)^2 2.2 \cdot 10^{-5} 6.75 \cdot 10^{-8} = 1.33 \ kJ.$$

The water requests 2269 kJ/kg for evaporation. Consequently, we need 0.6 gram/Hz of water for cooling of a needle edge in every cartridge.

8. Cost of the thermonuclear fuel.

Deuterium. The sea water contains deuterium about $1.55 \cdot 10^{-4}$ %. The World produces about tens thousand tons in year. Cost 1 \$/g.

Tritium. The special nuclear reactors can produced it. Now the cost is 30,000 \$/g. In future an expected cost will be from 100K÷200K \$/g.

Helium-3. Very rare isotope. The Helium-4 contains $1.3 \cdot 10^{-6}/1$ of the Helium-3. Cost is 30K \$/g. One project offers to extract it on Moon and delivery to Earth.

Lithium 6 -7. Nature mixture cost 150 \$/kg.

Uranium-238 contains 0.7% of Uranium-235. It cost 90÷250 \$/kg.

Plutonium-239. Cost 5600 \$/g.

As you see the thermonuclear fuel D+D is the cheapest, but D+T has the lowest temperature for thermonuclear reaction. All the current experimental thermonuclear installations are using the D+T. Look your attention, the offered method allow to get very high thermonuclear temperature. We take $U = 15$ kV, but no limit take $U = 50, 100, 200$ kV. The 200 kV produce the temperature $T = 200 \cdot 10^3 \cdot 1.18 \cdot 10^4 = 2.36 \cdot 10^9$ K (two billions!). As you see in fig. 4 and estimations over, that significantly increase the probability of thermonuclear reaction and produce a fuel for the other reactor. We can use the cheap fuel produced small neutrons, large protons, expensive elements, which can be a fuel for thermonuclear reactors.

Below in Table 7 the properties of some material suitable for the offer installation.

Table 7. Properties of some material suitable for the offer installation.

Material	Tensile strength kg/mm^2	Density g/cm^3	Fibers	Tensile strength kg/mm^2	Density g/cm^3
Steel A514	76	7.8	S-Glass	471	2.48
Aluminum alloy	45.5	2.7	Basalt fiber	484	2.7
Titanium alloy	90	4.51	Carbon fiber	565	1,75
Steel Piano wire	220-248	7.8	Carbon nanotubes	6200	1.34

Issue: [7] p.370.

Conclusion

The author offers a new small cheap electric cumulative and impulse inertial thermonuclear reactors, which increases the pressure and temperature of a nuclear fuel in thousands times, reaches the ignition and full thermonuclear reaction. Electric Cumulative and Impulse AB Reactor, herein offered by its originator, contains several innovations and inventions.

Main of them is using a electric explosive, which allows to accelerate the thermonuclear fuel to very high speed (up and more than 1000 km/s) which (as it is shown by computations) compresses the fuel in million times and heating up the hundreds million degrees of temperature. The second main innovation is the additional heating the fuel by electric impulse to up temperature in 15keV and more (hundreds millions of degrees). Important innovation is compressed gas fuel at room temperature, instellation for electric and mechanical energy and thermonuclear rocket.

The offered reactor is small, cheap, may be used for cheap electricity, as engine for Earth transportation (train, truck, sea-going ships, aircraft), for space apparatus and for producing small and cheap and powerful weapons. Closed ideas are in [1]-[10].

RÉFÉRENCES

(READER CAN FIND PART OF THESE ARTICLES IN WEBS: HTTPS://ARCHIVE.ORG/DETAILS/LIST5OFBOLONKINPUBLICATIONS,HTTP://BOLONKIN.NAROD.RU/P65.HTM, HTTP://ARXIV.ORG/FIND/ALL/1/AU:+BOLONKIN/0/1/0/ALL/0/1, HTTP://VIXRA.ORG).

[1] Bolonkin,A.A., "Inexpensive Mini Thermonuclear Reactor". International Journal of Advanced Engineering Applications, Vol.1, Iss.6, pp.62-77 (2012). http://viXra.org/abs/1305.0046 .

http://archive.org/details/InexpensiveMiniThermonuclearReactor,

[2] Bolonkin A.A. , Cumulative Thermonuclear AB-Reactor.. Vixra 7/ 8/2015,
http://viXra.org/abs/1507.0053 ,
https://archive.org/details/ArticleCumulativeReactorFinalAfterCathAndOlga7716

[3] Bolonkin A.A., Ultra-Cold Thermonuclear Synthesis: Criterion of Cold Fusion. 7 18 2015.
http://viXra.org/abs/1507.0158, **GSJornal**: http://gsjournal.net/Science- Journals/%7B$cat_name%7D/View/6140
.

[4] Bolonkin A.A., Cumulative and Impulse Mini Thermonuclear Reactors. 3 30 16,
http://viXra.org/abs/1605.0309, https://archive.org/download/ImpulseMiniThermonuclearReactors ,

[5] Bolonkin, A.A., "Non Rocket Space Launch and Flight". Elsevier, 2005. 488 pgs. ISBN-13:
978-0-08044-731-5, ISBN-10: 0-080-44731-7 . http://vixra.org/abs/1504.0011 v4,

[6] Bolonkin, A.A., "New Concepts, Ideas, Innovations in Aerospace, Technology and the Human
Sciences", NOVA, 2006, 510 pgs. ISBN-13: 978-1-60021-787-6. http://viXra.org/abs/1309.0193,

[7] Bolonkin, A.A., Femtotechnologies and Revolutionary Projects. Lambert, USA, 2011. 538 p. 16 Mb.
ISBN:978-3-8473-0839-0.http://viXra.org/abs/1309.0191,

[8] Bolonkin, A.A., Innovations and New Technologies (v2).Lulu, 2014. 465 pgs. 10.5 Mb, ISBN:
978-1-312-62280-7. https://archive.org/details/Book5InnovationsAndNewTechnologiesv2102014/

[9] Bolonkin, A.A., Stability and Production Super-Strong AB Matter.
International Journal of Advanced Engineering Applications. 3-1-3, February 2014, pp.18-33.
http://fragrancejournals.com/wp-content/uploads/2013/03/IJAEA-3-1-3.pdf
The General Science Journal, November, 2013, #5244.

[10] Bolonkin, A.A., Converting of Any Matter to Nuclear Energy by AB-Generator.
American Journal of Engineering and Applied Science, Vol. 2, #4, 2009, pp.683-693.
http://viXra.org/abs/1309.0200,

[11] Kikoin I.K., Tables of Physical Values, Moscow, Atomizdat, 1975 (Russian).

[12] AIP, Physics Desk Reference, 3-rd Edition. Springer, AIP PRESS.

17 July 2016

Chapter 8
Cumulative-Rocket Reactor with additional Electric Heating
(Version 1)

Abstract

The author offers two types Chemical Explosive Pressing Cumulative Reactor with additional Electric Heating. Reactors increase the pressure and temperature of its nuclear fuel by chemical explosive and electric charge. Offered reactors contain several innovations to achieve its product.

Chief among them in an chemical cumulative reactor is using moving explosives (by rocket thrust) and an electric discharge, which allows to accelerate the explosive to very high speed (more 20 km/s) which (as shown by integral computations) compresses the fuel capsule a million times and additional heating fuel by electric impulse up hundreds millions degrees of temperature.

In impulse version of AB thermonuclear reactor the gas fuel, primary high compressed into a pellet by a chemical explosive. Than the fuel pellet is heating by an electric impulse up the needed temperature in hundred of millions degrees, produces the thermonuclear reaction. Author gives theory and estimations of the suggested reactors.

Description and Innovations of Cumulative and Impulse AB reactors
Description.

Cumulative method. Author offered three the new methods [1 – 3], which is cheaper by thousands of times, more efficiency and does not have many disadvantages of the laser and magnetic methods. In given article the author offers two (cumulative and impulse) improved reactors. Detailed consideration of advantages the new methods and computation proofs are in next paragraph.

Description of new cumulativr reactor and method.

The improved version 1 of the Cumulative AB thermonuclear reactor is presented in figures 1 – 4.

The new thermonuclear reactor contains (Fig.1): strong spherical body of reactor 1; cartridge (holder) of fuel pellet (or pellet in Version 2) 2; holders (electric conductor) of fuel cartridge (pellet) 3; enter of compressed air (gas) 4; exit of a hot compressed air after thermonuclear heating, 5; contacts of voltage 6 for electric condenser.

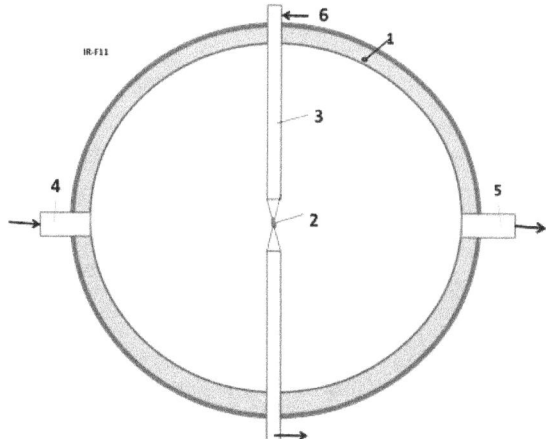

Fig. 1. AB thermonuclear cumulative reactor (Version 1). *Notations*: 1- strong spherical body of reactor, diameter about 0.5 – 5 m ; 2 – cartridge (holder) of fuel pellet, diameter 1÷2 cm (or pellet in Version 2, diameter 2 ÷ 3 mm); 3 – holder (electric conductor) of fuel pellet; 4 – enter of compressed air (gas); 5 – exit of a hot compressed air (gas) after thermonuclear heating; 6 – electric voltage.

The fuel cartridge 6 has diameter about 10 – 20 mm and the next design (fig.2): strong sphere 1; net fuse (electric net 2 for ignition of explosive 3, fig 2), explosive 3, film (piston) of heavy material 4, ampoule of nuclear fuel (pellet) 5, electric conductors 6.

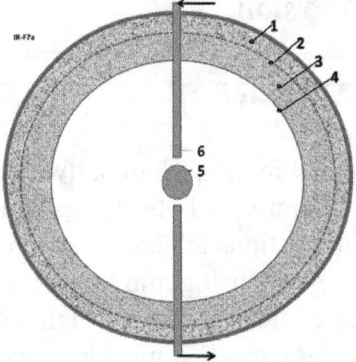

Fig.2. Cartridge of Cumulative AB-thermonuclear Reactor (Version 1). *Notations*: 1 –strong sphere, 2 – net fuse (electric net), 3 – explosive, 4 - layer (piston) of heavy material, 5 – nuclear fuel (pellet), 6 - electric conductor.

In shortly the reactor works the following way: The lasting sphere 1of reactor (fig.1) is filled the compressed gas (for example, air). Electric signal is sending across the electric conductor 3 (fig.1) and blow up the explosive 3 (fig.2) into fuel cartridge (fig.3a). The cumulative explosive works as rocket engine (fig.3b) and presses the layer (piston) of heavy material 4 around the fuel pullet 5 (fig.3) and high presses (and heating) the fuel pullet 5. The strong electric impulse from condenser, sending across the insulated conductors 3 (fig.1) and 6 (fig.2) into pellet, additional heats the nuclear fuel up the need thermonuclear temperature. The nuclear fuel explodes.

Note: It is very important to *simultaneously* ignite the *all* outer surface of explosive 3 (fig.2). In only this case the explosive begin to move towards mass center and works as rocket engine, accelerate and compress the explosive and layer 4 (having small mass) for high speeds in tens times more that in conventional explosive. That is main innovation in offered method.

We can reach the simultaneously ignition all outer surface of explosive by an electric net having a small cells of net. The electric impulse will ignite the entire outer surface of explosive.

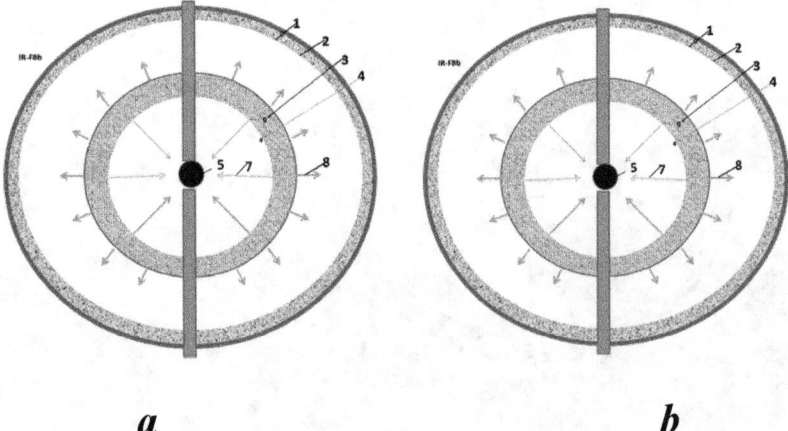

a   *b*

Fig. 3. Cumulative cartridge of thermonuclear fuel. Work of Cumulative thermonuclear cartridge (Version 1). *Notations*: *a* – initial layer explosive, *b* – rocket part of explosive. 1 - 5 are same with Fig.3; 6 is pressure of initial layer of explosion, 7 is moving of explosive (force from jetted gas rocket thrust) , 8 is flow of reactive explosive gas.

Simultaneously (or early) into the big sphere of reactor body may be injected the water 7 (fig.4a) (optional).

The compressed air (or injected water) is heating by the thermonuclear explosive 8 (fig.4a) and go out across a hole 9 (fig.4a) into MHG (or gas-, steam turbine) and produces the electric or mechanical energy.

In rocket engine the gas flow out across the nozzle and creates the rocket trust (fig.4b).

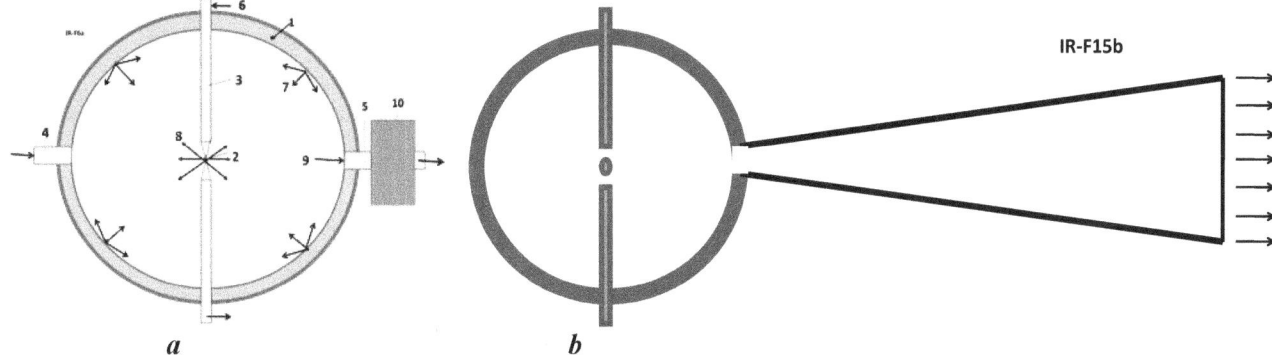

a   *b*

Fig. 4. Final (industrial) work of Cumulative and Impulse AB thermonuclear reactors. *a*) Hot compressed gas from sphere runs to the magneto-hydrodynamic generator (MHG) 10 and produces electric energy or runs to gas turbine and produces an useful work (Fig. 2a). *b*) Hot compressed gas runs to rocket nozzle and produces the rocket trust. *Notation*: 1 – 6 are same Fig.1; 7 – injection the cooling liquid (for example, water)(option); 8 – thermonuclear explosive of fuel pellet; exit of hot gas; 10 – MHG or gas (steam) turbine.

The main difference the offered reactor from cumulative reactors (versions 1, 2) [1] is location of a compressed explosive. In [1] the explosive is located into main spherical body 1 (fig.1)(or gun in [2]). In current version 1 (fig.2) the explosive 3 is small and located in *the special fuel cartridge* (fig.2). It easer and it is more comfortable in using.

The versions 1, 2 of current reactors the fuel pullet is filling by the compressed gas fuel (up 1000 atmosphere or more) and version 2 not has the explosive for an additional compressing of fuel. The fuel pellet is heating only by strong electric charge. The pullet has an additional features (fig.5). The computation shows that is possible. This method needs in additional research. In cumulative version 1 we can use the conventional pellet with frozen fuel.

AB Reactors are cooled well-known methods between explosives or by an injection of water into sphere (fig. 4a).

In more details the Cumulative AB Reactor works the following way (figs. 3 – 5):

The net fuse 2 (Fig.3a) simultaneously blows an outer layer of the explosive 3 and an explosive gas 6 is pushing the explosive 3 to center of sphere fuel cartridge. The outer surface of explosive 3 is burning 8 (fig.3b) and create jet rocket thrust 7 which is moving, accelerating, compressing the explosive and thin layer 4 in direction to center 5 of sphere fuel pellet.

As the result the layer 4 (piston) made from heavy material bumps with high speed (about 20 km/s) and produced a high pressure (millions atmospheres). This pressure is acting more time than laser pressure and reaches to center of fuel capsule (Fig.3a). The strong electric impulse increases the fuel temperature up need value. The thermonuclear fuel capsule explodes (Fig.3b). The heavy material of layer breaks the nuclear explosion, increases the conformation time and efficiency of the thermonuclear reaction. If installation is used as reactor for MHG or turbine, the cooling liquid (for example, water) is injected into strong sphere. One is converted to hot gas (steam) and rotates the turbine blades (Fig.4a).

Note: It is very important to *simultaneously* ignite the *all* outer surface of explosive 3 (fig.2). In only this case the explosive mass begin to move towards to center and works as rocket engine, accelerate and compress the explosive and layer 4 (having small mass) for high speeds in tens times more that in conventional explosive. That is main innovation in offered method.

Compressed gas pellet for Cumulative and Impulse thermonuclear reactors is shown in fig.5. One has a compressed up 300 – 1000 atm thermonuclear gas fuel 5, the insulator spherical cover 2, the contacts connect the inside and outside electric contacts 2-3; and thin internal electric conductor 4 (only for Impulse reactor) which produces the initial ion canal for an electric currency.

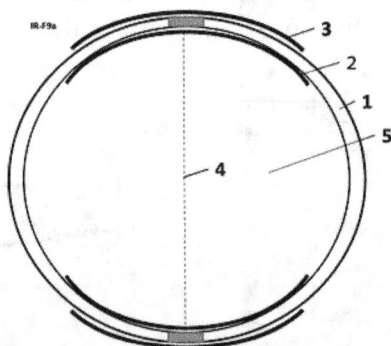

Fig.5. Compressed gas pellet for Cumulative and Impulse Thermonuclear reactors. *Notations*: 1- isolator cover; 2, 3 – internal and external electric contacts; 4 – initial hot ionizer; 5 – compressed gas thermonuclear fuel.

Cumulative-Rocket Reactor (Version 2)
Description and Innovations of Cumulative-Rocket reactors

Description.

Cumulative-Rocket method. Author offers the new method, which is cheaper more efficiency and does not have many disadvantages of the laser and magnetic methods. Detailed consideration of advantages the new method and computation proofs will be in below paragraph.

Description of Cumulative-Rocket reactor and method.

The most comfortable version 1 of the Cumulative AB thermonuclear reactor is presented in figures 6 – 8. The new thermonuclear reactor contains (Fig.6): strong sphere 1, fuse 2, six (or more) explosive injectors 3, piston 4 from heavy appropriate material, fuel pellet 5 and tube 6 for loading and primary acceleration.

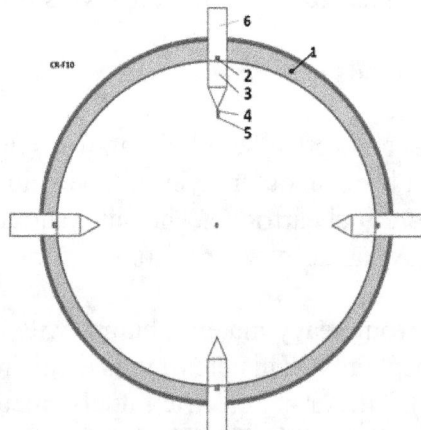

Fig.6. Impulse Cumulative-Rocket Thermonuclear Reactor (Version 2). *Notations:* 1 – strong sphere of reactor body, 2 – fuse, 3– explosive, 4 - piston from heavy material, 5 – capsule (pellet) of nuclear fuel; 6 – tube for loading and additional initial acceleration.

The Cumulative AB reactor works in the following way (figs. 7):

The net fuses 2 (six or more) (Fig.6) simultaneously blows an nearest layer of the explosive (shoot, fire off) and an explosive gas is pushing the explosive 3 to center of sphere (Fig.7a). The explosive 3 is burning, produces gas jet 8 (fig.7a) and create jet rocket thrust 7 which is moving, accelerating, pistons 4 in direction to center of sphere 1.

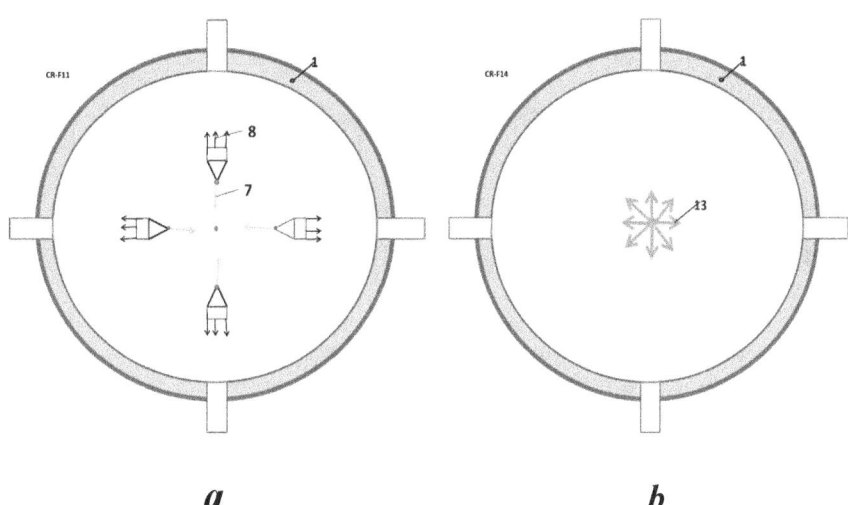

a   **b**

Fig. 7. Work of cumulative-rocket thermonuclear reactor (Version 2). *Notations*: *a* – rocket acceleration of pistons, *b* – nuclear explosion; 7 is moving of explosive and pistons (force from rocket thrust), 8 is flow of rocket explosive gas. 13 – Thermonuclear explosion.

Note: It is very important to *simultaneously* ignite *all* fuses 2 of explosive 3. In only this case, the explosive begin to move at the same time into target mass and reach simultaneously its center. They work like rocket engine, accelerate the pistons 4 (having small mass from heavy material) for high speeds in tens times more that in conventional cumulative explosive. That is main innovation in offered method, version 2.

Estimation of Chemical Cumulative-Rocket AB thermonuclear reactor.

The proposed Chemical Cumulative AB Reactor is an internal rocket engine, which accelerates the small layer from heavy material by cumulative chemical explosion (Figs. 1-3). This piston bumps into pellet of contained nuclear fuel, compresses and heats the pellet up to very high values, producing a nuclear reaction. Most important innovation is in design, the cumulative explosion which works as rocket engine and produces a final speed of the small piston in 10 (and more, from 2 km/s to 20 km/s) times and piston energy in $(10)^2 = 100$ times more than a convention explosion. The second important innovation is the additional heating the fuel by the strong electric impulse. Below is not project, below is the estimations of the typical parameters of AB reactors.

1. **Final speed of the piston (heavy layer).** Let us to estimate the offered design. It is well known the final speed V of rocket is

$$V = -W_e \ln \frac{M_k}{M} = -W_e \ln \mu_k$$

(1)

where W_e is speed exhaust gas of rocket, m/s; M_k is final mass of rocket, kg; M is initial mass of rocket, kg; $\mu_k = M_k/M$ is ratio the final and initial mass of rocket.

The distance L (acceleration path) of rocket is

$$L = g_0^{-1} W_e^2 v_0 [1 - \mu_k (1 - \ln \mu_k)] \quad \text{if} \quad \mu_k < 0.05 \quad \text{then} \quad L \approx g_0^{-1} W_e^2 v_0$$

(2)

In (2) it is used the notations: $g_o = 9.81$ m/s$^2 \approx 10$ m/s^2 is Earth acceleration, $v_o = Mg_o/P_0$ is an initial weight-to-thrust, N/N.

The rocket engine has the solid fuel $W_e = 2400 \div 2800$ m/s, liquid fuel $W_e = 3000 \div 3400$ m/s, hydrogen-oxygen up $W_e = 4000$ m/s.

The explosive matters have:

TNT: specific energy $E_s = 4.184$MJ/kg ≈ 4.2 MJ/kg, density $\rho = 1,650$ kg/m^3, speed of detonation 6900 m/s;

Dynamite: specific energy up E_s= 7 MJ/kg, standard$s$ = 5.3 MJ/kg, density ρ= 1,400 kg/m^3 , speed of detonation 6000 m/s.

From $E = mV^2/2$ we get the average speed of exhaust gas for TNT:

$$W_e = (2E_s)^{1/2} = (2 \cdot 4.184 \cdot 10^6)^{1/2} = 2893 \; m/s \tag{3}$$

Maximum pressure of explosive is

$$p = \frac{E}{v} = \frac{Em}{mv} = E_s \rho = 4.184 \cdot 10^6 \cdot 1650 = 6.9 \cdot 10^9 \; \frac{N}{m^2} = 6.9 \cdot 10^4 \; atm \;, \tag{4}$$

where v is volume of explosive, m^3; E_s= E/m is specific energy of explosion, J/kg.

Density of particles and temperature of TNT explosion initial moment of explosion (exhause gas has average $\mu = 20$, TNT reaction) is:

$$n = \frac{\rho}{\mu m_p} = \frac{1650}{20 \cdot 1.67 \cdot 10^{-27}} = 4.94 \cdot 10^{28} \; m^{-3},$$

$$T = \frac{p}{nk} = \frac{6.9 \cdot 10^9}{4.94 \cdot 10^{28} 1.38 \cdot 10^{-23}} = 10.1 \cdot 10^3 \; K, \tag{5}$$

Let us estimate the final speed of the layer for data: mass of explosive is $M = 10^{-3}$ kg, mass of layer $M_k = 1$ mg $= 10^{-6}$ kg, $W_e = 2650$ m/s.

$$V = -W_e \ln \frac{M_k}{M} = -2650 \ln \frac{10^{-6}}{10^{-3}} = 2650 \cdot 3 \cdot 2.3 \approx 18300 \; m/s \approx 18 \; km/s \;. \tag{6}$$

Let us to find the minimal acceleration distance L of layer (minimal distance from lower part of explosive to center of sphere cartridge). For Version 1 we receive:

From $L = g_0^{-1} W_e^2 v_o$, $v_o = Mg_0/F$, $m_c = \rho V_d S$,

$$F = ma = m_c W_e = \rho V_d S W_e, \; v_o = Mg_0/\rho V_d S W_e, \; V_d = W_e. \tag{7}$$

we get $\quad L = \dfrac{M}{\rho S} = \dfrac{10^{-3}}{1.65 \cdot 10^3 \cdot 6 \cdot 10^{-4}} = 0.1 \cdot 10^{-2} m = 1 \; mm$

where g_0= 9.81 m/s^2 is gravitation; W_e is speed of rocket exhaust gas, m/s; ρ is rocket fuel density, kg/m^3; V_d is rate (speed) of combustion of rocket fuel, m/s; S is initial area of the combustion created rocket thrust, m^2; F is initial rocket thrust, N/m^2, M is mass of rocket fuel, kg. We can change speed V_d to add special additives.

2. Temperature T and pressure p in pellet after compressing by piston is (for piston speed $V = 1.8 \cdot 10^4$ m/s and density of piston $\rho = 2 \cdot 10^4$ kg/m^3, $\mu = 200$):

$$p = \frac{\rho(2V)^2}{2} = \frac{2 \cdot 10^4 (2 \cdot 1.8 \cdot 10^4)^2}{2} = 1.36 \cdot 10^{13} \; \frac{N}{m^2} = 1.36 \cdot 10^8 \; atm \;. \tag{8}$$

We take here layer speed $2V$ because we have two opposed layers.

Temperature

$$\text{From} \quad E = \frac{\mu m_p V^2}{2} = \frac{3kT}{2} \quad \text{we have} \quad T = \frac{\mu m_p V^2}{3k} \;. \tag{9}$$

The mixture D+T has $\mu = 2.5$, piston about $\mu = 200$ (for example: tungsten has $\rho = 19.34 \cdot 10^3$ kg/m^3, μ =184; uran-238 has $\rho = 19.1 \cdot 10^3$ kg/m^3 , $\mu = 238$; lead has $\rho = 11.35 \cdot 10^3$ kg/m^3 , $\mu = 207$).

The temperature of mixture D+T is:

$$\text{For fuel D+T} \quad T = \frac{\mu m_p V^2}{3k} = \frac{2.5 \cdot 1.67 \cdot 10^{-27}(1.8 \cdot 10^4)^2}{3 \cdot 1.38 \cdot 10^{-23}} = 34.5 \cdot 10^3 K \approx 3.14 eV,$$
$$\tag{10}$$
$$\text{For layer} \quad T = \frac{\mu m_p V^2}{3k} = \frac{200 \cdot 1.67 \cdot 10^{-27}(1.8 \cdot 10^4)^2}{3 \cdot 1.38 \cdot 10^{-23}} = 2.61 \cdot 10^6 K = 225 eV,$$

The mass of the layer is in $5 \div 20$ times more then mass of fuel and the layr has direct contact to fuel. That means the fuel will has temperature about 3 millions degree. That is less the need value 10 keV but in thousands of time more than in a laser method. In offered method we have also very more pressure that in laser method. The high-pressure significantly decreases the need temperature because one decreases the need distance between nuclear particles.

3. Estimation the Criterion of ignition the Cumulative-Rocket AB Reactor.

The process of compression converted the solid-liquid fuel into gas. In according Ch.1 the initial pressure this gas is $p_0 = 26.5$ atm. The final pressure is about $p = 1.44 \cdot 10^8$ atm. The rate of fuel compression is

$$\xi = \frac{p}{p_0} = \frac{1.44 \cdot 10^8}{26.5} \approx 5.43 \cdot 10^5 \qquad (11)$$

(compare this reached value with maximum $\xi = 600$ in laser method. The Cumulative AB Reactor has compression many times more than laser method).

That means the liner size of fuel pellet will be in $(\xi)^{1/3} = 82$ times less. If capsule has initial diameter $D = 0.1$ cm (fuel mass $= 21 \cdot 10^{-6}$ kg, $\rho_o = 0.2$ g/cm^3), one has $R = 0.05/82 = 6.1 \cdot 10^{-4}$ cm. The offered Cumulative AB thermonuclear reactor produced *direct* compression almost a thousands times greater than the usual shock wave laser compression machines at the center of a fuel pullet. The density of the fuel will be $\rho = \rho_o \xi = 0.2 \cdot 5.43 \cdot 10^5 = 1.1 \cdot 10^5$ g/cm^3.

The criterion of the inertial ignition is

$$\rho R = 1.1 \cdot 10^5 \cdot 6.1 \cdot 10^{-4} = 67 > 1 \qquad (12)$$

One is in 67 times MORE than needed ($67 > 1 > 0.28$). That means we can use instead of very expensive tritium the deuterium which is the thousands times cheaper. The corresponding reactions are:

D + T $\rightarrow ^4$He (3.5MeV) + n (14.1MeV);
D + D $\rightarrow$ T (1.01MeV) + p (3.02MeV) 50%
D + D $\rightarrow ^3$He(0.82MeV) + n (2.45MeV) 50%

The deuterium cannot be used in the laser reactor because one requests in 100 times more ignition criterion then D + T. But, as you see in (12), one may be used in AB reactor (Fig.5, Ch.1) with an additional heating by electric charge.

The ^{3}He is received in deuterium reaction may be used in next reactions:

D + ^{3}He $\rightarrow ^4$He (3.6MeV) + p (14.7MeV);
^{3}He + ^{3}He $\rightarrow ^4$He +2p (12.9MeV).

They produce only high-energy protons which can be directly converted in electric energy. Last reactions do not produce radio isotopic matters (no neutrons).

Reaction D + D has the other distinct advantages:
1. One produces the protons which energy can be converted directly to electric energy.
2. One produces the tritium which is expensive and may be used for thermonuclear reaction.
3. One produces less and low energy neutrons which create radioactive matters.

The other important advantage is using the pellets with compression gas fuel. Let us take a micro-balloon (pellet) having fuel gas with $p_0 = 100$ atm., radius 0.1 cm., temperature 300K. The mass fuel will be 4.19 mg.

The compression rate is $\xi = p / p_0 = 1.44 \cdot 10^8 / 100 = 1.44 \cdot 10^6$. Liner size decreases by $(\xi)^{1/3} = 113$ times. The radius of compressed fuel pellet will be $R = 0.1/113 = 0.88 \cdot 10^{-3}$ cm. The initial density is

$$\rho_0 = \frac{\mu m_p p_0}{k T_0} = \frac{2.5 \cdot 1.67 \cdot 10^{-27} 10^7}{1.38 \cdot 10^{-23} 300} = 10 \frac{kg}{m^3} = 10^{-2} \frac{g}{cm^3}, \qquad (13)$$

and inertial criterion is

$$\rho R = \rho_0 R_0 \xi^{2/3} = 10^{-2} 0.1 \cdot (1.44 \cdot 10^6)^{2/3} = 12.7 > 1. \qquad (14)$$

Criterion is good for compressed fuel D+T, but it is small for fuel D+D. For fuel D+D we must decrease pressure in pellet up 400÷1000 atm or increase diameter (and power) our installation or use the additional heating of fuel by strong electric impulse.

Compressed micro-balloon (pellet) is more comfortable for working because it is unnecessary to store the fuel at lower temperature.

Estimation of other parameters the Cumulative-Rocket AB Reactor.

1. *Thermonuclear energy.* One mg (10^{-6} kg) of thermonuclear fuel D+T has energy:
Number of nucleus:

$$n_1 = \frac{M}{\mu m_p} = \frac{10^{-6}}{2.5 \cdot 1.67 \cdot 10^{-27}} = 2.4 \cdot 10^{20}$$

(15)

One pair of nuclear D+T produces energy $E_1 = 17.6$ MeV. The n_1 nuclear particles contain the energy

$$E = 0.5 \, n_1 E_1 = 0.5 \cdot 2.4 \cdot 10^{20} 17.6 \cdot 10^6 = 21.1 \cdot 10^{26} \, eV = 21.1 \cdot 10^{26} 1.6 \cdot 10^{-19} = 3.38 \cdot 10^8 \text{ J}$$

(16)

If coefficient efficiency of the Cumulative AB Reactor is $\eta = 0.3$, one mg of fuel produces the energy 100 millions joules. If we make one explosion per sec, installation has the power of 100 million watts. The part of this energy will be produced inside fuel micro-capsule fuel pellet (3.5 MeV from ^{4}He, $E = 6.72 \cdot 10^7$J) the most of energy (14.1 MeV from neutrons) will be produced into the big containment sphere.

Conventional coefficient of nuclear reactor efficiency is about $0.3 \div 0.5$, the steam (gas) turbine is about 0.9.

2. *Energy is delivered by piston to fuel capsule* is $E = mV^2/2$. For $m = 5$ mg, piston speed $V = 2 \cdot 10^4$ m/s final piston energy is $E = 2 \cdot 10^3$ J. That is less then typical energy $20 \div 50$ kJ delivered by laser installation. But laser energy is spent in vaporizing the cover of the fuel pellet and only small part as shock wave reaches the center of fuel pellet mass. In Cumulative AB Reactor, all piston energy passes directly into the target fuel pellet. The piston energy is easy to increase up 20 kJ by increasing the layer/piston (see Ch.5) mass and piston speed (also by using more explosive). The layer/piston mass hinders the fuel micro-balloon and increases the nuclear reaction time in many times.

Part of this energy will be used for ionization of the fuel. One mg of fuel, for its ionization, requests $E = n_1 \cdot 13.6$ eV = 522 J, compression of solid fuel about $E = 624$ J, compression of gas fuel from $p = 100$ atm $E = 105$ J. That is a part of the derived piston energy.

3. *Reaction of explosive* TNT is $2C_7H_5N_3O_6 = 3N_2 + 5H_2O + 7CO + 7C$.

4. *Estimation of pressure and temperature after nuclear explosion.*

Let us to find the pressure and temperature after thermonuclear explosive the one mg fuel D+T.
Number of nuclear particles in sphere 1 m^3 is

$$n_n = \frac{M}{\mu m_p} = \frac{10^{-6}}{2.5 \cdot 1.67 \cdot 10^{-27}} = 2.4 \cdot 10^{20} \quad \frac{1}{m^3}$$

(17)

Full thermonuclear energy

$$E_n = 0.5 n_n E_1 = 0.5 \cdot 2.4 \cdot 10^{20} 17.6 \cdot 10^6 = 21.1 \cdot 10^{26} \text{ eV} = 3.38 \cdot 10^8 \text{ J}$$

(18)

Number of air particles with air density $\rho = 1.225$ kg/m^3 in pressure $p = 1$ atm is

$$n_o = \frac{M}{\mu m_p} = \frac{1.225}{28 \cdot 1.67 \cdot 10^{-27}} = 2.6 \cdot 10^{25} \quad \frac{1}{m^3} \, .$$

(19)

If coefficient efficiency of thermonuclear reaction is $\eta = 0.3$ in volume 1 m^3:

$$p = \frac{\eta E_n}{v} = \frac{0.3 \cdot 3.38 \cdot 10^8}{1} = 1 \cdot 10^8 \approx 10^8 \, \frac{N}{m^2} = 1000 \text{ atm}$$

(20)

Total pressure– nuclear explosive together with chemical explosive - is $p \approx 1000$ atm.
Temperature of gas mixture of explosive plus nuclear fuel is

$$T = \frac{p}{(n_0 + n_n)k} = \frac{10^8}{(2.6 \cdot 10^{25} + 2.4 \cdot 10^{20}) \cdot 1.38 \cdot 10^{-23}} = 279 \cdot 10^3 \text{ K}$$

(21)

If we increase the initial pressure into reactor body up 100 atm, that the temperature desreases to 2790K. The same temperature is in a combustion chamber of conventional engine of the internal combustion.

We can use the conventional cooling system.

The same method may be used for estimation of injection water into instellation body or any garbige material in a space ship (or asteroid).

5. *Thickness of sphere cover*. Assume the spherical cover is made from conventional steel having safety tensile stress $\sigma = 50$ kg/mm$^2 = 5 \cdot 10^8$ N/m^2. The full tensile force is $F = \pi r^2 p = 3.14 \cdot 0.5^2 \cdot 10^8 = 0.785 \cdot 10^8$ N. Requested area of steel is $S_r = F/\sigma = 0.785 \cdot 10^8/5 \cdot 10^8 = 0.157$ m^2. The thickness of sphere wall is $\delta = S_r/2\pi r = 0.157/2 \cdot 3.14 \cdot 0.5 = 0.05$ m. Mass of sphere is $M_c \approx \gamma S_s \delta = 7800 \cdot 4.536 \cdot 0.05 = 1769$ kg. Here S_s is average surface of sphere.

If we use the more strong material for sphere wall, for example: 1µm iron whisker having safety tensile stress $\sigma \approx 400$ kg/mm$^2 = 4 \cdot 10^9$ N/m^2, we decrease the sphere's mass by 4 – 8 times. We can also make the sphere wall from composite materials (example: an artificial fiber carbon or glass having safety stress $\sigma \approx 100 \div 150$ kg/mm^2 and density $\gamma = 1500 \div 2700$ kg/m^3).

6. *Cooling the sphere by water*. If explosions are very frequent, we then can decrease the wall or/and gas temperature by injection of the chilled or room-temperature water. The water also protects our installation from high-energy neutrons in other words, it behaves as a shielding materials.

Let us estimate the amount of water which decreases the temperature and pressure of gas (at most steam H_2O) into sphere for magnitudes acceptable for current steam turbines: $T = 400°C = 672$ K. The critical point of water (triple point) is $T = 273°C$, $p = 22$ MPa.

Heating 1 kg water from 20°C to 100°C requests energy $E = C_p \Delta T = 4.19 \cdot 80 = 333$ kJ, evaporation $- r = 2260$ kJ, heating of steam up 400°C - $E = C_p \Delta T = 1.05 \cdot 300 = 315$ kJ. Total amount of water heat energy is $E_w = 333 + 2260 + 315 = 2908$ kJ/kg. Total mass of water for nuclear efficiency $\eta = 1$ equals $M_w = E/E_w = 3.4 \cdot 10^8/2.9 \cdot 10^6 = 117$ kg. For $\eta = 0.3$ $M_w = 35$ kg. The 2 – 3 cm of water thickness protects the instellation fron high energy of neutrons produced by reaction D+T.

Unfortunately, the injection of water before decompressing strongly decreases the efficiency of installation.

7. *Run protons and heavy nuclear particles.* (see p. 27).

7a. *Converting the nuclear energy of AB reactors to electric, mechanical energy or a rocket thrust.*

The best means for converting a Cumulative AB Reactor nuclear energy is magneto hydrodynamic electric generator (MHD-generator) which converts with high efficiency the high temperature and high pressure plasma directly in electric energy. Together with capacitors one can produces continuous electric currency. Impulse work of reactor allows to cool the reactor by injection the cooler (or conventional cooling) and protect the Cumulative AB Reactor installation from very high temperature.

The second way for converting an Cumulative AB Reactor nuclear energy is conventional heat exchanger and gas turbine. As cooler may be used the FLiBe – melted mix of fluoride salts of lithium and beryllium.

The third way is injection of water inside sphere and steam turbine as description over.

8. *Using the Cumulative AB reactor as an impulse space rocket engine.*

There are good prospects (possibility) to use the suggested Cumulative AB Reactor as an impulse rocket engine.

If plasma will flow from sphere to space the average speed V of jet is

$$\text{From} \quad E = \frac{mV^2}{2} \quad \text{we get} \quad V = \left(\frac{2E}{m}\right)^{1/2} = \left(\frac{2 \cdot 10^8}{5 \cdot 10^{-3}}\right)^{1/2} = 4 \cdot 10^5 \ \frac{m}{s}. \quad (22)$$

Here E is nuclear energy in one impulse of one mg nuclear fuel, J; m is the mass injected to outer space (fuel catridge together with conventional explosive), kg.

Received speed $V = 400$ km/s is in many times more than a current exhaust chemical speed 3 km/s. If of space apparatus has mass $m_2 = 1$ tonn the ship speed changes in $V_2 = (m/m_2) V_1 = 2$ m/s in one impulse. If we spend 10 kg of fuel cartridges, the apparatus get speed 10 km/s.

More importantly, the next possibility is of the rocket powered by the Cumulative AB Reactor. Any matter from any planets, asteroids, space body may be used as fuel used for increasing the derivation of impulses. For example, assume the captured solid object moving through space is composed of some water, and we filled rocket tanks using that mined planet, comet or asteroid water. From (35-2) and Law of equal impulse we have from every impulse

$$V_1 = (2Em_1)^{1/2} / m_2 = (2 \cdot 10^8 \cdot 16)^{1/2} / 10^3 = 56.6 \quad m/s \quad . \tag{23}$$

Here V_1 is add speed m_1 mass jet kg, $m_1 = 16$ kg of water; m_2 is mass of space apparatus.

9. *Estimation of the fast neutron penetration*

$$l = 1/n\sigma, \tag{24}$$

Where l is path of penetration, cm; n is density of material, $1/cm^3$; $\sigma = 10^{-24}$ cm^2 is cross section of the nuclear. For steel $l = 12$ cm, for compressed air up 100 atm the $l = 410$ cm.

10. *Requested thickness of the spherical shell* is

$$\frac{D}{d} = \left(\frac{p}{\sigma} + 1 \right)^{0.5}, \tag{25}$$

Where D is outer diameter of spherical shell, d is inner diameter of spherical shell, p is pressure, atm; σ is safety tensail stress kg/cm^2. Exemple, if $p = 10$ kg/mm^2, $\sigma = 50$ kg/mm^2, then $D/d \approx 1.1$.

Detail Estimation of Cumulative-Rocket reactors for transportation engine

1. **Estimation of nuclear energy (power).** Let us make more detail estimation the Cumulative and Impulse reactors for engine of transport vinicle having the fuel pellet 0.1 mg ($M_f = 10^{-7}$ kg) with fuel D+T or D+D. The Impulse reactor has pressure into pellet 300 atm.

Estimation of energy (power) this D+T pellet if the coefficient efficiency is $\eta = 0.5$.

The couple nuclei T+D produses nuclear energy $E_1 = 17.6$ MeV

Number N_f of nucleis in pellet is:

$$N = \frac{M_f}{\mu m_p} = \frac{10^{-7}}{2.5 \cdot 1.67 \cdot 10^{-27}} = 2.4 \cdot 10^{19} \quad .$$

Here μ is averege moliar mass of D+T; m_p is mass of proton, kg.

The nucler energy of 1 mg D+T fuel in 1 Hz is

$$E = 0.5 E_1 N \eta = 0.5 \cdot 17.6 \cdot 10^6 \cdot 1.6 \cdot 10^{-19} 2.4 \cdot 10^{19} \cdot 0.5 = 16.9 \cdot 10^6 \approx 17 \; MJ / Hz \quad .$$

That is power energy of the 2-5 power aviation turbo-engines. If one cycle in second (1 Hz) is not enough, we can decrease the friquency. The piston engine has up 50-70 revolution per second, the high speed aviation gun up 30 shots in second.

If we use the D+D fuel having single energy $E_1 = 3.65$ MeV , $\mu = 2$, the nucler energy is aproximetely in 5 times less because E_1 is less.

2. **Size of cartridge and pellet**. Let us estimate the size of the cumulative cartrige for mass the explosive TNT $M_e = 1$ gram ($M_e = 10^{-3}$ kg, energy $E_e = 4.2$ NJ, density $\rho = 1650$ kg/m^3) and internel diameter cartridge is $d = 10$ mm. The thickness δ of explosive is:

$$\delta \approx \frac{M_e}{4\pi r^2 \rho} = \frac{1}{4 \cdot 3.14 \cdot 0.5^2 \cdot 1.65} = 0.2 \; cm$$

Outer diameter of cartidge for safety tensile stress 100 kg/mm^2 is D = 16 mm.

Let us estimate the compressed **pellet** having gas mass $M = 10^{-7}$ kg, pressure $p = 300$ atm $= 3 \cdot 10^7$ N/m^2 and T = 300 K. Specific denfity the gas D+D, D+T in compession $p = 1$ atm is $\rho_o = 0.1$ kg/m^3,atm. The internel radius of gas pellet is:

$$r = \left(\frac{3M}{4\pi p \rho_0}\right)^{1/3} = \left(\frac{3 \cdot 10^{-7}}{4 \cdot 3.14 \cdot 3 \cdot 10^2 0.1}\right)^{1/3} \approx 0.926 \cdot 10^{-3} m \approx 1\, mm$$

The reletive outer diameter of pellet for pressure $p = 3$ kg/mm^2 and the safety tensile stress of the pellet cover $\sigma = 50$ kg/mm^2 with according (38-2) is

$$\frac{D}{d} = \left(\frac{p}{\sigma}+1\right)^{0.5} = \left(\frac{3}{50}+1\right)^{1/2} = 1.03$$

Nuclear prosses in to pellet. After cumulative explosive into cartridge the density of fuel D+T into pellet after cumulative compressing is

$$n = \frac{3p}{\mu m_p V_p^2} = \frac{3 \cdot 1.36 \cdot 10^{13}}{2.5 \cdot 1.67 \cdot 10^{-27}(18 \cdot 10^3)^2} = 3 \cdot 10^{25} cm^{-3},$$

Where p is pressure after cumulative compressing, N/m^2, V_p is final speed of piston, m/s , m_p is mass of proton, kg. Density of D+D fuel is $n = 3.75 \cdot 10^{25}$ cm^{-3}.
Time of fuel combustion for $T = 15$ keV is

For D + T $\quad t = \dfrac{0.5\eta E_1}{5.6 \cdot 10^{-13} n <\sigma v>} = \dfrac{0.5 \cdot 0.5 \cdot 2.82 \cdot 10^{-12}}{5.6 \cdot 10^{-13} 3 \cdot 10^{25} 2.65 \cdot 10^{-16}} = 1.58 \cdot 10^{-10} \, s,$

For D + D $\quad t = \dfrac{0.5\eta E_1}{3.3 \cdot 10^{-13} n <\sigma v>} = \dfrac{0.5 \cdot 0.5 \cdot 0.58 \cdot 10^{-12}}{5.6 \cdot 10^{-13} 3.75 \cdot 10^{25} 3.2 \cdot 10^{-18}} = 3.7 \cdot 10^{-9} \, s,$

Where η is coefficient efficiency, E_1 is energy couple nucleis (for D+T $E_1 = 17.6$MeV$\cdot 1.6 \cdot 10^{-19} = 2.82 \cdot 10^{-12}$ J; for D+D $E_1 = 3.65$MeV$= 0.58 \cdot 10^{-12}$ J). Here we used Eq. (8-1) and Table 5.
For primary compressed gas fuel pellet $p = 300$ atm without cumulative compressing, the density of the fuel gas into pellet for $T = 300$K is

$$n = \frac{p}{kT} = \frac{3 \cdot 10^7}{1.38 \cdot 10^{-23} 300} = 2.25 \cdot 10^{21} cm^{-3},$$

Where $k = 1.38 \cdot 10^{-23}$ - is Boltzmann constant.
The time of nucler fuel combustion for $T = 15$ keV is

For D + T $\quad t = \dfrac{0.5\eta E_1}{5.6 \cdot 10^{-13} n <\sigma v>} = \dfrac{0.5 \cdot 0.5 \cdot 2.82 \cdot 10^{-12}}{5.6 \cdot 10^{-13} 2.25 \cdot 10^{21} 2.65 \cdot 10^{-16}} = 2.1 \cdot 10^{-6} \, s,$

For D + D $\quad t = \dfrac{0.5\eta E_1}{3.3 \cdot 10^{-13} n <\sigma v>} = \dfrac{0.5 \cdot 0.5 \cdot 0.58 \cdot 10^{-12}}{5.6 \cdot 10^{-13} 2.25 \cdot 10^{21} 3.2 \cdot 10^{-18}} = 6.14 \cdot 10^{-5} \, s,$

As you see, the combustion time significantly is increased but it is enough for reaction. We can decrease it if we increases density of fuel.
Estimation of electric condenser. For heating of fuel we use the short strong electric impulse. For impulse the electric conderser may be used. Let us to estimate the condenser parameters for getting the fuel temperature $T = 15$ keV.
If fuel mass is $M = 1$ mg $= 10^{-7}$ kg, the number of nucleis for D+T is

$$N = \frac{M}{\mu m_p} = \frac{10^{-7}}{2.5 \cdot 1.67 \cdot 10^{-27}} = 2.4 \cdot 10^{19}.$$

For D+D the $N = 3 \cdot 10^{19}$.
The energy is needed for heating the fuel D+T up $T = 15$ keV is

$$W = NT \cdot 1.6 \cdot 10^{-19} = 2.4 \cdot 10^{19} 15 \cdot 10^3 1.6 \cdot 10^{-19} \approx 60\, kJ.$$

For heating D+D fuel is W = 72 kJ.

The minimal specific weight of conventional conductor according [7] p. 368 is $\gamma = 2$ kJ/kg.

Consequently, the requeted mass of condenser is about 30 – 36 kg. But if we can use the advanced supercapacitor ($\gamma = 10$ kJ/kg) or ultracapacitor ($\gamma = 20$ kJ/kg) or capacitor EEStor, having claimed capasity $\gamma = 1000$ kJ/kg, we can decreased the capacitor mass. In any case, the capacitor mass is small part of themonuclear engine.

Estimation of capacitor discharge.

a) Need condenser *after cumulative compressing* of pellet for heating fuel up $T = 15$ keV.

Assume the initial temperature of cumulative compressed gas fuel is $T = 3.14$ eV, mass of fuel $M = 10^{-7}$ kg, initial pressure $p = 300$ atm, initial diameter of pellet $d = 0.2$ cm..

The specific electric Spitzer resistance of plasma is

$$\rho = \eta_\perp = 1{,}03 \cdot 10^{-2} Z \ln \Lambda \cdot T^{-3/2}.$$

Where Z is rate of charge, $\ln\Lambda = (5 \div 15)$ is Columbus logariphm. For $\ln\Lambda = 10$ we have

$$\rho = 1.03 \cdot 10^{-2} 1 \cdot 10 / 3.14^{3/2} = 1.85 \cdot 10^{-3}\ \Omega \cdot cm.$$

Diameter of the cumulative compressed pellet having initial gas pressure $p_o = 300$ atm and $l_0 = 0.2$ cm is

$$l = l_0 \left(\frac{p_0}{p}\right)^{1/3} = 0.2 \left(\frac{300}{3.24 \cdot 10^7}\right)^{1/3} = 4.2 \cdot 10^{-3}\ cm, \quad s = \frac{3.14}{4} l^2 = 1.4 \cdot 10^{-5}\ cm^2$$

Electric resistance is

$$R = \rho \frac{l}{s} = 1.85 \cdot 10^{-3} \frac{4.2 \cdot 10^{-3}}{1.4 \cdot 10^{-5}} = 0.555\ \Omega \cdot cm$$

Where l is diameter of pellet, cm; s is cross-section area, sm^2.

Needed inisial voltage and currency of condenser for time of recharde $t = 10^{-5}$ second is

$$U = \left(\frac{RW}{t}\right)^{1/2} = \left(\frac{0.555 \cdot 6 \cdot 10^4}{10^{-5}}\right)^{1/2} = 57.7\ kV, \quad I = \left(\frac{W}{Rt}\right)^{1/2} = \left(\frac{6 \cdot 10^4}{0.555 \cdot 10^{-5}}\right)^{1/2} = 104\ kA.$$

Capacity of condenser

$$C = \frac{t}{R} = \frac{10^{-5}}{0.555} = 18 \cdot 10^{-6}\ F$$

b) Need condenser *without* cumulative compressing of pellet. Initial data: Initial temperature is $T = 0.1$eV, mass of fuel $M = 10^{-7}$ kg, pressire $p = 300$ atm, diameter of pellet $d = 0.2$ cm, final temperature T = 15 keV.

$$\rho = 1.03 \cdot 10^{-2} 1 \cdot 10 / 0.1^{3/2} = 3.16\ \Omega \cdot cm.$$

Cross-section area of the pellet having fuel gas pressure $p_o = 300$ atm and diameter of pellet $d_o = l_o = 0.2$ cm is

$$s = \pi d_0^2 / 4 = 0.0314\ cm^2 .$$

Electric resistance is

$$R = \rho \frac{l_0}{s} = 3.16 \frac{0.2}{0.0314} = 20\ \Omega \cdot cm$$

Where l is diameter of pellet, cm; s is cross-section area, sm^2.

Needed initial voltage and currency of condenser for time of recharge $t = 10^{-5}$ second is

$$U = \left(\frac{RW}{t}\right)^{1/2} = \left(\frac{20 \cdot 6 \cdot 10^4}{10^{-5}}\right)^{1/2} = 346\ kV, \quad I = \left(\frac{W}{Rt}\right)^{1/2} = \left(\frac{6 \cdot 10^4}{20 \cdot 10^{-5}}\right)^{1/2} = 54.8\ kA$$

Capacity of condenser

$$C = \frac{t}{R} = \frac{10^{-5}}{20} = 7 \cdot 10^{-7} \ F \ .$$

The specific energy weight γ_c [J/kg] of the condenser my be estimate by formules

$$\gamma_c = \frac{\varepsilon_0 \varepsilon E_q^2}{\gamma} \ .$$

Where $\varepsilon_0 = 8.85 \cdot 10^{-12}$ F/m is electric constant; $\varepsilon \approx 3$ dielectric constant; $E_q \approx 160 \div 640$ MV/m is safety electric stress of isolator; $\gamma \approx 1000 \div 3000$ kg/m^3 is spesific weight of isolator.

Initial magnetic pressure from charged currency.

a) Pellet having cumulative compressing has initial currency $I = 10^4$ kA, radius of pellet after compressing $r = 2.1 \cdot 10^{-5}$ m has magnetic intensity H and magnetic pressure p:

$$H = \frac{I}{2\pi r} = \frac{104 \cdot 10^3}{2 \cdot 3.14 \cdot 2.1 \cdot 10^{-5}} = 7.9 \cdot 10^8 \ \frac{A}{m},$$

$$p = \frac{\mu_0 H^2}{2} = \frac{4\pi \cdot 10^{-7} (7.9 \cdot 10^8)^2}{2} = 4 \cdot 10^{11} \ \frac{N}{m^2} = 4 \cdot 10^6 \ atm \ .$$

That is closed to piston pressure $3.4 \cdot 10^7$ atm.

b) Without cumulative pressure $H = 0.87 \cdot 10^7$ A/m and $p = 477$ atm.

Table 7. Properties of some material suitable for the offer installation.

Material	Tensile strength kg/mm²	Density g/cm³	Fibers	Tensile strength kg/mm²	Density g/cm³
Steel A514	76	7.8	S-Glass	471	2.48
Aluminum alloy	45.5	2.7	Basalt fiber	484	2.7
Titanium alloy	90	4.51	Carbon fiber	565	1,75
Steel Piano wire	220-248	7.8	Carbon nanotubes	6200	1.34

Source: [7] p.370.

Conclusion

The author offers a new small cheap cumulative and impulse inertial thermonuclear reactors, which increases the pressure by chemical explosive and temperature by electric charge. Cumulative and Impulse AB Reactor, herein offered by its originator, contains several innovations and inventions.

Main of them is using a moved explosive, which allows to accelerate the special piston/layer to very high speed (more than 20 km/s) which (as it is shown by computations) compresses the fuel capsule in million times and heating up the million degrees of temperature. The second main innovation is the additional heating the fuel pellet by electric impulse to up temperature in 15keV and more (hundreds millions of degrees). Important innovtion is compressed pellet at room temperature, installation for electric and mechanical energy and thermonuclear rocket.

The offered reactor is small, cheap, may be used for cheap electricity, as engine for Earth transportation (train, truck, sea-going ships, aircraft), for space apparatus and for producing small and cheap and powerful weapons. Closed ideas are in [1]-[9].

92

RÉFÉRENCES

(READER CAN FIND PART OF THESE ARTICLES IN WEBS:
HTTPS://ARCHIVE.ORG/DETAILS/LIST5OFBOLONKINPUBLICATIONS,HTTP://BOLONKIN.NAROD.RU/P65.HTM,
HTTP://ARXIV.ORG/FIND/ALL/1/AU:+BOLONKIN/0/1/0/ALL/0/1, HTTP://VIXRA.ORG).

[1] Bolonkin, A.A., "Inexpensive Mini Thermonuclear Reactor". International Journal of Advanced Engineering Applications, Vol.1, Iss.6, pp.62-77 (2012) . http://viXra.org/abs/1305.0046 http://archive.org/details/InexpensiveMiniThermonuclearReactor,

[2] Bolonkin A.A. , Cumulative Thermonuclear AB-Reactor.. Vixra 7 8 2015, http://viXra.org/abs/1507.0053 https://archive.org/details/ArticleCumulativeReactorFinalAfterCathAndOlga7716

[3] Bolonkin A.A., Ultra-Cold Thermonuclear Synthesis: Criterion of Cold Fusion. 7 18 2015. http://viXra.org/abs/1507.0158, GSJornal: http://gsjournal.net/Science- Journals/%7B$cat_name%7D/View/6140

[4] Bolonkin, A.A., "Non Rocket Space Launch and Flight". Elsevier, 2005. 488 pgs. ISBN-13: 978-0-08044-731-5, ISBN-10: 0-080-44731-7 . http://vixra.org/abs/1504.0011 v4,

[5] Bolonkin, A.A., "New Concepts, Ideas, Innovations in Aerospace, Technology and the Human Sciences", NOVA, 2006, 510 pgs. ISBN-13: 978-1-60021-787-6. http://viXra.org/abs/1309.0193,

[6] Bolonkin, A.A., Femtotechnologies and Revolutionary Projects. Lambert, USA, 2011. 538 p. 16 Mb. ISBN:978-3-8473-0839-0.http://viXra.org/abs/1309.0191,

[7] Bolonkin, A.A., Innovations and New Technologies (v2).Lulu, 2014. 465 pgs. 10.5 Mb, ISBN: 978-1-312-62280-7. https://archive.org/details/Book5InnovationsAndNewTechnologiesv2102014/

[8] Bolonkin, A.A., Stability and Production Super-Strong AB Matter. International Journal of Advanced Engineering Applications. 3-1-3, February 2014, pp.18-33. http://fragrancejournals.com/wp-content/uploads/2013/03/IJAEA-3-1-3.pdf The General Science Journal, November, 2013, #5244.

[9] Bolonkin, A.A., Converting of Any Matter to Nuclear Energy by AB-Generator. American Journal of Engineering and Applied Science, Vol. 2, #4, 2009, pp.683-693. http://viXra.org/abs/1309.0200,

[10] Kikoin I.K., Tables of Physical Values, Moscow, Atomizdat, 1975 (Russian).

[11] AIP, Physics Desk Reference, 3-rd Edition. Springer, AIP PRESS.

7 July 2015

Expenses in R&D nuclear energy

Organization:	Technology	Found/End	Participants	Est. Amount:
EMCC	Polywell-ish	1987 / -	22-30	$ 41,728,353
CSI	Polywell-ish	2010 / -	4	$ 120,000
Radiant Matter (Guess)	Polywell-ish	2010 / -	2	$ 30,000
Lockheed Martin (Guess)	Polywell-ish	2007 / -	6-8	5 to 10 Million
Iran Polywell (est)	Polywell-ish	2010 / -	8	$ 8,000,000
Pomethius Fusion Perfection	Polywell-ish	2008 / 2013	2	$ 40,000
SHINE Medical	Fusor-ish	2010 / -	15-25	$ 134,000,000
Phenoix Nuclear Labs	Fusor-ish	2005 / -	25-30	$ 34,874,182
Tri Alpha Energy (est)	Field Reverse	1998 / -	150	$ 150,000,000
MSNW	Field Reverse	1994 / -	7-11	$ 5,000,000
Helion	Field Reverse	2010 / -	4-11	$ 6,500,000
LPPX	Focus Fusion	1974 / -	7	$ 4,480,279
General Fusion	Mag. Target	2002 / -	60	$ 55,000,000
FP Generation	Beam Fusion	2009 / 2011	6	$ 3,000,000
		Total:	334	$ 450,272,814

Chapter 9
Multi-Reflection Thermonuclear Reactors*

ABSTRACT

The author offers several innovations that he first suggested publicly early in 1983 for the AB multi-reflex engine, space propulsion, getting energy from plasma, etc. (see: A. Bolonkin, Non-Rocket Space Launch and Flight, Elsevier, London, 2006, Chapters 12, 3A). It is the micro-thermonuclear Reactors. That is new micro-thermonuclear reactor with very small fuel pellet that uses plasma confinement generated by multi-reflection of laser beam or its own magnetic field. The Lawson criterion increases by hundreds of times. The author also suggests a new method of heating the power-making fuel pellet by outer electric current as well as new direct method of transformation of ion kinetic energy into harvestable electricity. These offered innovations dramatically decrease the size, weight and cost of thermonuclear reactor, installation, propulsion system and electric generator. Non-industrial countries can produce these researches and constructions. Currently, the author is researching the efficiency of these innovations for two types of the micro-thermonuclear reactors: multi-reflection reactor (ICF) and self-magnetic reactor (MCF).

Keywords: *Small thermonuclear reactor, Multi-reflex AB-thermonuclear reactor, Self-magnetic AB-thermonuclear reactor, aerospace thermonuclear engine.*

----

* Presented as Bolonkin's paper AIAA-2006-8104 in 14th Space Plane and Hypersonic Systems Conference, 6-8 November, 2006, USA.

INNOVATION

As you can see in the Equation for thermonuclear reaction (reaction's "ignition", Ch,1) it is necessary to rapidly and greatly increase the target-enveloping temperature, the density of target proper and to shorten the time of the operation in order to keep the fuel in these precisely induced conditions. In ICF the density of plasma is very high (10^{28} m^{-3}, it increases in 20-30 times in target), the temperature reaches tens of millions $^\circ$K, but time is measured in nanoseconds. As a result, the Lawson criterion is tens to hundreds of times lower than is required. In a tokomak, the time is mere parts of second and the ambient temperature is tens of millions of degrees, but density of plasma is very small (10^{20} m^{-3}). The Lawson criterion is also tens to hundreds of times lower than needed.

The author offers some innovations and names these reactors as AB-reactors.. The main innovations are below.

Multi-reflect reactor (MRR). The first innovation suggested early in 1983 [14] and developed later in [1]-[26] for multi-reflex engine and space propulsion. Conventional ICF has conventional inside surface of the combustion chamber. This surface absorbs part of the heat radiation emanating from the pellet and plasma, the rest of the radiation reflects in all directions and is also absorbed by walls of combustion chamber. As result the target loses energy expensively delivered by lasers. This loss is so huge that we need very powerful lasers and we cannot efficiently heat the target to reach ignition temperature (Lawson's criterion). In all current ICF installations this loss is tens of times more than acceptable.

The innovative ICF has, on the inside surface of combustion chamber, a covering of small Prism Reflectors (PR) (figs. 1, 2) (or multi-layer reflector. Note: Multi-layer reflector can only reflect the laser beam). The system of prism reflectors has great advantages in comparison with conventional mirror and especially with conventional combustion chamber. The advantages are listed:

(1) The prism reflector has very high efficiency. The coefficient of its radiate absorption is less about million times the rate of the conventional mirror.

(2) The prism mirror reflects the radiation in widely diapason of continuous spectrum. A conventional mirror reflects the radiation only in narrow diapason of continuous spectrum. That means that any conventional mirror has big absorption of radiation energy even if it has high reflectivity (up 99%) in narrow interval of the continuous spectrum. The prism reflector allows to use the thermal plasma radiation.

(3) The prism reflector bounces the heat radiation exactly to a point where heat beam comes up, even if it has defect at position. The conventional mirror having small defect in position (or the pellet is not located exactly in center of sphere) destroys the pellet.

(4) According with Point 3 above, the prism reflector may be used in cylindrical (toroidal) camera (Figure5) (tokomak, stellarator). A conventional mirror cannot be employed because reflected ray will be sent in the other direction.

(5) The prism reflector can uniformly distribute the beam energy in pellet surface. The small spherical plasma pellet reflected the scattered radiation. That means the laser beam after the first reflection reflects on semi-sphere, after two reflections that presses on full sphere, after 3 - 4 reflection the pressure is almost uniformly. That decreases a number of needed laser beams, simplify, and decreases cost of laser installation.

In particular, this innovation may be used in already built current reactors for their improvement.
Self-Magnetic reactor (SMR) (Figures 3, 4). The magnetic pressure is proportional to the inverse value of electric conductor diameter. (The conventional magnetic reactor has a diameter of plasma flex some meters). The high temperature plasma has excellent conductivity which does not depend from plasma density. If the diameter is decreased to 0.1 mm and electric currency is high, the magnetic pressure is increased by hundreds or thousand times and that can keep the high-density plasma. Thus, the plasma is confined by self-generated magnetic field (and by pinch-effect) and it does not need in powerful outer magnetic field created very complex, high cost super-conductivity system! This innovation in MCF is dramatically decreasing the size of reaction zone and using of gaseous compression fuel pellet (micro-capsule) in magnetic confinement reactor.

8

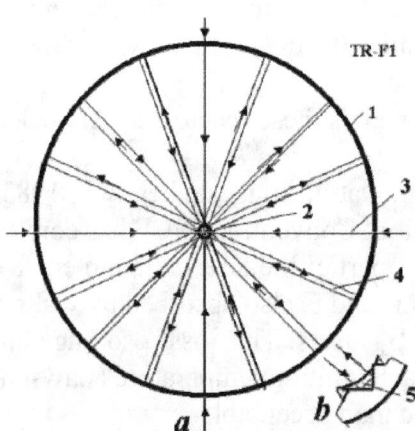

Figure 1. Multi-reflex Reactor. (*a*) Cross-section of ICF; (*b*) Cross-section of spherical combustion chamber and prism reflectors. Notations: 1 - spherical shell; 2 - target (pellet); 3- ignition laser beams; 4 - reflected laser beams; 5 - prism reflector.

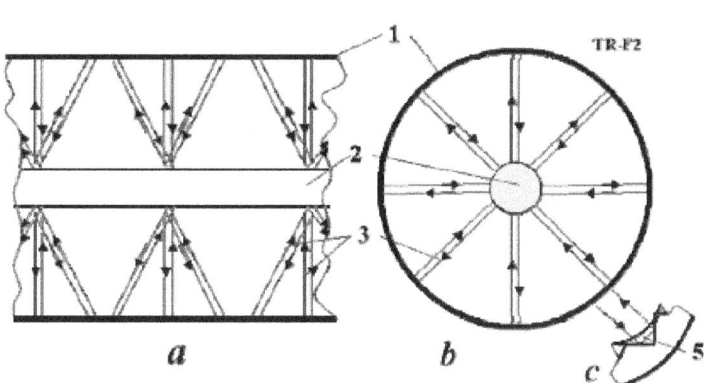

Figure 2. Multi-reflective in cylindrical tube chamber or tokomak. (*a*) Cross-section along long axis. (*b*) Cross-section along transverse axis; (c) Cross-section of toroidal or cylindrical combustion and prism reflectors. Notations: 1 - combustion chamber; 2 - plasma (fuel capsule); 3 - reflected laser beams; 5 - prism reflectors.

The other innovation in SMR is uses the electric current (electric impulse) for initial heating of microcapsule targets. That means we don't need a large, very complex and expensive laser (or ion beam) system for inertial confinement reactor or induce system in magnetic confinement reactor. That is possible in special design of the fuel microcapsule. The energy for heating of the microcapsule to thermonuclear temperature is small and conventional condensers may be used for it.

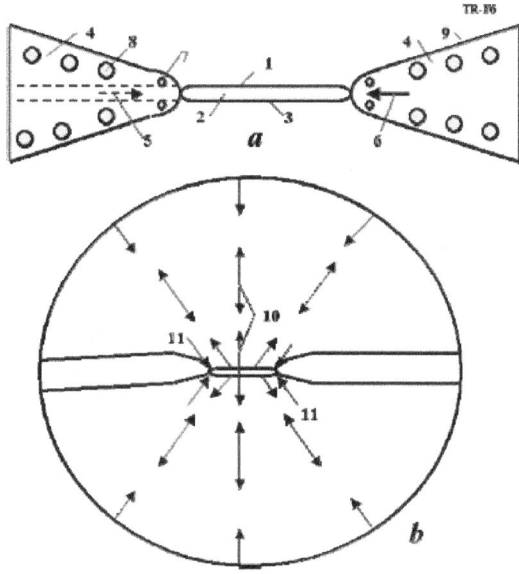

Figure 3. Micro AB-reactor with self-magnetic confinement and radiation support of plasma. (a) – fuel micro-capsule and electrodes; (b) - Reflective camera. Notation: 1 - micro fuel capsule; 2 - thermonuclear fuel into capsule; 3 - capsule shell and covering; 4 - electrodes; 5 - feeding of capsule; 6 - electric currency (electron injector); 7 - magnetic stopper; 8 - cooling system; 9 - thermo protection; 10 - radiation; 11 - additional radiation pressure to fuel capsule ends.

The self-magnetic reactor uses very small capsule diameter when the magnetic intensity is very high. The magnetic intensity has good distribution (decreases to plasma center, Figure 4c) and magnetic pressure is big (it is enough to keep the kinetic plasma pressure which is not so much for low density plasma).

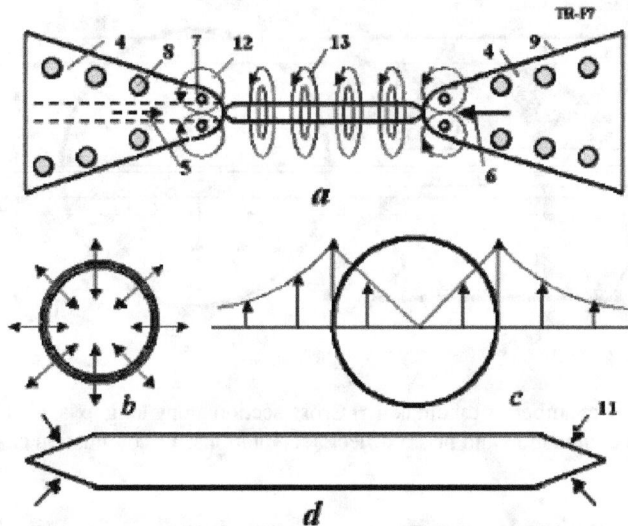

Figure 4. Micro AB-thermonuclear reactor with self-magnetic confinement. (*a*) Self-magnetic field around fuel capsule, (*b*) - Explosion initial compression, (*c*) Magnetic intensity distribution in cross-section plasma flex, (*d*) Fuel capsule. Notations: 4 - electrodes; 5 - feeding of capsule; 6 - electric currency (electron injector); 7 - magnetic stopper; 8 - cooling system; 9 - thermo protection; 11 - additional end capsule pressure of self-magnetic field and radiation; 12 - magnetic corks. 13 - self-magnetic field of capsule (and pinch-effect).

The some innovations are magnetic and radiation stoppers (confinements) of plasma in ends of the fuel capsule. It is suggested two methods:

(1) The capsule has conic ends (Figure 4d). The capsule radius decreases at ends. That means the magnetic intensity will increase at the capsule ends (up 10 and more times). They will work as magnetic mirror (plasma stopper) (Figure 4).

(2) Our magnetic field has other direction and form then it is in conventional tokomak. In tokomak the magnetic lines are parallel to a toroid (or cylinder) axis. In our SMR the magnetic lines are circles around cylindrical capsule. That means there is an axis magnetic force which put obstacles a heat transfer from plasma to electric electrodes. That works as radial and axial magnetic stopper.

There are others innovations which reader can apprehend in comparison the offered micro AB-reactors and current and under construction reactors.

These innovations decrease the size, weight and the monetary cost of thermonuclear reactors by thousands of times and allow the widespread future construction of thermonuclear electric stations, engines, and space propulsions.

The offered self-magnetic reactor has the following differences in comparison with Z -machine of Sandia Laboratory (USA). Z-machine used a set of very fine tungsten wires running around the fuel would be "dumped" with the current instead. The wires would quickly vaporize into a plasma, which is conductive, and the current flow would then cause the plasma to pinch. The key difference is that the plasma would not be the fuel, as in our SMR, but used solely to generate very high-energy X-rays as the metal plasma compressed and heated. The X-rays would compress a tiny fuel cylinder containing deuterium-tritium mix, in the same fashion that the X-rays generated from a nuclear bomb compress the fuel load in an H-bomb. The superpower X-ray output pulse (up 2.7 megajouls!) generated by heavy tungsten metal plasma ($_{184}^{74}W$) is very danger for people. In additional, the powerful fluctuation in the magnetic field (an "electromagnetic pulse") also generates strong electric current in all of the metallic objects in the room and demises electronic devices. In our machine the small fuel cylinder has a thin conductive layer from light metall. The capsule can also have conductivity filaments into fuel. They help to produce initial heating of fuel plasma (up $10^4 \div 10^5$ °K) and initial the plasma compression (rocket and/or inertial). The father plasma heating and confinement is produced by voltage curve of Figure 13 which create self-magnetic field equal (or more) plasma gas pressure.

Summary. This work offers two types of micro-AB-thermonuclear reactors: by multi-reflex radiation confinement of plasma and self-magnetic confinement. They can use high and low density fuel (compressed gas or liquid/frizzed fuel) and they can work in pulse or continuous regimes.

The offered micro-AB-reactor with self-magnetic confinement includes: micro fuel capsule with compressed gaseous or liquid (frizzed) fuel; two electric electrodes, and a combustion chamber. Internal surface of combustion chamber is covered by prism or multi-layer reflectors.

The capsule contains thermonuclear fuel (it conventionally has two component, example D + T), and conducting capsule shell. Fuel may be composed by conducting fiber for quick heating. The capsule has the conic ends.

The electric electrodes have windings for creating magnetic locks, canals for feeding of fuel capsule (or injector for liquid fuel), and electron injector (electric currency). Last may be electron (currency) emitter. Electrodes also contain a cooling system and thermo-protection.

The suggested reactor works the following way. The strong impulse electric current passes through capsule. The capsule shell explodes, creating an initial plasma flux and compressed, heating, and creating initial fuel plasma fuse. The plasma radiation erupts and part of them returns and compresses the plasma, helping the electric current to heat the plasma to its ignition temperature.

We spoke about micro AB-reactors. But it does not mean that power of them is small. In next articles I will discuss the methods for transformation and utilization of the thermonuclear energy into other types of energy and propulsion. These completed research show the power of micro AB-reactor can reach some thousands of kW.

ESTIMATIONS OF OFFERED REACTOR

We consider two new Micro-AB-Reactors having the innovative: multi-reflect radiation and self-magnetic confinement features.

In multi-reflect radiation confinement of AB-Reactor the offered innovation is the special prisms, a high reflectivity mirror that returns the laser beam exactly to its point of origination. As a result, the all energy absorbs by plasma, the laser radiation multi-times presses the plasma and impedes it or, at least, it does not allow its expansion. The plasma has high reflectivity and this press effect may be increased hundreds to thousands of times. Practically speaking, we are weakly limited and can use the cheap and solid fuel.

The uniformly heating of target by laser beam is very big and important problem in ICF. The non-equal rocket forces on target shell destroy the capsule before thermonuclear ignition. If the first ICF reactor had some laser beams, the second generation had 10 laser beams (NOVA), the third generation has 60 beams (OMEGA), and the next generation will have 192 beams (NIF). All laser beams must be equal and work in coordination - that is complex, difficult and expensive problem. The prism reflector is easy designed such the reflected beam runs round the target and presses it uniformly from all sides after 2 - 4 reflections.

The second innovation is the special form microcapsule that is filled by compression gas or liquid (frozen) fuel.

In self-magnetic confinement Micro-AB-Reactor main innovation is super thin microcapsule and electric heating which produce high-intensity magnetic field, keeping the plasma pressure and conic (or close to conic) ends of ampoule capsule which work as plasma stoppers. The important innovation is the using an electric currency for straight heating of capsule. The magnetic lines in our reactor are circles located into and around plasma channel. The magnetic intensity increases from central axis to maximal plasma radius. That pushes plasma into center of plasma channel. In the ends of plasma channel the magnetic forces put obstacles in plasma leakage.

For estimation possibilities of these innovations in the first AB-reactor we compute the multi-reflection pressure, the condition of plasma reflection and compare them with dynamic pressure of plasma. In the second AB-reactor we consider the equilibrium the magnetic and kinetic pressures.

Capsules. For multi-reflex conformation is more suitable the *spherical capsule*. Let us consider the gas compressed fuel capsule. The shell thickness and relative weight of gas compressed fuel spherical capsule can be computed by equations:

$$\frac{\delta}{r} = \frac{P}{2\sigma}, \quad M_R = \frac{3}{2}\frac{P}{\sigma}\frac{\gamma_s}{\gamma_f}, \tag{1}$$

where δ is shell thickness of capsula, [m]; r is capsule radius, [m[; δr is relative thickness of fuel shell; P is fuel gas pressure into capsule (over the atmospheric pressure), [N/m²]; σ is safety tensile stress, [N/m²]; M_R is relative capsule mass; γ_s is density of capsule shell, [kg/m³]; γ_f is density of fuel, [kg/m³].
Example, for gas pressure $P = 100$ atm $= 10^7$ N/m², $\sigma = 200$ kgf/mm² $= 2\times10^9$ N/m², $\gamma_s = 1800$ kg/m³; $\gamma_f = 11$ kg/m³ (at $P = 100$ atm) we get $\delta r = 2.5\times10^{-3}$, $M_R = 0.6$.

Cylindrical capsule ($l >> r$). For our estimations we take the capsule having the length 1 mm, radius $r = 0.05$ mm, cross section area $S \approx 8\times10^{-3}$ mm³, fuel volume $V_c = 8\times10^{-3}$ mm³ $= 8\times10^{-12}$ m³. That is very small. It is a microcapsule.

If the gas in a microcapsule is pressed the relative thickness and relative mass of cylindrical shell may be computed by equations:

$$\frac{\delta}{r} \approx \frac{P}{\sigma}, \quad M_R \approx 2\frac{P}{\sigma}\frac{\gamma_s}{\gamma_f}, \tag{2}$$

For $P = 100$ atm ($P = 10^7$ N/m²) and $\sigma = 200$ kg/mm² ($\sigma = 2\times10^9$ N/m²) ratio $\delta r = 5\times10^{-3}$, $M_S = 1.3$.

Fuel density. The particles (ions) density n of fuel in 1 m³ and number of particles n_C in microcapsule equal

$$n = \frac{1}{2m_p}\left(\frac{\gamma_{f1}}{\mu_1} + \frac{\gamma_{f2}}{\mu_2}\right) = \frac{\gamma_{fa}}{\mu_a m_p} = \frac{\gamma_{fa}}{m_{ia}} \approx \frac{\gamma_{f1}}{\mu_1 m_p}, \quad n_c = nV_C$$

$$\text{where} \quad \mu_1 = \frac{m_{i1}}{m_p}, \quad \mu_2 = \frac{m_{i2}}{m_p} \tag{3}$$

Here $m_{ia} = 2.5\times1.672\times10^{-27}$ kg for D+T is average mass of fuel ion; $m_p = 1.672\times10^{-27}$ kg is proton mass; low indexes "$_{1,2}$" means the first and the second fuel component.

The $n \approx 10^{20}$ 1/m³ in the present magnetic confinement fusion reactor; $n \approx 2.6\times10^{25}$ 1/ m³ for gas D+T in a pressure 1 atm, $T = 288$ °K (the density of deuterium D is $\gamma_{f1} = 0.0875$ kg/m³, $\mu_1 = 2$, the density of Tritium is 1.5 more). For D and other pressure the n must be changed in same times, for example, if $P = 100$ atm then $\gamma_f = 8.75$ kg/m³, $n \approx 2.6\times10^{27}$ 1/m³, $n_c = 20.8\times10^{15}$; $n \approx 2.1\times10^{28}$ 1/m³, $n_c = 1.7\times10^{17}$ ($\gamma_f = 71$ kg/m³) for liquid hydrogen at a pressure of 1 atm. (In conventional inertial confinement fusion reactor, the fuel density may be more in 10 - 30 times, under a rocket pressure of a fuel capsule cover. For hydrogen the frizzed temperature is -259.34 °C, the boiling point is -252.87 °C). For D+T average $\mu_a = (2+3)/2 = 2.5$.

Fuel mass M_f [kg] and thermonuclear *energy* E_c into microcapsule are computed by equations:

$$E = 1.6\cdot10^{-19}\frac{M_f}{\mu_N m_p} E_r = 0.96\cdot10^8 M_f\frac{E_r}{\mu_N}, \tag{4}$$

$$M_f = \mu_N m_p nV_C, \quad E_C = 0.5\times1.6\cdot10^{-19}nV_C E_R$$

where E_r is reaction energy of one couple particles, eV; $\mu_N = \mu_1 + \mu_2$ is number of nucleons which take part in reaction, for D + T $\mu_N = 2+3 = 5$. If we want compute energy of one type of particles, the E_r is reaction energy for given type of particles, for D+T the energy $E_r = 17.5$ MeV;. For example, our capsule (1×0.1 mm) filled by liquid fuel D+T has fuel mass $M_f = 0.71×10^{-3}$ µg and will produce energy $E_c = 240$ kJ. If we burn out 10 capsules per 1 s, the engine power will be 2400 kW. An estimated 20% this energy gives the charged particles and 80% of neutrons. The fuel capsule having $M = 10$ µg $= 10^{-5}$ kg of a mixture D+T produces 3.34×10⁹ J if all atoms take part in reaction. That is equivalent to the energy derived from 84 liters of benzene. Computations are presented in Figure 5.

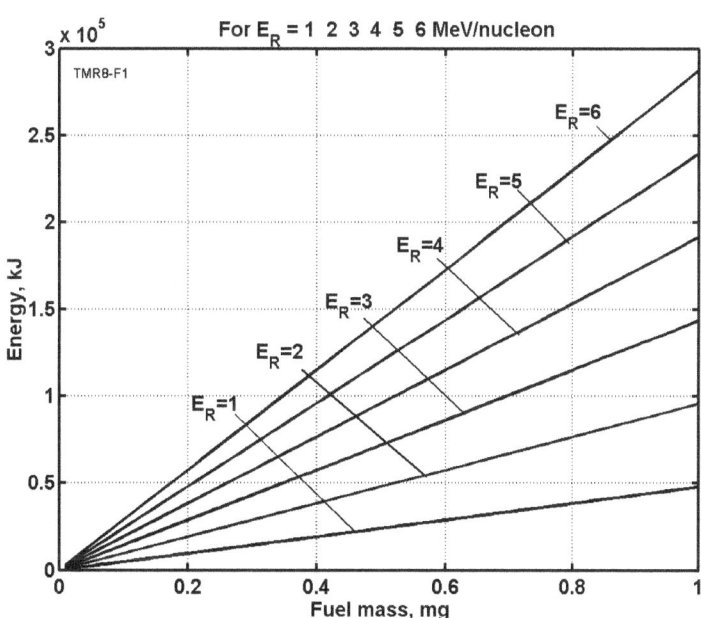

Figure 5. Energy of thermonuclear reactor versus the fuel mass and energy per one nucleon. $E_R = Er / \mu N$.

Distribution of thermonuclear energy between particles. In most cases the result of thermonuclear reaction is two components. As you see in Table 1 that may be "He" and neutron or proton. The thermonuclear energy distributes between them the following manner:

$$\text{From} \quad E = E_1 + E_2 = \frac{m_1 V_1^2}{2} + \frac{m_2 V_2^2}{2}, \quad m_1 V_1 = m_2 V_2, \tag{5}$$

$$\text{we have} \quad \frac{E_1}{E} = \frac{m_2}{m_1 + m_2} = \frac{\mu_2}{\mu_1 + \mu_2}, \quad E_2 = E - E_1$$

where m is particle mass, kg; V is particle speed, m/s; E is particle energy, J; $\mu = m_i/m_p$ is relative particle mass. Lower indexes "$_{1,2}$" are number of particles.

Unfortunately, as you can see (in Table 1, Ch1), most particle energy catches the neutron as the lightest particle. But its emission has high penetrating capability, creating radioactive isotopes, causing damage to the main construction, very dangerous for living beings, and that can be converted only in heat energy.

Energy is needed for fuel heating. This energy can be estimated by equation:

$$E_h = \frac{c}{\mu_a} M T_k, \quad c = \frac{k}{2m_p} = 4.13 \cdot 10^3, \quad \mu_a = \frac{m_{ia}}{m_p} \tag{6}$$

where c is plasma thermal capacity, J/kg·°K; T_k is temperature in °K; k is Boltzmann constant, m_{ia} is average mass of ions, kg; M is fuel mass, kg. This computation is presented in Figure 6. Our capsule filled by liquid mixture D+T requests ignition energy about 124 J for its heating up to 100 million °K. That is energy of electric condenser having size about 10×10×10 cm for $\beta = 10^8$ V/m (see below).

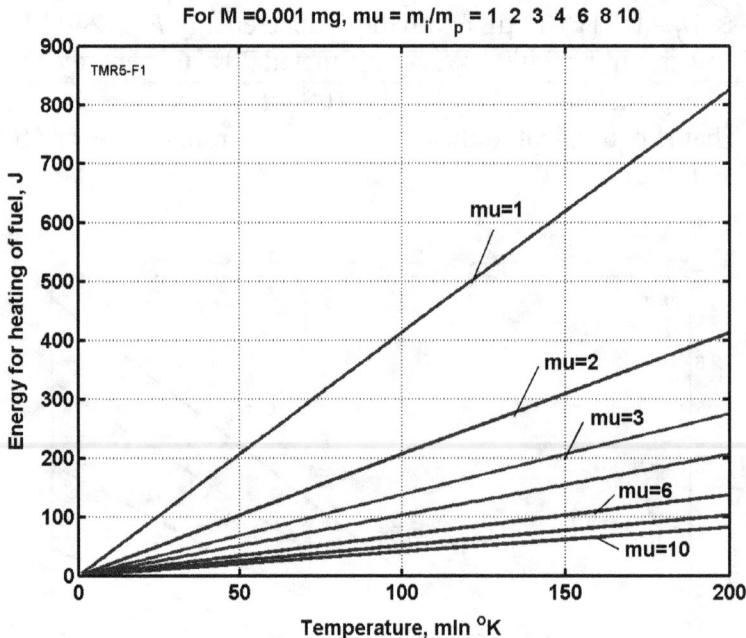

For M =0.001 mg, mu = m_i/m_p = 1 2 3 4 6 8 10

Figure 6. Energy requested for fuel D+T heating.

Capacitor for thermonuclear ignition. Condenser requested as storage of energy for fuel thermonuclear ignition my be estimated by equation

$$\frac{W}{V} = \frac{1}{2}\varepsilon_0 \varepsilon \beta^2, \quad \frac{W}{M} = \frac{\varepsilon_0 \varepsilon \beta^2}{2\gamma} \qquad (7)$$

where W is condenser energy, J; V is condenser volume, m³; M is condenser mass, kg; $\varepsilon_0 = 8.85\times10^{-12}$ Φ/m is the electrostatic constant; ε is dielectric constant; β is dielectric strength, V/m; γ is specific density of dielectric, kg/m³. Energy from capacitor is about one joule from one centimeter cub.

Electron plasma frequency. Electron frequency of plasma is computed by equation (8). For $n \approx 10^{20}$ 1/m³ that is equals $\omega_{pe} = 5.64\times10^{11}$ rad/s, for $n \approx 10^{28}$ 1/m³ that equals $\omega_{pe} = 5.64\times10^{15}$ rad/s . That is more then the laser frequency ($\lambda = 0.3\times10^{-9}$ m, $\omega = 2.1\times10^{10}$ rad/s). That means the plasma will reflect the laser beam.

Plasma skin depth. The depth in plasma to which an electromagnetic radiation can penetrate (Eq. (9)) is: For $n \approx 10^{20}$ 1/m³ that is equals $d_p = 5.31\times10^{-2}$ cm, for $n \approx 10^{28}$ 1/m³ that equals $d_p = 5.31\times10^{-6}$ cm. As you see, the depth is small.

Coefficient reflectivity of plasma. No data about plasma reflectivity. However, from general theory of reflectivity it is known the reflectivity depends from conductivity. Silver has the best conductivity from solid body and best reflectivity. It is about $q = 0.78 \div 0.99$ (it depends from frequency of radiation: for ultra-violet radiation $q \approx 0.78$, for thermal radiation $q \approx 0.99$). The plasma for $T > 15\times10^6$ °K has better conductivity then silver. The plasma conductivity increases as $T^{3/2}$. That means the plasma having the $T \approx 10^8$ °K has reflectivity in 17.2 times better then silver. That means the plasma reflectivity is more 0.999. We take in our computation $q = 0.999$. The efficiency of offered innovation very strong depends from reflectivity of plasma. The reflectivity of the prism mirror is very high [23]. We neglect the loss in it.

Bremsstrahlung (brake) radiation. That is proportional the energy spectra E and has Maxwell distribution:

$$f_E dE = f_p \left(\frac{dP}{dE}\right) dE = 2\sqrt{\frac{E}{\pi(kE)^3}} \exp\left[\frac{-E}{kT_k}\right] dE,$$

$$\lambda = \frac{c}{\nu}, \quad \nu = \frac{c}{\lambda} = \frac{E}{h} \tag{8}$$

where $k = 1,38\times10^{-23}$ is Boltzmann's constant, J/°K; T - temperature, °K; E - energy, J; ν - frequency, 1/s; λ length of wave, m; $h = 6.525\times10^{-34}$ is Planck's constant, J·s. Assume the brake radiation has same specter.
   Computations are presented in figures 7 - 8.

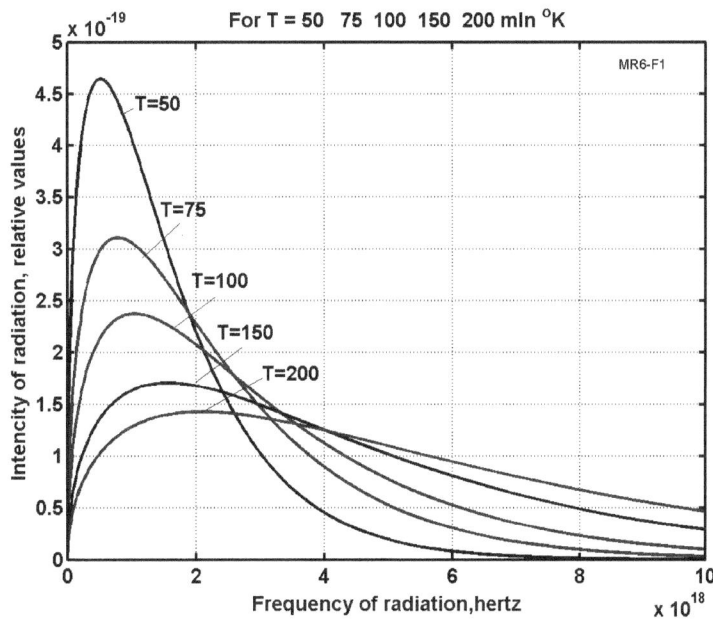

Figure 7. Spectra of brake radiation for plasma temperature 60 - 200 millions degrees (°K).

The ultra-violet rays are below approximately $< 3\times10^{17}$ hertz ($\lambda > 10^{-9}$ m), the soft x-rays are below $< 3\times10^{19}$ hertz ($\lambda > 10^{-11}$ m). That means the brake radiation can be reflected by special methods. For example, the silver having high electro-conductivity has average reflectivity 0.99 in heat region, 0.95 in light region, and 0.78 in ultra-violet region. Some metals has reflectivity up 0.2 for $\lambda = 40\times10^{-9}$ m. But plasma having the temperature more than 15×10^6 °K has more electro-conductivity then silver and it must, therefore, have better reflectivity. The reflectivity coefficient of prism mirror is very high and we can neglect its losses. However, the reflectivity of prism mirror for brake radiation is needed in a detailed test.

The average energy of Bremsstrahlung photon equals average energy of plasma electron. The formula for average wavelength is:

From $\quad E = kT_k = h\nu, \quad \lambda = \dfrac{c}{\nu}$

we receive $\quad \lambda_e = \dfrac{ch}{kT_k} = \dfrac{0.0144}{T_k} \tag{9}$

where E is electron energy, $c = 3\times10^8$ is light speed, m/s; T_k is temperature in °K; λ_e is wave length, m ; ν is wave frequency, 1/s;.
   For example, for $T_k = 10^8$ °K the $\lambda_e = 1.44\times10^{-10}$ m. That value is the lower ultra-violet diapason $\lambda > 10^{-9}$ m.
   For very high temperature the most part of this spectrum is in the soft x-ray region, but soft x-ray can be reflected and retracted by special methods.

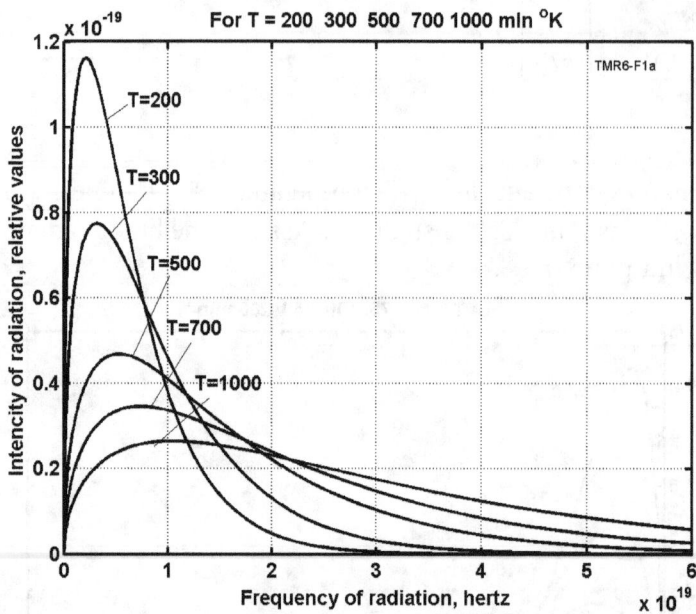

Figure 8. Spectra of brake radiation for plasma temperature 200 - 1000 millions degrees (°K).

The reactive pressure. We can estimate that the ion speed for $T = 10^8$ °K. That is approximately $V = 600$ km/s. If $M = 0.1$ $\mu g = 10^{-7}$ kg of a mixture D+T is increased their speed to this value in time $\tau = 10^{-9}$ s, the reactive force will be $F = MV/\tau \approx 5 \times 10^7$ N. If the fuel capsule has surface $s = 5$ mm$^2 = 5 \times 10^{-6}$ m^2 the capsule cover pressure is $p = F/s = 10^{13}$ N/m$^2 = 10^8$ atm. This pressure produces the shockwave which compresses the fuel microcapsule and create high ion temperature.

RADIATION CONFINEMENT

Radiation confinement is suitable for multi-reflex laser beam support.

Equilibrium of Multi-Reflex Laser and Kinetic Pressures
From equations (6), (10) we receive

$$P_k = 2nkT_k, \quad P_R = \frac{2E}{c}\left(\frac{2}{1-q}\right), \quad P_k = P_R,$$

$$P_L = SE = 0.5nkT_k cS(1-q)$$

(10)

where P_L is impulse power of laser, W; S is surface of capsule or plasma; q is plasma reflection. The additional number 2 appears because we neglect the prism reflection loss. The computations for $n = (0 \div 1) \times 10^{28}$ 1/m^3, $S = (1 \div 4) \times 10^{-6}$ m^2, $q = 0.999$, $T_k = 10^8$ are presented in Figure 13.

Look your attention that power of laser pulse for multi-reflection confinement is in tens - hundreds times less then one is in the current ICF reactors (OMEGA has 60×10^{12} W, Z-machine will have 350×10^{12} W). That shows, the multi-reflect conformation is more efficiency then rocket conformation for small targets.

We can increase the initial multi-reflex pressure in millions times if we cover the outer capsule surface the small reflective prism as internal surface of the combustion chamber. As it is shown in [26] p. 378, Figure A3.4, the pressere from 1 kW of laser power can reach more 10^4 N. If laser pulse power has $P_L = 10^{13}$ W, the pressure will be $F = 10^{17}$ N. The surface of a spherical capsule having diameter 1 mm is about $S = 3$ mm$^2 = 3 \times 10^{-6}$ m^2. Hence the pressure on target is $P = F/S = 3.3 \times 10^{22}$ N/m$^2 = 3.3 \times 10^{17}$ atm! That is in 10^9 times more then a gas dynamic pressure of the plasma at temperature 10^8 K.

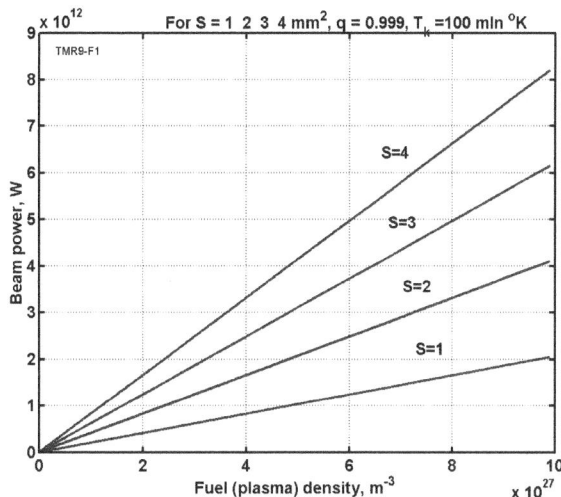

Figure 9. Equilibrium of multi-reflex laser pulse power versus plasma density and fuel capsule surface S for coefficient plasma reflectivity $q = 0.999$, plasma temperature $T_k = 10^8$ °K.

Note. The rocket force used for compressing and heating pullet at present time cannot keep the big pressure and temperature for very small capsules at need time because gas extension. For example, let us to estimate the extension time for two capsules having diameter $d = 0.3$ mm and $d = 3$ mm respectively at temperature 10^8 K. The average ion speed at this temperature is about 6×10^5 m/s. For typical pulse time 10^{-9} s the plasma radius is increased in 6×10^{-4} m $= 0.6$ mm. That means the volume of the first capsule increases in $(0.75/0.15)^3 = 125$ times, the volume of the second capsule is increased only in $(2.1/1.5)^3 = 2.7$ times. In our multi-reflex reactor the beam pressure does not allow to expansion the plasma and increases the reaction time and possibility thermonuclear reaction in hundreds times.

Requested minimum time of laser pulse. Duration of laser pulse needed for heating of capsule can be computed by equation

$$\tau = \frac{c_1 M_f T_k}{P_L} \tag{11}$$

where $c_1 = 4.13 \times 10^3$ is thermal capacity coefficient, J/kg.K; T_k is plasma temperature, K. The computations are presented in Figure 10.

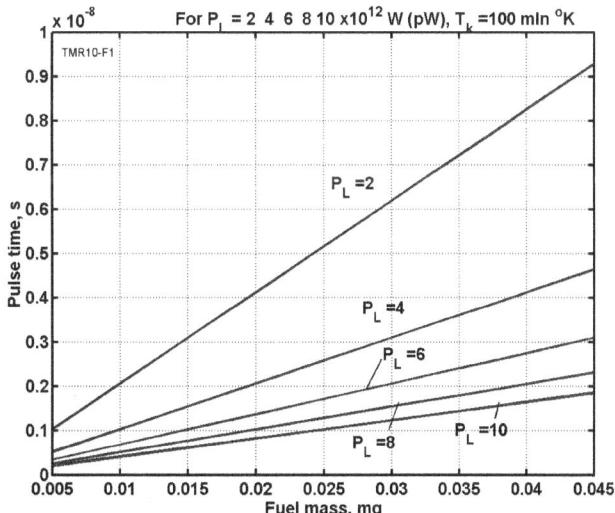

Figure 10. Request laser pulse time for capsule heating via capsule mass and laser pulse power for capsule temperature 10^8 K.

As you see, the pulse is same with current laser $(1 \div 10) \times 10^{-12}$ s, (ps). In conventional ICF reactor the most part beam energy is reflected by plasma and heating the shell of combustion chamber. In our reactor the nearly all beam energy is used for capsule heating.

Equilibrium of Brake Radiation and Kinetic Pressures

From equations (10), (7) we receive

$$P_k = 2nkT_k, \quad P_{BP} = \frac{2V}{c(1-q)S} 5.34 \cdot 10^{-37} n^2 T_e^{1/2} Z_{eff}, \quad P_k = P_{BP},$$

$$n = \frac{k\,c\,(1-q)S}{5.34 \cdot 10^{-37} V Z_{eff}} \frac{T_k}{T_e^{1/2}} = 9 \cdot 10^{28}(1-q)T_e^{1/2}\frac{S}{VZ_{eff}} \qquad (12)$$

where T_e is temperature in keV; V is plasma volume, m³; S is plasma surface, m²; q is average coefficient reflectivity of x-rays produced by brake radiation. The equilibrium of brake radiation and kinetic pressure can be reached for ratio $V/S \approx 1$. For reflection of the brake radiation one can be used a plasma mirror.

THE SELF-MAGNETIC CONFINEMENT

The self-magnitude confinement is suitable for low-density plasma. Your attention is called for to the big difference between a present conventional reactor magnetic field and the offered self-magnetic field. For creating of the present magnetic field, the large powerful superconductivity very expensive magnets are used. Our self-magnetic does not request anything. The self-magnetic field is produced by capsule electric current and that is more powerful by hundreds of times. Why? The magnetic intensity and pressure of electric current in inverse proportion of plasma radius (see equations (29) below). The present thermonuclear reactor has plasma camera of some meters. Our capsule has radius only 0.05 mm.

Equilibrium of Self-Magnetic and Kinetic Plasma Pressure

From equation (10) and (12) we receive

$$P_k = P_m, \quad P_k = 2nkT_k, \quad P_m = \frac{\mu_0 H^2}{2}, \quad H = \frac{I}{2\pi r}, \quad P_m = \frac{\mu_0}{8\pi^2}\left(\frac{I}{r}\right)^2,$$

$$I = 4\pi r\sqrt{\frac{knT_k}{\mu_0}} = 4.16 \cdot 10^{-8} r\,(nT_k)^{0.5}, \quad U = RI, \quad B = \frac{\mu_0 I}{2\pi r} \qquad (13)$$

where r is radius of capsule (plasma flux), m; I is electric currency, A; R is capsule (plasma) resistance, Ohm; U is capsule (electrodes) voltage, V; H is magnetic intensity, A/m; B is magnetic intensity, Tesla; T_k is plasma temperature , °K.

The computations for several n are presented in figures 11 - 14.

The present magnetic confinement reactor having superconductivity magnets has maximum magnetic intensity 5 - 6 Tesla. As you see in Figure 14 the offered AB-self-magnetic reactor has more magnetic intensity in hundreds times.

PROJECT

Below there are estimations of some projects, which show parameters of suggested AB-reactors. These are not optimal reactors. They demonstrate the methods of computations and possible technical data of new micro reactors.

1. Multi-reflection AB-reactor. Let us to take the spherical fuel capsule diameter $d = 1$ mm. Its surface is 3.14 mm², the volume is $v = 0.52$ mm³=5,2×10⁻¹⁰ m³. If gaseous fuel (D+T) has pressure 1, 10, 100 atm, the specific fuel density are $\rho = 0.11$ kg/m³, 1.1 kg/m³, 11 kg/m³ respectively. The fuel mass are $M_f = \rho v = 5.7 \times 10^{-11}$,

57×10^{-11}, 570×10^{-11} kg respectively. Particle densities are $n_1 = \rho/\mu_{\bar{a}} m_p = 2.63 \times 10^{25}$ 1/m³, 2.63×10^{26} 1/m³, 2.63×10^{27} 1/m³ respectively. Numbers of particles in the capsule are $n = n_1 v = 1.37 \times 10^{16}$, 1.37×10^{17}, 1.37×10^{18} respectively.

Thermonuclear energy in capsule are $E = 0.5 n E_1 = 1.9 \times 10^4$ J, 1.9×10^5 J, 1.9×10^6 J respectively. Here $E_1 = 17.6 \times 10^6 \times 1.6 \times 10^{-19} = 2.8 \times 10^{-12}$ J is the energy in single reaction of couple particles. Where 17.6×10^6 MeV is thermonuclear energy of reaction D+T. If we burn 1 capsule in 1 second, the thermonuclear power will be $W = 1.9 \times 10^4$ W, 1.9×10^5 W, 1.9×10^6 W respectively.

Fuel heating energy are $E_f = c_1 M_f T_k = 24$ J, 240 J, 2400 J respectively. Here $c_1 = 4.13 \times 10^3$ is average thermal capacity of plasma, $T_k = 10^8$ K is maximal plasma temperature. These heating energy must be increased some $(3 \div 6)$ times because we must to heat the capsule shell and coefficient of energy efficiency is less then 1. The current condensers have energy storage capability about 1 J/cm³.

Requested minimum (equilibrium) pulse laser power equal $N = 17.1 \times 10^9$ W, 17.1×10^{10} W, 17.1×10^{11} W respectively (Eq. (22)) for $q = 0.999$. Pulse time is $\tau = E_f/N = 1.4 \times 10^{-9}$ s.
We can use the liquid fuel. All parameter significantly will improvement (approximately in 10 times with comparison of the 100 atm capsule), but we get a problem with storage of capsules into a liquid helium.

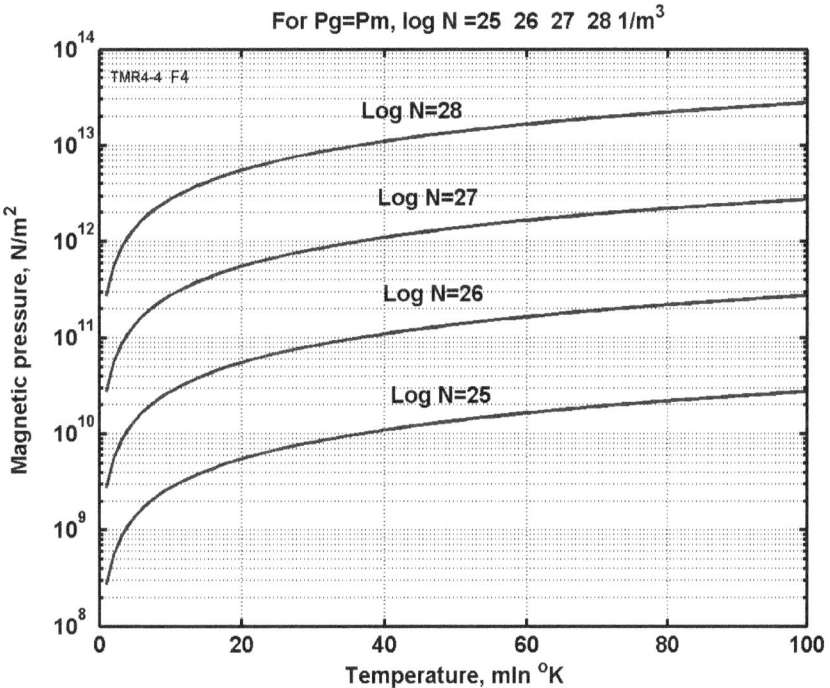

Note: This pressure is same for multi-reflex and plasma pressure.

Figure 11. Equilibrium self-magnetic and kinetic pressures versus plasma temperature and plasma densities. Capsule 0.1×1 mm. N is plasma density, 1/m³.

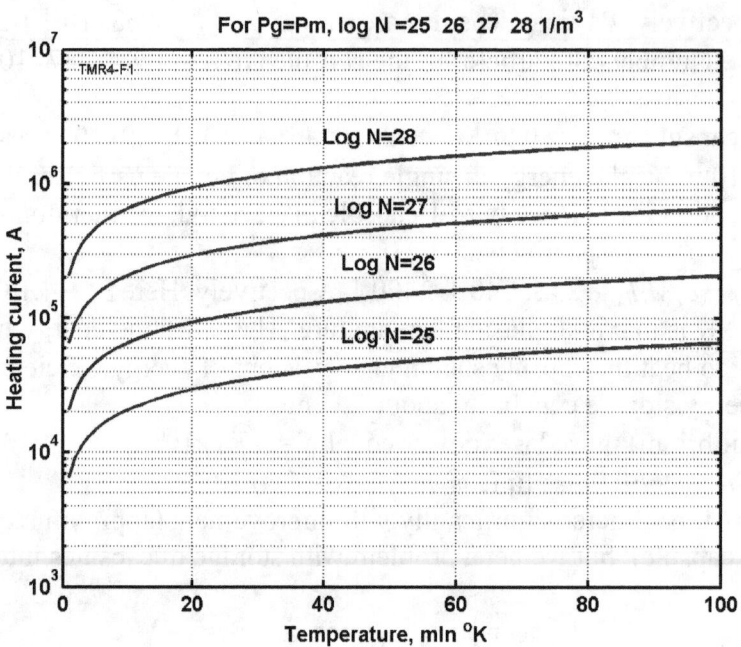

Figure 12. Electric currency needed for equilibrium kinetic and magnetic pressure for several plasma densities. Microcapsule has size 0.1×1 mm. *N* is plasma density, 1/m³.

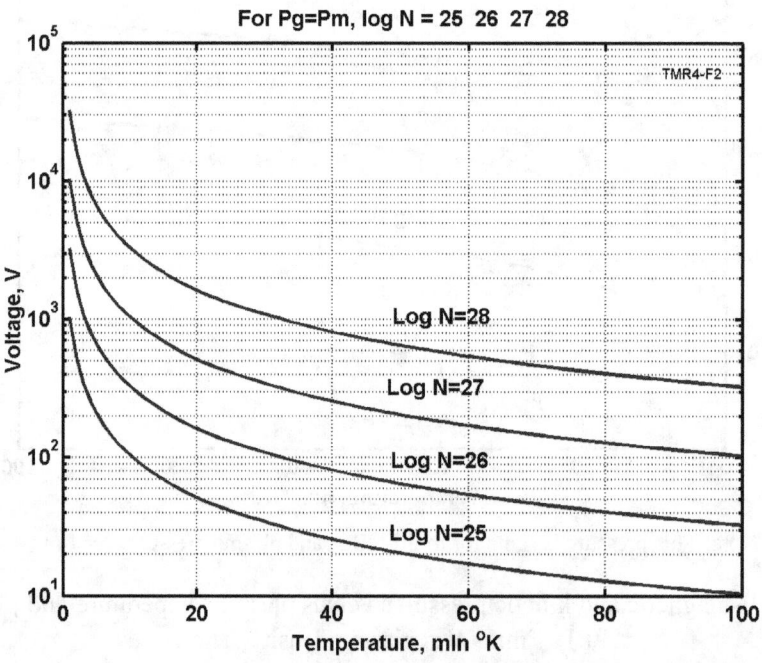

Figure 13. Electric voltage needed for equilibrium kinetic and magnetic pressures for several plasma densities. Capsule has size 0.1×1 mm. *N* is plasma density, 1/m³.

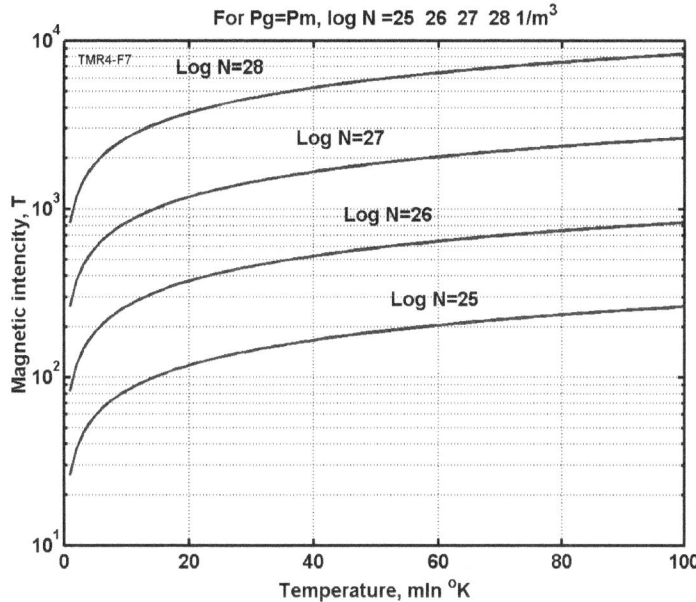

Figure 14. Equilibrium self-magnetic intensity on microcapsule surface via plasma temperature for several plasma densities. Capsule has size 0.1×1 mm. N is plasma density, $1/m^3$.

2. Self-magnetic AB-reactor. Let us to take the fuel capsule of the length $L = 1$ mm, diameter $2r = 0.1$ mm and gaseous fuel (D+T) pressure $p = 100$ atm . The cross-section of capsule is $S = 7.85 \times 10^{-3}$ mm^2, volume $v = 7.85 \times 10^{-12}$ m^3, fuel mass is $M_f = \rho v = 9.5 \times 10^{-11}$ kg, particle density is $n_1 = \rho/\mu_a m_p = 2.63 \times 10^{27}$ 1/m^3, number of particle into capsule is $n = n_1 v = 2.06 \times 10^{16}$. Heating fuel energy is $E_f = c_1 M_f T_k = 39$ J, for $T_k = 10^8$ K. If we burn 1 capsule in 1 second the thermonuclear power will be $W = 3 \times 10^4$ W.

Requested minimum (equilibrium) electric currency equal $I = 1.07 \times 10^6$ A (for $T_k = 10^8$ K),(Eq. (29)). The plasma electric specific resistance at $T_k = 10^8$ K is $\rho = 0.1Z/T^{3/2} = 1.23 \times 10^{-7}$ Ω sm. The electric plasma resistance is $R_f = \rho L/S = 1.5 \times 10^{-4}$ Ω (all for $T_k = 10^8$ K). Voltage $U = IR_f = 160$ V. Pulse power $N = IU = 17.1 \times 10^7$ W. Pulse time is $\tau = 2.3 \times 10^{-6}$ s. Maximum intensity of magnetic field is $B = \mu_o I/2\pi r = 4280$ T.

DISCUSSION

The offered thermonuclear AB-Reactors, as with any innovations, are needed in further more detailed laboratory research, product development and testing. However, theses new Reactors have gigantic advantages over present-day thermonuclear reactors:

(1) They are cheaper by many hundreds of times. That means not only non-industrial countries but middle-size companies can undertake RandD and production of perfected new thermonuclear reactors.
(2) They have a small weight and size but they have enough power (up 10,000 kW). That means they can be used as engine of land vehicles, small ships, aircraft, manned and unmanned spacecraft propulsion and community power utilities.
(3) They are not limited in high temperature regime as are all existing reactors. That means they can use inexpensive fuel (not tritium, helium-3, uranium as do extant reactors).

The parameters of AB-Reactors are considered and computed in given article very far from optima. They are only examples utilized to vividly illustrate the large possibilities of the innovative reactors.

The suggested AB-thermonuclear reactor has Lawson criterion in some order more then conventional current (2005) thermonuclear reactors (ICF). That strongly increases either of three multipliers in Lawson criterion. That increases the density n up to two-three orders. It increases the temperature T. It returns the laser and thermal radiation back to fuel pellet. (This emission is lost in present reactors). It increases the time of reaction τ. The suggested AB thermonuclear reactors may be a revolutionary jump in energy industry.

Note: In conventional ICF the initial (internal into plasma) radiation does not compress the plasma. Plasma is transparency for internal radiation. That emission influences only to an emitted particle. When radiation came out of source (fuel pellet) and reflected or adsorbed by chamber surface that does not press on pellet surface. By that means, the conventional inertial thermonuclear reactor has only losses from radiation. The offered AB Reactor has the big desirable benefits from thermal radiation. The more radiation, the more benefits.

The offered AB-Reactor can also have problems. The radiation mirror can have a bad reflectivity for ultra-violent rays or experimenters may have problems with fast high-intensity electric impulse through small capsule. However, if mirror will be reflect only conventional ultra-violet, light, and thermal radiations that would be enough for ignition of a thermonuclear reaction. As any innovation the offered reactor needs further perfecting RandD.

The offered AB-self magnetic reactor is different from present magnetic confinement reactor. That is smaller because AB-self-magnetic reactor works a small fuel capsule. In present-day reactor, the rare fuel gas (D+T) fills all volume of large chamber. In AB-Reactor the fuel is located into small capsule under high pressure (or, as solid, liquid or frizzed fuel under conventional pressure). In this case the fuel density can reach $n = 10^{-26} \div 10^{-27}$ 1/m^3 (or solid, liquid, frozen fuel may be inside conductive matter, $n = 10^{-28} \div 10^{-29}$ 1/m^3). If the plasma reflectivity is high ($q > 0{,}99$), that is enough for thermonuclear ignition and keeping plasma under the radiation pressure and magnetic pressure. For current MCF the magnetic intensity is 5 T. For AB-Self-MCF the magnetic intensity may be about 10^4 T. For AB-radiation reactor the radiation pressure is about $10^{10} \div 10^{13}$ N/m^2 (millions atm) (Figure15). We can neglect the outer magnetic force in AB-Reactor and we may design AB-Self-MCF reactor without very complex and expensive superconductivity magnetic system.

The article below is usefull for this topic.

Références

(Reader can find part of these articles in WEBs: http://Bolonkin.narod.ru/p65.htm, http://arxiv.org, search: Bolonkin, and in the book Bolonkin A.A., *"New Concepts, Idea and Innovation in Aerospace, Technology and Human Science"*, NY, Nova, 2006. Part B, Ch,1.)

[1] Bolonkin, A.A., (1982a), *Installation for Open Electrostatic Field*, Russian patent application #3467270/21 116676, 9 July, 1982 (in Russian), Russian PTO.

[2] Bolonkin, A.A., (1982b), *Radioisotope Propulsion*. Russian patent application #3467762/25 116952, 9 July 1982 (in Russian), Russian PTO.

[3] Bolonkin, A.A., (1982c), *Radioisotope Electric Generator*. Russian patent application #3469511/25 116927. 9 July 1982 (in Russian), Russian PTO.

[4] Bolonkin, A.A., (1983a), *Space Propulsion Using Solar Wing and Installation for It*, Russian patent application #3635955/23 126453, 19 August, 1983 (in Russian), Russian PTO.

[5] Bolonkin, A.A., (1983b), *Getting of Electric Energy from Space and Installation for It*, Russian patent application #3638699/25 126303, 19 August, 1983 (in Russian), Russian PTO.

[6] Bolonkin, A.A., (1983c), *Protection from Charged Particles in Space and Installation for It*, Russian patent application #3644168 136270, 23 September 1983, (in Russian), Russian PTO.

[7] Bolonkin, A. A., (1983d), *Method of Transformation of Plasma Energy in Electric Current and Installation for It*. Russian patent application #3647344 136681 of 27 July 1983 (in Russian), Russian PTO.

[8] Bolonkin, A. A., (1983e), *Method of Propulsion using Radioisotope Energy and Installation for It.* Russian patent application #3601164/25 086973 of 6 June, 1983 (in Russian), Russian PTO.

[9] Bolonkin, A. A.,(1983f), *Transformation of Energy of Rarefaction Plasma in Electric Current and Installation for it.* Russian patent application #3663911/25 159775, 23 November 1983 (in Russian), Russian PTO.

[10] Bolonkin, A. A., (1983g), *Method of a Keeping of a Neutral Plasma and Installation for it.* Russian patent application #3600272/25 086993, 6 June 1983 (in Russian), Russian PTO.

[11] Bolonkin, A.A.,(1983h), *Radioisotope Electric Generator.* Russian patent application #3620051/25 108943, 13 July 1983 (in Russian), Russian PTO.

[12] Bolonkin, A.A., (1983i), *Method of Energy Transformation of Radioisotope Matter in Electricity and Installation for it.* Russian patent application #3647343/25 136692, 27 July 1983 (in Russian), Russian PTO.

[13] Bolonkin, A.A., (1983j), *Method of stretching of thin film.* Russian patent application #3646689/10 138085, 28 September 1983 (in Russian), Russian PTO.

[14] Bolonkin A.A., (1983k), *Light Pressure Engine,* Patent (Author Certificate) No. 11833421, 1985 USSR (priority on 5 January 1983).

[15] Bolonkin, A.A., (1987), "*New Way of Thrust and Generation of Electrical Energy in Space*". Report ESTI, 1987, (Soviet Classified Projects).

[16] Bolonkin, A.A., (1990), "*Aviation, Motor and Space Designs*", Collection *Emerging Technology in the Soviet Union*, 1990, Delphic Ass., Inc., pp.32–80 (English).

[17] Bolonkin, A.A., (1991), *The Development of Soviet Rocket Engines*, 1991, Delphic Ass.Inc., 122 p., Washington, (in English).

[18] Bolonkin, A.A., (1992a), "*A Space Motor Using Solar Wind Energy (Magnetic Particle Sail)*". The World Space Congress, Washington, DC, USA, 28 Aug. – 5 Sept., 1992, IAF-0615.

[19] Bolonkin, A.A., (1992b), "*Space Electric Generator, run by Solar Wing*". The World Space Congress, Washington, DC, USA, 28 Aug. – 5 Sept. 1992, IAF-92-0604.

[20] Bolonkin, A.A., (1992c), "*Simple Space Nuclear Reactor Motors and Electric Generators Running on Radioactive Substances*", The World Space Congress, Washington, DC, USA, 28 Aug. – 5 Sept., 1992, IAF-92-0573.

[21] Bolonkin, A.A. (1994), "*The Simplest Space Electric Generator and Motor with Control Energy and Thrust*", 45th International Astronautical Congress, Jerusalem, Israel, 9–14 Oct.,1994, IAF-94-R.1.368.

[22] Bolonkin A.A. (2004a), *Light Multi-reflex Engine, JBIS, vol.* 57, No. 9/10, 2004, pp. 353-359.

[23] Bolonkin A.A.(2004b), *Multi-reflex Space Propulsion, JBIS, Vol.* 57, No. 11/12, 2004, pp. 379 -390.

[24] Bolonkin A.A. (2006a), *Non-Rocket Space Launch and Flight,* Elsevier, London, 2006, 488 pgs.

[25[Bolonkin A.A. (2006b), *New Thermonuclear Reactor,* AIAA-2006-7225, Conference "Space-2006", USA.

January 2005

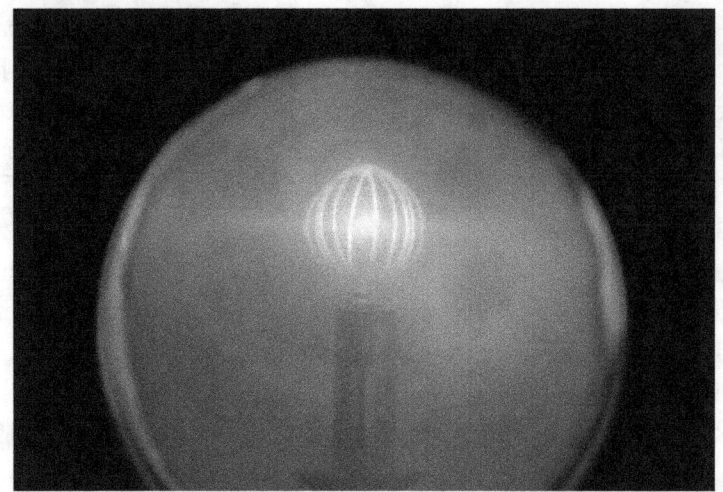

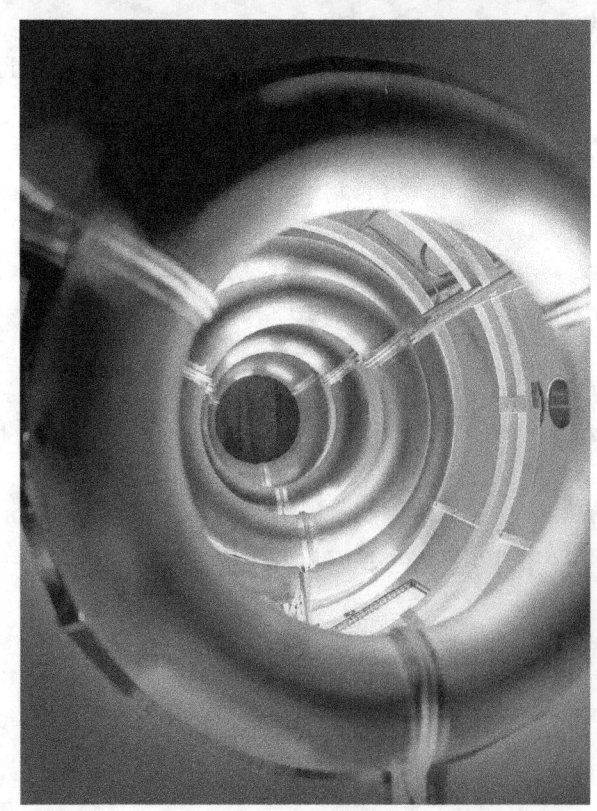

Chapter 10

Converting of any Matter to Nuclear Energy by AB-Generator and its Application*

Abstract

Author offers a new nuclear generator which allows to convert any matter to nuclear energy in accordance with the Einstein equation $E=mc^2$. The method is based upon tapping the energy potential of a Micro Black Hole (MBH) and the Hawking radiation created by this MBH. As is well-known, the vacuum continuously produces virtual pairs of particles and antiparticles, in particular, the photons and anti-photons. The MBH event horizon allows separating them. Anti-photons can be moved to the MBH and be annihilated; decreasing the mass of the MBH, the resulting photons leave the MBH neighborhood as Hawking radiation. The offered nuclear generator (named by author as AB-Generator) utilizes the Hawking radiation and injects the matter into MBH and keeps MBH in a stable state with near-constant mass. The AB-Generator can not only produce gigantic energy outputs but should be hundreds of times cheaper than a conventional electric generation processes. The AB-Generator can be used in aerospace as a photon rocket or as a power source for numerous space vehicles. Many scientists expect the Large Hadron Collider at CERN will produce one MBH every second and the technology to capture them may be used for the AB-Generator.

Key words: Production of nuclear energy, Micro Black Hole, energy AB-Generator, photon rocket.
* Presented as Paper AIAA-2009-5342 in 45 Joint Propulsion Conferences, 2–5 August, 2009, Denver, CO, USA.

Introduction

Black hole. In general relativity, a black hole is a region of space in which the gravitational field is so powerful that nothing, including light, can escape its pull. The black hole has a one-way surface, called the event horizon, into which objects can fall, but out of which nothing can come out. It is called "black" because it absorbs all the light that hits it, reflecting nothing, just like a perfect blackbody in thermodynamics. Despite its invisible interior, a black hole can reveal its presence through interaction with other matter. A black hole can be inferred by tracking the movement of a group of stars that orbit a region in space which looks empty. Alternatively, one can see gas falling into a relatively small black hole, from a companion star. This gas spirals inward, heating up to very high temperature and emitting large amounts of radiation that can be detected from earthbound and earth-orbiting telescopes. Such observations have resulted in the general scientific consensus that, barring a breakdown in our understanding of nature, that black holes do exist in our universe. Although it is impossible to directly observe a black hole, its existence is inferred by its gravitational action on the surrounding environment, particularly with microquasars and active galactic nuclei, where material falling into a nearby black hole is significantly heated and emits a large amount of X-ray radiation. The only objects that agree with these observations and are consistent within the framework of general relativity are black holes.

A black hole has only three independent physical properties: mass, charge and angular momentum. In astronomy black holes are classed as:

- Supermassive - contain hundreds of thousands to billions of solar masses and are thought to exist in the center of most galaxies, including the Milky Way.

- Intermediate - contain thousands of solar masses.

- Micro (also *mini black holes*) - have masses much less than that of a star. At these sizes, quantum mechanics is expected to take effect. There is no known mechanism for them to form via normal processes of stellar evolution, but certain inflationary scenarios predict their production during the early stages of the evolution of the universe.

According to some theories of quantum gravity they may also be produced in the highly energetic reaction produced by cosmic rays hitting the atmosphere or even in particle accelerators such as the Large Hadron Collider. The theory of Hawking radiation predicts that such black holes will evaporate in bright flashes of gamma radiation. NASA's Fermi Gamma-ray Space Telescope satellite (formerly GLAST) launched in 2008 is searching for such flashes.

Fig 1. Artist's conception of a stellar mass black hole. Credit NASA.

The defining feature of a black hole is the appearance of an *event horizon*; a boundary in spacetime beyond which events cannot affect an outside observer. Since the event horizon is not a material surface but rather merely a mathematically defined demarcation boundary, nothing prevents matter or radiation from entering a black hole, only from exiting one.

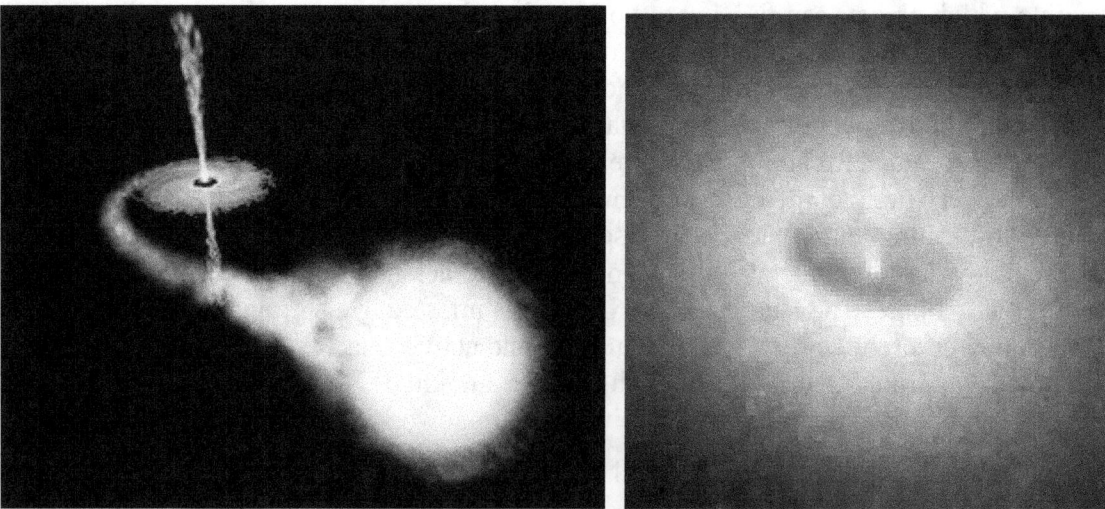

Fig.2 (left). Artist's impression of a binary system consisting of a black hole and a main sequence star. The black hole is drawing matter from the main sequence star via an accretion disk around it, and some of this matter forms a gas jet.

Fig.3 (right). Ring around a suspected black hole in galaxy NGC 4261. Date: Nov.1992. Courtesy of Space Telescope Science

For a non rotating (static) black hole, the *Schwarzschild radius* delimits a spherical event horizon. The Schwarzschild radius of an object is proportional to the mass. Rotating black holes have distorted, nonspherical event horizons. The description of black holes given by general relativity is known to be an approximation, and it is expected that quantum gravity effects become significant near the vicinity of the event horizon. This allows observations of matter in the vicinity of a black hole's event horizon to be used to indirectly study general relativity and proposed extensions to it.

Fig.4. Artist's rendering showing the space-time contours around a black hole. Credit NASA.

Though black holes themselves may not radiate energy, electromagnetic radiation and matter particles may be radiated from just outside the event horizon via *Hawking radiation*. At the center of a black hole lies the *singularity*, where matter is crushed to infinite density, the pull of gravity is infinitely strong, and spacetime has infinite curvature. This means that a black hole's mass becomes entirely compressed into a region with zero volume. This zero-volume, infinitely dense region at the center of a black hole is called a *gravitational singularity*. The singularity of a non-rotating black hole has zero length, width, and height; a rotating black hole's is smeared out to form a ring shape lying in the plane of rotation. The ring still has no thickness and hence no volume.

The *photon sphere* is a spherical boundary of zero thickness such that photons moving along tangents to the sphere will be trapped in a circular orbit. For non-rotating black holes, the photon sphere has a radius 1.5 times the Schwarzschild radius. The orbits are dynamically unstable, hence any small perturbation (such as a particle of infalling matter) will grow over time, either setting it on an outward trajectory escaping the black hole or on an inward spiral eventually crossing the event horizon.

Rotating black holes are surrounded by a region of spacetime in which it is impossible to stand still, called the *ergosphere*. Objects and radiation (including light) can stay in orbit within the ergosphere without falling to the center. Once a black hole has formed, it can continue to grow by absorbing additional matter. Any black hole will continually absorb interstellar dust from its direct surroundings and omnipresent cosmic background radiation. Much larger contributions can be obtained when a black hole merges with other stars or compact objects.

Hawking radiation. In 1974, Stephen Hawking showed that black holes are not entirely black but emit small amounts of thermal radiation.[1]He got this result by applying quantum field theory in a static black hole background. The result of his calculations is that a black hole should emit particles in a perfect black body spectrum. This effect has become known as Hawking radiation. Since Hawking's result many others have verified the effect through various methods. If his theory of black hole radiation is correct then black holes are expected to emit a thermal spectrum of radiation, and thereby lose mass, because according to the theory of relativity mass is just highly condensed energy ($E = mc^2$). Black holes will shrink and evaporate over time. The temperature of this spectrum (Hawking temperature) is proportional to the surface gravity of the black hole, which in turn is inversely proportional to the mass. Large black holes, therefore, emit less radiation than small black holes.

On the other hand if a black hole is very small, the radiation effects are expected to become very strong. Even a black hole that is heavy compared to a human would evaporate in an instant. A black hole the weight of a car ($\sim 10^{-24}$ m) would only take a nanosecond to evaporate, during which time it would briefly have a luminosity more than 200 times that of the sun. Lighter black holes are expected to evaporate even faster, for example a black hole of mass 1 TeV/c^2 would take less than 10^{-88} seconds to evaporate completely. Of course, for such a small black hole quantum gravitation effects are expected to play an important role and could even – although current developments in quantum gravity do not indicate so – hypothetically make such a small black hole stable.

Micro Black Holes. Gravitational collapse is not the only process that could create black holes. In principle, black holes could also be created in high energy collisions that create sufficient density. Since classically black holes can take any mass, one would expect micro black holes to be created in any such process no matter how low the energy. However, to date, no such events have ever been detected either directly or indirectly as a deficiency of the mass balance in particle accelerator experiments. This suggests that there must be a lower limit for the mass of black holes. Theoretically this boundary is expected to lie around the Planck mass ($\sim 10^{19}$ GeV/c^2, $m_p = 2.1764 \cdot 10^{-8}$ kg), where quantum effects are expected to make the theory of general relativity break down completely. This would put the creation of black holes firmly out of reach of any high energy process occurring on or near the Earth. Certain developments in quantum gravity however suggest that this bound could be much lower. Some braneworld scenarios for example put the Planck mass much lower, maybe even as low as 1 TeV. This would make it possible for micro black holes to be created in the high energy collisions occurring when cosmic rays hit the Earth's atmosphere, or possibly in the new Large Hadron Collider at CERN. These theories are however very speculative, and the creation of black holes in these processes is deemed unlikely by many specialists.

Smallest possible black hole. To make a black hole one must concentrate mass or energy sufficiently that the escape velocity from the region in which it is concentrated exceeds the speed of light. This condition gives the Schwarzschild radius, $r_o = 2GM / c^2$, where G is Newton's constant and c is the speed of light, as the size of a black hole of mass M. On the other hand, the Compton wavelength, $\lambda = h / Mc$, where h is Planck's constant, represents a limit on the minimum size of the region in which a mass M at rest can be localized. For sufficiently small M, the Compton wavelength exceeds the Schwarzschild radius, and no black hole description exists. This smallest mass for a black hole is thus approximately the Planck mass, which is about 2×10^{-8} kg or 1.2×10^{19} GeV/c^2.

Any primordial black holes of sufficiently low mass will Hawking evaporate to near the Planck mass within the lifetime of the universe. In this process, these small black holes radiate away matter. A rough picture of this is that pairs of virtual particles emerge from the vacuum near the event horizon, with one member of a pair being captured, and the other escaping the vicinity of the black hole. The net result is the black hole loses mass (due to conservation of energy). According to the formulae of black hole thermodynamics, the more the black hole loses mass the hotter it becomes, and the faster it evaporates, until it approaches the Planck mass. At this stage a black hole would have a Hawking temperature of $T_P / 8\pi$ (5.6×10^{32} K), which means an emitted Hawking particle would have an energy comparable to the mass of the black hole. Thus a thermodynamic description breaks down. Such a mini-black hole would also have an entropy of only 4π nats, approximately the minimum possible value.

At this point then, the object can no longer be described as a classical black hole, and Hawking's calculations also break down. Conjectures for the final fate of the black hole include total evaporation and production of a Planck mass-sized *black hole remnant*. If intuitions about quantum black holes are correct, then close to the Planck mass the number of possible quantum states of the black hole is expected to become so few and so quantised that its interactions are likely to be quenched out. It is possible that such Planck-mass black holes, no longer able either to absorb energy gravitationally like a classical black hole because of the quantised gaps between their allowed energy levels, nor to emit Hawking particles for the same reason, may in effect be stable objects. They would in effect be WIMPs, weakly interacting massive particles; this could explain dark matter.

Creation of micro black holes. Production of a black hole requires concentration of mass or energy within the corresponding Schwarzschild radius. In familiar three-dimensional gravity, the minimum such energy is 10^{19} GeV, which would have to be condensed into a region of approximate size 10^{-33} cm. This is far beyond the limits of any current technology; the Large hadron collider (LHC) has a design energy of 14 TeV. This is also beyond the range of known collisions of cosmic rays with Earth's atmosphere, which reach center of mass energies in the range of hundreds of TeV. It is estimated that to collide two particles to within a distance of a Planck length with currently achievable magnetic field strengths would require a ring accelerator about 1000 light years in diameter to keep the particles on track.

Some extensions of present physics posit the existence of extra dimensions of space. In higher-dimensional spacetime, the strength of gravity increases more rapidly with decreasing distance than in three dimensions. With certain special configurations of the extra dimensions, this effect can lower the Planck scale to the TeV range. Examples of such extensions include large extra dimensions, special cases of the Randall-Sundrum model, and String theory configurations. In such scenarios, black hole production could possibly be an important and observable effect at the LHC.

Virtual particles. In physics, a virtual particle is a particle that exists for a limited time and space, introducing uncertainty in their energy and momentum due to the Heisenberg Uncertainty Principle. Vacuum energy can also be thought of in terms of virtual particles (also known as vacuum fluctuations) which are created and destroyed out of the vacuum. These particles are always created out of the vacuum in particle-antiparticle pairs, which shortly annihilate each other and disappear. However, these particles and antiparticles may interact with others before disappearing.

  The net energy of the Universe remains zero so long as the particle pairs annihilate each other within Planck time. Virtual particles are also excitations of the underlying fields, but are detectable only as forces. The creation of these virtual particles near the event horizon of a black hole has been hypothesized by physicist Stephen Hawking to be a mechanism for the eventual "evaporation" of black holes. Since these particles do not have a permanent existence, they are called *virtual particles* or vacuum fluctuations of vacuum energy. An important example of the "presence" of virtual particles in a vacuum is the Casimir effect. Here, the explanation of the effect requires that the total energy of all of the virtual particles in a vacuum can be added together. Thus, although the virtual particles themselves are not directly observable in the laboratory, they do leave an observable effect: their zero-point energy results in forces acting on suitably arranged metal plates or dielectrics. Thus, virtual particles are often popularly described as coming in pairs, a particle and antiparticle, which can be of any kind.

The evaporation of a black hole is a process dominated by photons, which are their own antiparticles and are uncharged. The uncertainty principle in the form $\Delta E \Delta t \geq h$ implies that in the vacuum one or more particles with energy ΔE above the vacuum may be created for a short time Δt. These *virtual particles* are included in the definition of the vacuum.

Vacuum energy is an underlying background energy that exists in space even when devoid of matter (known as free space). The vacuum energy is deduced from the concept of virtual particles, which are themselves derived from the energy-time uncertainty principle. Its effects can be observed in various phenomena (such as spontaneous emission, the Casimir effect, the van der Waals bonds, or the Lamb shift), and it is thought to have consequences for the behavior of the Universe on cosmological scales.

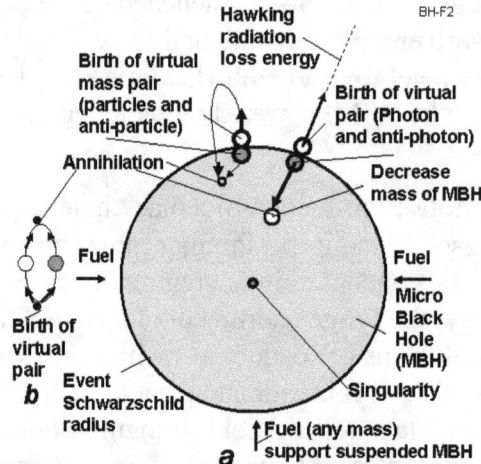

Fig.5. Hawking radiation. *a*. Virtual particles at even horizon.
b. Virtual particles out even horizon (in conventional space).

AB-Generator of Nuclear Energy and some Innovations

Simplified explanation of MBH radiation and work of AB-Generator (Fig.5). As known, the vacuum continuously produces, virtual pairs of particles and antiparticles, in particular, photons and anti-photons. In conventional space they exist only for a very short time, then annihilate and return back to nothingness. The MBH event horizon, having very strong super-gravity, allows separation of the particles and anti particles, in particular, photons and anti-photons. Part of the anti-photons move into the MBH and annihilate with photons decreasing the mass of the MBH and return back a borrow energy to vacuum. The free photons leave from the MBH neighborhood as Hawking radiation. That way the MBH converts any conventional matter to Hawking radiation which may be converted to heat or electric energy by the AB-Generator. This AB- Generator utilizes the produced Hawking radiation and injects the matter into the MBH while maintaining the MBH in stable suspended state.

Note: The photon does NOT have rest mass. Therefore a photon can leave the MBH's neighborhood (if it is located beyond the event horizon). All other particles having a rest mass and speed less than light speed *cannot* leave the Black Hole. They cannot achieve light speed because their mass at light speed equals infinity and requests infinite energy for its' escape—an impossibility.

Description of AB- Generator. The offered nuclear energy AB- Generator is shown in fig. 6. That includes the Micro Black Hole (MBH) 1 suspended within a spherical radiation reflector and heater 5. The MBH is supported (and controlled) at the center of sphere by a fuel (plasma, proton, electron, matter) gun 7. This AB- Generator also contains the 9 – heat engine (for example, gas, vapor turbine), 10 – electric generator, 11 – coolant (heat transfer agent), an outer electric line 12, internal electric generator (5 as antenna) with customer 14.

Work. The generator works the following way. MBH, by selective directional input of matter, is levitated in captivity and produces radiation energy 4. That radiation heats the spherical reflector-heater 5. The coolant (heat transfer agent) 11 delivers the heat to a heat machine 9 (for example, gas, vapor turbine). The heat machine rotates an electric generator 10 that produces the electricity to the outer electric line 12. Part of MBH radiation may accept by sphere 5 (as antenna) in form of electricity.

The control fuel guns inject the matter into MBH and do not allow bursting of the MBH. This action also supports the MBH in isolation, suspended from dangerous contact with conventional matter. They also control the MBH size and the energy output.

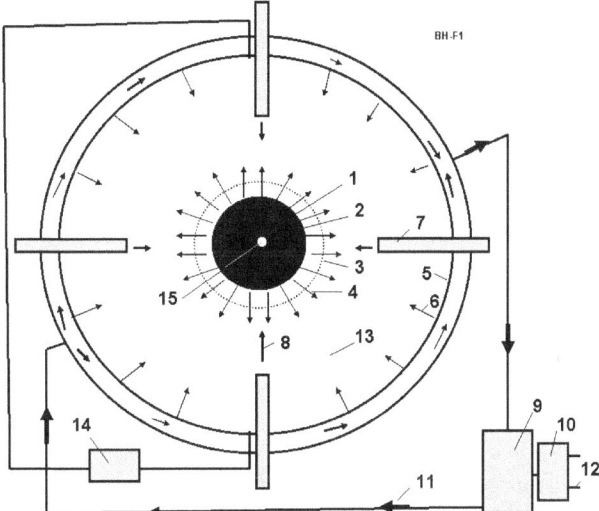

Fig.6. Offered **nuclear-vacuum energy AB- Generator**. *Notations*: 1- Micro Black Hole (MBH), 2 - event horizon (Schwarzschild radius), 3 - photon sphere, 4 – black hole radiation, 5 – radiation reflector, antenna and heater (cover sphere), 6 – back (reflected) radiation from radiation reflector 5, 7 – fuel (plasma, protons, electrons, ions, matter) gun (focusing accelerator), 8 – matter injected to MBH (fuel for Micro Black hole), 9 – heat engine (for example, gas, vapor turbine), 10 – electric generator connected to heat engine 9, 11 – coolant (heat transfer agent to the heat machine 9), 12 – electric line, 13 – internal vacuum, 14 – customer of electricity from antenna 5, 15 – singularity.

Any matter may be used as the fuel, for example, accelerated plasma, ions, protons, electrons, micro particles, etc. The MBH may be charged and rotated. In this case the MBH may has an additional suspension by control charges located at the ends of fuel guns or (in case of the rotating charged MBH) may have an additional suspension by the control electric magnets located on the ends of fuel guns or at points along the reflector-heater sphere.

Innovations, features, advantages and same research results

 Some problems and solutions offered by the author include the following:

1) A practical (the MBH being obtained and levitated, details of which are beyond the scope of this paper) method and installation for converting any conventional matter to energy in accordance with Einstein's equation $E = mc^2$.
2) MBHs may produce gigantic energy and this energy is in the form of dangerous gamma radiation. The author shows how this dangerous gamma radiation Doppler shifts when it moves against the MBH gravity and converts to safely tapped short radio waves.
3) The MBH of marginal mass has a tendency to explode (through quantum evaporation, very quickly radiating its mass in energy). The AB- Generator automatically injects metered amounts of matter into the MBH and keeps the MGH in a stable state or grows the MBH to a needed size, or decreases that size, or temporarily turns off the AB- Generator (decreases the MBH to a Planck Black Hole).
4) Author shows the radiation flux exposure of AB- Generator (as result of MBH exposure) is not dangerous because the generator cover sphere has a vacuum, and the MBH gravity gradient decreases the radiation energy.
5) The MBH may be supported in a levitated (non-contact) state by generator fuel injectors.

Theory of AB- Generator

Below there are main equations for computation the conventional black hole (BH) and AB-Generator.

General theory of Black Hole.

1. Power produced by BH is

$$P = \frac{hc^6}{15360\pi G^2}\frac{1}{M} \approx 3.56 \cdot 10^{32}\frac{1}{M}, \quad W, \tag{1}$$

where $h = h/2\pi = 1.0546 \cdot 10^{-34} \; J/s$ is reduced Planck constant, $c = 3 \cdot 10^8 \; m/s$ - light speed, $G = 6.6743 \cdot 10^{-11} \; m^3/kg.s^2$ is gravitation constant, M – mass of BH, kg.

2. Temperature of black body corresponding to this radiation is

$$T = \frac{hc^3}{8\pi Gk_b}\frac{1}{M} \approx 1.23 \cdot 10^{23}\frac{1}{M}, \quad K, \tag{2}$$

where $k_b = 1.38.10^{-23}$ J/k is Boltzmann constant.

3. Energy E_p [J] and frequency ν_0 of photon at event horizon are

$$E_p = \frac{hc^3}{16\pi G}\frac{1}{M}, \quad \nu_0 = \frac{E_p}{h} = \frac{c^3}{16\pi G}\frac{1}{M} = 8.037 \cdot 10^{33}\frac{1}{M}, \quad \lambda_0 = \frac{c}{\nu_0} = 3.73 \cdot 10^{-26} M. \tag{3}$$

where $c = 3 \cdot 10^8$ m/s is light speed, λ_0 is wavelength of photon at even radius, m. h is Planck constant.

4. Radius of BH event horizon (Schwarzschild radius) is

$$r_0 = \frac{2G}{c^2}M = 1.48 \cdot 10^{-27} M, \quad m, \tag{4}$$

5. Relative density (ratio of mass M to volume V of BH) is

$$\rho = \frac{M}{V} = \frac{3c^2}{32\pi G^3}\frac{1}{M^2} \approx 7.33 \cdot 10^{79}\frac{1}{M^2}, \quad kg/m^3. \tag{5}$$

6. Maximal charge of BH is

$$Q_{max} = 5 \cdot 10^9 eM \approx 8 \cdot 10^{-10} M, \quad C, \tag{6}$$

where $e = -1.6 \cdot 10^{-19}$ is charge of electron, C.

7. Life time of BH is

$$\tau = \frac{5120\pi G^2}{hc^4}M^3 = 2.527 \cdot 10^{-8} M^3, \quad s. \tag{7}$$

8. Gravitation around BH (r is distance from center) and on event horizon

$$g = \frac{GM}{r^2}, \quad g_0 = \frac{c^4}{4G}\frac{1}{M} = 3 \cdot 10^{42}\frac{1}{M}, \quad m \, s^{-2}. \tag{8}$$

Developed Theory of AB- Generator

Below are research and the theory developed by author for estimation and computation of facets of the AB-Generator.

9. Loss of energy of Hawking photon in BH gravitational field. It is known that a red shift allows estimating the frequency of photon in central gravitational field when it moves TO the gravity center. In this case the photon increases its frequency because photon is accelerated the gravitational field (wavelength decreases). But in our case the photon moves FROM the gravitational center, the gravitational field brakes it and the photon loses its energy. That means its frequency decreases and the wavelength increases. Our photon gets double energy because the black hole annihilates two photons (photon and anti-photon). That way the equation for photon frequency at distance $r > r_0$ from center we can write in form

$$\frac{\nu}{\nu_0} \approx 1 + \frac{2\Delta\varphi}{c^2}, \tag{9}$$

Where $\Delta\varphi = \varphi - \varphi_0$ is difference of the gravity potential. The gravity potential is

$$\Delta\varphi = \varphi - \varphi_0, \quad \varphi = \frac{GM}{r}, \quad \varphi_0 = \frac{GM}{r_0}, \quad r_0 = \frac{2GM}{c^2}. \tag{10}$$

Let us substitute (10) in (9), we get

$$\frac{\nu}{\nu_0} \approx 1 + \frac{r_0}{r} - \frac{r_0}{r_0}, \quad \text{or} \quad \frac{\nu}{\nu_0} = \frac{\lambda_0}{\lambda} \approx \frac{r_0}{r}. \tag{11}$$

It is known, the energy and mass of photon is

$$E_f = h\gamma, \quad E_f = m_f c^2, \quad m_f = E_f / c^2, \tag{12}$$

The energy of photon linear depends from its frequency. Reminder: The photon does not have a rest mass.

The relative loss of the photon radiation energy ξ at distance r from BH and the power P_r of Hawking radiation at radius r from the BH center is

$$\xi = \frac{r_0}{r}, \quad \nu = \xi\nu_0, \quad P_r = \xi P. \tag{13}$$

The r_0 is very small and ξ is also very small and $\nu \ll \nu_0$.

The result of an energy loss by Hawking photon in the BH gravitational field is very important for AB-Generator. The energy of Hawking radiation is very big; we very need to decrease it in many orders. The initial Hawking photon is gamma radiation that is dangerous for people and matter. In r distance the gamma radiation may be converted in the conventional light or radio radiation, which are not dangerous and may be reflected, focused or a straightforward way converted into electricity by antenna.

10. Reflection Hawking radiation back to MBH. For further decreasing the MBH produced energy the part of this energy may be reflected to back in MBH. A conventional mirror may reflect up $0.9 \div 0.99$ of radiation ($\xi_r = 0.01 \div 0.1$, ξ_r is a loss of energy in reflecting), the multi layers mirror can reflect up 0.9999 of the monochromatic light radiation ($\xi_r = 10^{-3} \div 10^{-5}$), and AB-mirror from cubic corner cells offered by author in [2], p. 226, fig.12.1g , p. 376 allows to reflect non-monochromatic light radiation with efficiency up $\xi_r = 10^{-13}$ strong back to source. In the last case, the loss of reflected energy is ([2] p.377)

$$\xi_r = 0.00023al, \quad l = m\lambda, \quad m \geq 1, \tag{14}$$

where l is size of cube corner cell, m; m is number of radiation waves in one sell; λ is wavelength, m; a is characteristic of sell material (see [2], fig.A3.3). Minimal value $a = 10^{-2}$ for glass and $a = 10^{-4}$ for KCl crystal.

The reflection of radiation to back in MBH is may be important for MBH stabilization, MBH storage and MBH 'switch off'.

11. Useful energy of AB- Generator. The useful energy P_u [J] is taken from AB- Generator is

$$P_u = \xi\xi_r P. \tag{15}$$

12. Fuel consumption is

$$\overset{\cdot}{M} = P_u / c^2, \quad \text{kg.} \tag{16}$$

The fuel consumption is very small. AB- Generator is the single method in the World now known which allows full converting reasonably practical conversion of (any!) matter into energy according the Einsteinian equation $E = mc^2$.

13. Specific pressure on AB- Generator cover sphere p [N/m²] and on the surface of MBH p_0 is

$$p = \frac{kP_r}{Sc} = \frac{kP_r}{4\pi r^2 c} = 2.65 \cdot 10^{-10} \frac{kP_r}{r^2}, \quad p_0 = \frac{P}{S_0 c} = \frac{hc^8}{15360 \cdot 16\pi^2 G^4} \frac{1}{M^4} = 1.44 \cdot 10^{28} \frac{1}{M^4}, \tag{17}$$

where $k = 1$ if the cover sphere absorbs the radiation and $k \approx 2$ if the cover sphere high reflects the radiation, S is the internal area of cover sphere, m^2; S_0 is surface of event horizon sphere, m^2; p_0 is specific pressure of Hawking radiation on the event horizon surface. Note, the pressure p on cover sphere   is small (see Project), but pressure p_0 on event horizon surface is very high.

14. Mass particles produced on event surface. On event horizon surface may be also produced the mass particles with speed $V < c$. Let us take the best case (for leaving the BH) when their speed is radially vertical. They cannot leave the BH because their speed V is less than light speed c. The maximal radius of lifting r_m [m] is

$$dV = -gdt, \quad dV = -\frac{g}{V}dr = -\frac{GM}{V}\frac{dr}{r^2}, \quad r_m = \frac{2GM}{c^2 - V_0^2} = \frac{r_0}{1 - (V/c)^2}, \quad (18)$$

where g is gravitational acceleration of BH, m/s^2; t is time, sec.; r_0 is BH radius, m; V_0 is particle speed on event surface, m/s^2. If the r_m is less than radius of the cover sphere, the mass particles return to BH and do not influence the heat flow from BH to cover sphere. That is in the majority of cases.

15. Explosion of MBH. The MBH explosion produces the radiation energy

$$E_e = Mc^2. \quad (19)$$

MBH has a small mass. The explosion of MBH having $M = 10^{-5}$ kg produces 9×10^{11} J. That is energy of about 10 tons of good conventional explosive (10^7 J/kg). But there is a vacuum into the cover sphere and this energy is presented in radiation form. But in reality only very small part of explosion energy reaches the cover sphere, because the very strong MBH gravitation field brakes the photons and any mass particles. Find the energy which reaches the cover sphere via:

$$dE = \xi c^2 dM, \quad \xi = \frac{r_0}{r}, \quad r_0 = \frac{2G}{c^2}M, \quad dE = \frac{2G}{r}MdM, \quad E = \frac{G}{r}M^2 = 6{,}674 \cdot 10^{-11}\frac{M^2}{r}. \quad (20)$$

The specific exposure radiation pressure of MBH pressure p_e [N/m^2] on the cover sphere of radius $r < r_0$ may be computed by the way:

$$p_e = \frac{E}{V} = \frac{3G}{4\pi}\frac{M^2}{r^3} = 1.6 \cdot 10^{-11}\frac{M^2}{r^3}, \quad r > r_0, \quad (21)$$

where $V = 3/4\ \pi r^3$ is volume of the cover sphere.

That way the exposure radiation pressure on sphere has very small value and presses very short time. Conventional gas balloon keeps pressure up 10^7 N/m^2 (100 atm). However, the heat impact may be high and AB- Generator design may have the reflectivity cover and automatically open windows for radiation.

Your attention is requested toward the next important result following from equations (20)-(21). Many astronomers try to find (detect) the MBH by a MBH exposure radiation. But this radiation is small, may be detected but for a short distance, does not have a specific frequency and has a variably long wavelength. This may be why during more than 30 years nobody has successfully observed MBH events in Earth environment though the theoretical estimation predicts about 100 of MBH events annually. Observers take note!

16. Supporting the MBH in suspended (levitated) state. The fuel injector can support the MBH in suspended state (no contact the MBH with any material surface).

The maximal suspended force equals

$$F = qV_f, \quad q = \frac{P_u}{c^2}, \quad F = \frac{P_u V_f}{c^2}, \quad (22)$$

where q is fuel consumption, kg; V_f is a fuel speed, m/s. The fuel (plasma) speed $0.01c$ is conventionally enough for supporting the MBH in suspended state.

17. AB- Generator as electric generator. When the Hawking radiation reaches the cover as radio microwaves they may be straightforwardly converted to electricity because they create a different voltage between different isolated parts of the cover sphere as in an antenna. Maximal voltage which can produces the radiation wave is

$$w = \frac{\varepsilon \varepsilon_0 E^2}{2} + \frac{\mu \mu_0 H^2}{2}, \quad w = \frac{P_r}{c}, \tag{23}$$

where w is density of radiation energy, J/m^3; E is electric intensity, V/m; H is magnetic intensity, T; $\varepsilon_0 = 8.85 \times 10^{-12}$ F/m is the coefficient of the electric permeability; $\mu_0 = 4\pi \times 10^{-7}$ N/A^2 is the coefficient of the magnetic permeability; $\varepsilon = \mu = 1$ for vacuum.

Let us take moment when $H = 0$, then

$$E = \sqrt{\frac{2w}{\varepsilon_0}} = \sqrt{\frac{2P_r}{\varepsilon_0 c}} = 2.73\sqrt{P_r} \quad U \approx bDE, \quad b = \frac{D}{\lambda} \le 1,$$

$$P_e \approx bP_r, \quad \lambda = \lambda_0 \frac{r}{r_0} = 16r, \quad b = \frac{2r}{16r} = \frac{1}{8}, \tag{24}$$

where E is electric intensity, V/m; U is voltage of AB-generator, V; b is relative size of antenna, D is diameter of the cover sphere if the cover sphere is used as a full antenna, m; P_e is power of the electric station, W.

As you see about 1/8 of total energy produced by AB- Generator we can receive in the form of electricity and 7/8ths reflects back to MBH; we may tap heat energy which convert to any form of energy by conventional (heat engine) methods. If we reflect the most part of the heat energy back into the MBH, we can have only electricity and do not have heat flux.

If we will use the super strong and super high temperature material AB-material offered in [3] the conversion coefficient of heat machine may be very high.

18. Critical mass of MBH located in matter environment. Many people are afraid the MBH experiments because BH can absorb the Earth. Let us find the critical mass of MBH which can begin uncontrollably to grow into the Earth environment. That will happen when BH begins to have more mass than mass of Hawking radiation. Below is the equation for the critical mass of initial BH. The educated reader will understand the equations below without detailed explanations.

$$dV = gdt, \quad g = \frac{GM}{r^2}, \quad dt = \frac{dr}{V}, \quad VdV = gdr, \quad \int_V^c VdV = \int_r^{r_o} \frac{GM}{r^2}dr, \quad r_0 = \frac{2G}{c^2}M, \quad V^2 = c^2 \frac{r_0}{r},$$

$$V = c\sqrt{\frac{r_0}{r}}, \quad dt = \frac{\sqrt{r}\,dr}{c\sqrt{r_0}}, \quad \int_t^0 dt = \frac{1}{c\sqrt{r_0}} \int_r^{r_o} \sqrt{r}\,dr, \quad t = \frac{2}{3c\sqrt{r_0}}\left(r^{3/2} - r_0^{3/2}\right) \approx \frac{2r^{3/2}}{3cr_0^{1/2}}, \quad r = \left(\frac{3c\sqrt{r_0}}{2}t\right)^{3/2},$$

$$r = 1.65G^{1/2}M^{1/3}t^{2/3}, \quad \overset{\bullet}{M} = \frac{P}{c^2} = \frac{hc^4}{15360\pi G^2} \frac{1}{M^2} = 4 \cdot 10^{15} \frac{1}{M^2}, \quad \text{for} \quad t = 1 \ s, \tag{25}$$

$$\overset{\bullet}{M}_e = \frac{4}{3}\pi r^3 \gamma = 6\pi \gamma G^{3/2}M \approx 10^{-4}\gamma M, \quad M = M_c e^{6\pi \gamma G^{3/2}t} \approx M_c e^{10^{-4}\gamma t}, \quad t = \frac{1}{6\pi \gamma G^{3/2}}\ln\frac{M}{M_c} \approx \frac{10^4}{\gamma}\ln\frac{M}{M_c},$$

where V is speed of environment matter absorbed by MBH, m/s; g is gravity acceleration of MBH, m/s; r is distance environment matter to MBH center, m; t is time, sec; $\overset{\bullet}{M}$ is mass loss by MBH, kg; $\overset{\bullet}{M}_e$ is mass taken from Earth environment by MBH, kg; γ is density of Earth environment, kg/m^3; M_c is critical mass of MBH when one begin uncontrollable grows, kg; t is time, sec.

Let us to equate the mass $\overset{\bullet}{M}$ radiated by MBH to mass $\overset{\bullet}{M}_e$ absorbed by MBH from Earth environment, we obtain the critical mass M_c of MBH for any environment:

$$M_c^3 = \frac{hc^4}{92160\pi^2 G^3}\frac{1}{\gamma} = 3.17 \cdot 10^{24}\frac{1}{\gamma}, \quad \text{or} \quad \gamma = 3.17 \cdot 10^{24}\frac{1}{M_c^3}, \tag{26}$$

If MBH having mass $M = 10^7$ kg (10 thousands tons) is put in water ($\gamma = 1000$ kg/m^3), this MBH can begin uncontrollable runaway growth and in short time (~74 sec) can consume the Earth into a black hole having diameter ~ 9 mm. If this MBH is located in the sea level atmosphere ($\gamma = 1.29$ kg/m^3), the initial MBH must has critical mass $M = 10^8$ kg (100 thousand tons). The critical radius of MBH is very small. In the first case ($M =$

10^7 kg) $r_0 = 1.48 \times 10^{-20}$ m, in the second case ($M = 10^8$ kg) $r_0 = 1.48 \times 10^{-19}$ m. Our MBH into AB-Generator is not dangerous for Earth because it is located in vacuum and has mass thousands to millions times less than the critical mass.

However, in a moment of extreme speculation, if far future artificial intelligence (or super-small reasoning) beings will be created from nuclear matter [3] they can convert the Earth into a black hole to attempt to access quick travel to other stars (Solar systems), past and future Universes and even possibly past and future times.

19. General note. We got our equations in assumption $\lambda/\lambda_o = r/r_o$. If $\lambda/\lambda_o = (r/r_o)^{0.5}$ or other relation, the all above equations may be easy modified.

AB-Generator as Photon Rocket

The offered AB- Generator may be used as the most efficient photon propulsion system (photon rocket). The photon rocket is the dream of all astronauts and space engineers, a unique vehicle) which would make practical interstellar travel. But a functioning photon rocket would require gigantic energy. The AB-Generator can convert any matter in energy (radiation) and gives the maximum theoretical efficiency.

The some possible photon propulsion system used the AB –Generator is shown in Fig.7. In simplest version (*a*) the cover of AB generator has window 3, the radiation goes out through window and produces the thrust. More complex version (*c*) has the parabolic reflector, which sends all radiation in one direction and increases the efficiency. If an insert in the AB- Generator covers the lens 6 which will focuses the radiation in a given direction, at the given point the temperature will be a billions degree (see Equation (2)) and AB-Generator may be used as a photon weapon.

The maximal thrust T of the photon engine having AB- Generator may be computed (estimated) by equation:

$$T = M \overset{\bullet}{c}, \quad \text{N}, \tag{26}$$

For example, the AB-generator, which spends only 1 gram of matter per second, will produce a thrust 3×10^5 N or 30 tons.

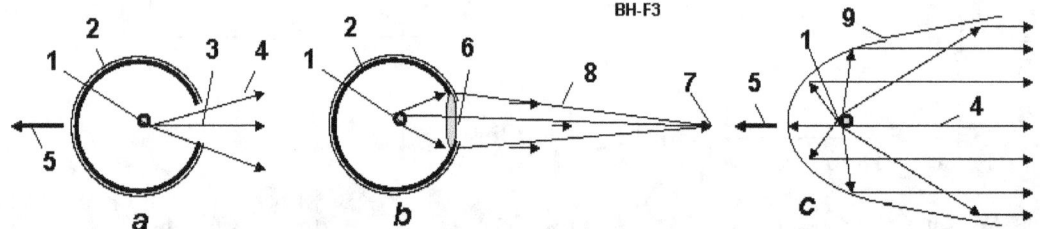

Fig.7. AB- Generator as Photon Rocket and Radiation (Photon) Weapon. (*a*) AB- Generator as a Simplest Photon Rocket; (*b*) AB- Generator as focused Radiation (photon, light or laser) weapon; (*c*) Photon Rocket with Micro-Black Hole of AB-Generator. *Notations*: 1 – control MBH; 2 – spherical cover of AB-Generator; 3 – window in spherical cover; 4 – radiation of BH; 5 – thrust; 6 – lens in window of cover; 7 – aim; 8 - focused radiation; 9 – parabolic reflector.

AB-Generator Energy Production

To estimate the energy production of an AB-Generator which is only by way of example of a computation and possible parameters. Let us take the MBH mass $M = 10^{-5}$ kg and radius of the cover sphere $r = $ 5m. No reflection. Using the equations (1)-(24) we receive:

$$P = 3.56 \cdot 10^{32} / M^2 = 3,56 \cdot 10^{42} \quad W,$$

$$r_0 = 1.48 \cdot 10^{-27} M = 1.48 \cdot 10^{-32} \quad m,$$

$$\xi = r_0 / r = 2.96 \cdot 10^{-33},$$

$$P_r = \xi P = 1.05 \cdot 10^{10}, \quad P_u = \xi \xi_r P = P_r, \quad W, \quad \xi_r = 1.$$

$$\lambda_0 = 3,73 \cdot 10^{-26} M = 3.73 \cdot 10^{-31} \quad m. \tag{27}$$

$$\lambda = 16 \cdot r = 80 \quad m.$$

$$p = \frac{P_r}{4\pi c r^2} = 0.111 \quad \frac{N}{m^2}, \quad c = 3 \cdot 10^8 \quad m/s,$$

$$\overset{\bullet}{M} = P_u / c^2 = 1.17 \cdot 10^{-7} \quad kg/s,$$

$$p_e = 1.6 \cdot 10^{-11} \frac{M^2}{r^3} = 1.28 \cdot 10^{-23} \quad N/m^2$$

Remaing main notations in equations (27): $P_r = P_u = 1.05 \times 10^{10}$ W is the useful energy (1/8 of this energy may be taken as electric energy by cover antenna, 7/8 is taken as heat); $\lambda = 80$ m is wavelength of radiation at cover sphere (that is not dangerous for people); $\overset{\bullet}{M} = 1.17 \times 10^{-7}$ kg/s is fuel consumption; $r_o = 1.48 \times 10^{-32}$ m is radius of MBH; $p_e = 1.28 \times 10^{-23}$ N/m² is explosion pressure of MBH.

Note that pressure of the explosion pressure is very small, less than a billion times of radiation pressure on the cover surface $p = 0.111$ N/m² which is not surprising because BH takes back the energy with that spent for acceleration the matter in eating the matter. As such, there is no danger of explosion of MBH.

Heat transfer and internal electric power are

$$q = \frac{P_u}{S} = \frac{P_u}{4\pi r^2} = 3.34 \cdot 10^7 \quad \frac{W}{m^2},$$

$$\text{For} \quad \delta = 2 \cdot 10^{-3} \text{ m}, \quad \lambda_h = 100, \quad \Delta T \approx q\delta / \lambda_h = 668^o K, \tag{28}$$

$$E = 2.73\sqrt{P_r} = 2.8 \cdot 10^5 \quad V/m, \quad U = E \cdot 2r = 2.8 \cdot 10^6 \quad V, \quad P_e = P_r / 8 = 1.31 \cdot 10^9 \quad W,$$

where q is specific heat transfer through the cover sphere, S is internal surface of the cover sphere, m²; δ is thickness of the cover sphere wall, m; λ_h is heat transfer coefficient for steel; ΔT is difference temperature between internal and external walls of the cover sphere; E is electric intensity from radiation on cover sphere surface, V/m; U is maximal electric voltage, V; P_e is electric power, W.

The power heat and electric output of a AB- Generator as similar to a very large complex of present day Earth's electric power stations ($P_r = 10^{10}$ W, ten billion of watts). The AB- Generator is a hundred times cheaper than a conventional electric station, especially since, heat energy can be reflected back to the MBH avoiding all the problems of conventional power conversion equipment (using only electricity from spherical cover as antenna). We hope the Large Hadron Collider at CERN can get the initial MBH needed for AB-Generator. The other way to obtain one is to find the Planck MBH (remaining from the time of the Big Bang and former MBH) and grow them to target MBH size.

Results

1. Author has offered the method and installation for converting any conventional matter to energy according the Einstein's equation $E = mc^2$, where m is mass of matter, kg; $c = 3 \cdot 10^8$ is light speed, m/s.
2. The Micro Black Hole (MBH) is offered for this conversion.
3. Also is offered the control fuel guns and radiation reflector for explosion prevention of MBH.

4. Also is offered the control fuel guns and radiation reflector for the MBH control.
5. Also is offered the control fuel guns and radiation reflector for non-contact suspension (levitation) of the MBH.
6. For non contact levitation of MBH the author also offers:
   a) Controlled charging of MBH and of ends of the fuel guns.
   b) Control charging of rotating MBH and control of electric magnets located on the ends of the fuel guns or out of the reflector-heater sphere.
7. The author researches show the very important fact: A strong gamma radiation produced by Hawking radiation loses energy after passing through the very strong gravitational MBH field. The MBH radiation can reach the reflector-heater as the light or short-wave radio radiation. That is very important for safety of the operating crew of the AB- Generator.
8. The author researches show: The matter particles produced by the MBH cannot escape from MBH and can not influence the Hawking radiation.
9. The author researches show another very important fact: The MBH explosion (hundreds and thousands of TNT tons) in radiation form produces a small pressure on the reflector-heater (cover sphere) and does not destroys the AB- Generator (in a correct design of AB-generator!). That is very important for safety of the operating crew of the AB-generator.
10. The author researches show another very important fact: the MBH cannot capture by oneself the surrounding matter and cannot automatically grow to consume the planet.
11. As the initial MBH can be used the Planck's (quantum) MBH which *may* be everywhere. The offered fuel gun may to grow them (or decrease them) to needed size or the initial MBH may be used the MBH produce Large Hadron Collider (LHC) at CERN. Some scientists assume LHC will produce one MBH every second (86,400 MBH in day). The cosmic radiation also produces about 100 MBH every year.
12. The spherical dome of MBH may convert part of the radiation energy to electricity.
13. A correct design of MBH generator does not produce the radioactive waste of environment.
14. The attempts of many astronomers find (detect) the MBH by a MBH exposure radiation will not be successful without knowing the following: The MBH radiation is small, may be detected only over a short distance, does not have specific frequency and has a variable long wavelength.

Discussion

Our equations are based upon the assumption $\lambda/\lambda_o = r/r_o$. If $\lambda/\lambda_o = (r/r_o)^{0.5}$ or other relation, the all above equations may be easy modified. The Hawking article was published 34 years ago (1974)[1] and since then hundreds of scientific works based in Hawking work appears and no known facts creates doubt in the possibility of Hawking radiation but neither it is not proven so that the Hawking radiation may not exist. The Large Hadron Collider has the main purpose to create the MBHs and detect the Hawking radiation [5].

Conclusion

The AB- Generator could create a revolution in many industries (electricity, car, ship, transportation, etc.) that allows designing photon rockets and flight to other star systems. The maximum possible efficiency is obtained and a full solution possible for the energy problem of humanity. These overwhelming prospects urge us to research and develop this achievement of science.
The articles usefull for thos topics are below in References.

Acknowledgement

The author wishes to acknowledge Joseph Friedlander (of Shave Shomron, Israel) for correcting the English and offering useful advice and suggestions.

References:

(The reader may find some of related articles at the author's web page http://Bolonkin.narod.ru/p65.htm; http://arxiv.org , http://www.scribd.com search "Bolonkin"; http://aiaa.org search "Bolonkin"; and in the Bolonkin's List 5.2 of Publications: http://viXra.org/abs/1604.0304; https://achive.org/details/List5.2OfBolonkinPublication42216 ,

1. Hawking, S.W. (1974), "Black hole explosions?", *Nature* 248: 30–31, doi:10.1038/248030a0, http://www.nature.com/nature/journal/v248/n5443/abs/248030a0.html.
2. Bolonkin A.A., Non-Rocket Space Launch and Flight, Elsevier, 2006, 488 pgs. http://www.archive.org/details/Non-rocketSpaceLaunchAndFlight ,
3. Bolonkin A.A., "New Concepts, Ideas, Innovations in Aerospace, Technology and the Human Sciences", NOVA, 2007, 502 pgs. http://www.archive.org/details/NewConceptsIfeasAndInnovationsInAerospaceTechnologyAndHumanSciences
4. Bolonkin A.A., Cathcart R., "Macro-Projects: Environments and Technologies", NOVA, 2008, 536 pgs. , http://www.archive.org/details/Macro-projectsEnvironmentsAndTechnologies ,
5. Bolonkin A.A., "New Technologies and Revolutionary Projects", Lulu, 2010, 324 pgs, http://www.archive.org/details/NewTechnologiesAndRevolutionaryProjects,
6. Bolonkin A.A., Femtotechnology: Nuclear AB-Matter with fantastic properties. American Journal of Engineering and Applied Science, Vol. 2, #2, 2009, pp.501-514. http://viXra.org/abs/1309.0201,
7. Bolonkin A.A., Converting of Matter to Nuclear Energy by AB-Generator. American Journal of Engineering and Applied Science, Vol. 2, #4, 2009, pp.683-693. Presented as paper AIAA-2009-5342 in 45 Joint Propulsion Conferencees, 2-5 Augest, 2009, Denver, CO. https://www.academia.edu/14515398/Converting_of_Matter_to_Nuclear_Energy , http://viXra.org/abs/1309.0200, https://www.scribd.com/search-documents?query=Bolonkin,
8. Femtotechnology: Design of the Strongest AB-Matter for Aerospace. Presented as paper AIAA-2009-4620 to 45 Joint Propulsion Conference, 2-5 August, 2009, Denver CO, USA. See also closed paper AIAA-2010-1556 in 48 Aerospace Meeting, New Horizons, 4 – 7 January, 2010, Orlando, FL, USA. Global Science Journal: http://gsjournal.net/Science-Journals/Research%20Papers-Engineering%20%28Applied%29/Download/5713 , Journal of Aerospace Engineering, Oct. 2010, Vol. 23, No. 4, pp.281-292. http://www.archive.org/details/FemtotechnologyDesignOfTheStrongestAb-matterForAerospace , http://viXra.org/abs/1401.0173 .
8. Wikipedia. Some background material in this article is gathered from Wikipedia under the Creative Commons license. http://wikipedia.org .

23 April 2009

Chapter 11
Ultra-Cold Thermonuclear Reactor. Criterion of Cold Fusion.

Abstract

All scientists know well: for reaching the nucleus fusion, we need the gigantic temperature in hundreds millions degrees. Only in this case the kinetic energy of nucleus overcomes the repulsive electric force of nucleus and connects two initial nucleuses to one new nucleus. In last sixty years, the goverment spent the tens billion dollars attempting to develop useful thermonuclear energy. But scientists cannot yet reach a stable thermonuclear reaction. They still are promising publically, after another 15 – 20 years, and more tens of billions of US dollars to finally design the expensive workable industrial installation, which possibly will produce electric energy more expensive than current heat, wind and hydro-electric stations can in 2015.

Author uses the well-known physical laws and shows the other opposed cheap way: very low temperatures (0.01 ÷ 10K) and high pressure (some thousands or millions of atmospheres) allow reaching the same results: themonuclear fusion. He uses not kinetic energy of nucleus again repulsive force of nucleus as in conventional methods. He uses the blocking the repulsive forces of nucleus by electrons (sphere Debya), very low temperature and high pressure. In current time to reach these temperature and pressure are easily than hundreds millions degrees by magnetic or inercial confinement. New method the thermonuclear fusion very cheap and allows to use other thermonuclear fuel which are cheaper and produce the aneutronic reaction. Author offers the new Criterion for Ultra Cold Thermonuclear Fusion.

Keywords: *Ultra-cold thermonuclear fusion, Micro-thermonuclear reactor, AB-thermonuclear reactor, transportation thermonucler reactor, aerospace thermonuclear engine, nucleus fusion.*

Main Idea. Theory, Estimations, Criteriuns of Cold Fusion.

Main Idea. Theory, Estimation. Plasma is the mixture the positive (nucleus) and negative (electrons) charges. The energy gives only the fusion of positive (nucleus) charges. The repulsive electric force overcome the fusion of nucleus. The plasma is used in current nuclear reactors is rare and scientists conventionally neglect its influence of electrons in fusion of nucleus.

The nucleos pulse one other but if between them is electron, one blocks the repulsive force. For example, the electrons in atoms and molecules are block the negative nucleus charge and atom (molecules) became neitral. The atoms can overcome one to other. The electrons connect them in molecules.

Decreasing the distance between nucleus by negative charges are used in muon catalase. The heave negative muon has orbite radius in 207 times less than conventional electron. One decreases the distance (and energy for association) nucleus and allows to connect nucleus. In conventional fusion, the distance between nucleuses the scientists try to overcome by high kinetic energy (temperature). The offer method tries to overcome by electrons and compression of a fuel.

Criterium of Cool Fusion. However, if temperature is very low and pressure is very high, electron effect become significant. In plasma physics there is Debye radius – distance the charge can come to other same charge (nucleus to nucleus) not fills its charges (the other electrons are blocked the positive charge of nucleus). Debye radius is (in SU)

$$\lambda_D \approx \sqrt{\frac{\varepsilon_0 k}{e^2}}\sqrt{\frac{T}{n}} = a\sqrt{\frac{T}{n}} = \left(\frac{8.85 \cdot 10^{-12} \cdot 1.38 \cdot 10^{-23}}{1.6^2 \cdot 10^{-19 \times 2}}\right)^{1/2}\left(\frac{T}{n}\right)^{1/2} = 69\left(\frac{T}{n}\right)^{1/2} \quad [\text{m}], \quad (1)$$

where ε_0 is electric constant, $C^2/N \cdot m^2$; $k = 1.38 \cdot 10^{-23}$ is Bolzmann constant, J/K; e is charge of electron $1.6 \cdot 10^{-19}$ C; T is temperature of electrons, K; n is number of electrons into 1 m^3.

In typical conditions (hydrogen at 1 atm, $\rho = 0.1$ kg/m^3, $T = 300$K) the $n = \rho/\mu \cdot m_p = 0.1/1 \cdot 1.67 \cdot 10^{-27} = 6 \cdot 10^{25}$ 1/m^3. $\lambda_D = 1.54 \cdot 10^{-10}$ m. This is usial radius of atom H (it is closed to electron radius of molecule H_2 $r = 1.25 \cdot 10^{-10}$ m).

The radius (length, sphere) of Debye is distance which the nucleus can approach (overcome) one to other without the repulsive force of same charges. The outer electrons blockade the repulsive forces of nucleus.

The strong nucleus attactive force of nucleus begine from distance less
$$d \approx 2 \cdot 10^{-15} \text{ m.} \tag{2}$$

If we can to bring together two nucleus in this distance, we can reach the fusion. Let us substitute this value (2) to equation (1) and estimate the ratio T/n requested for it.
$$\frac{T}{n} < \frac{\lambda_D^2}{69^2} = \frac{(2 \cdot 10^{-15})^2}{4761} \approx 8.4 \cdot 10^{-34}. \quad \text{Final} \quad B = \frac{T}{n} < 8.4 \cdot 10^{-34}. \tag{3}$$

Here B is new critereon, T is temperature of fuel electrons, K; n is number (density) of electrons into 1 m^3 the fuel.

Final equation B is the first version the ***criterion of the Ultra-Cool Fusion***. It is principal different from criterium of the inercial fusion $\rho R > 1$ (where ρ is density of fuel. g/cm^3, R is radius of fuel cupsule, cm.). The inercial criterium depends from density and RADIUS of capsule and request hundred millions of fuel temperature. The offer criterium depends from density and temperature (not from capsule size). The LOWER temperature is better for cool fusion!

It is more camfortable for estimation when n is presented throw the pressure of fuel:
$$n = \frac{p}{kT}, \quad p = 10^5 p_a, \tag{4}$$
where p is fuel pressure, N/m^2; p_a is fuel pressure in atmospheres; $k = 1.38 \cdot 10^{-23}$ J/K is Bolzmann constant. Substitute (4) to (3) we get the criterion (3) in form:
$$B = \frac{T^2}{p_a} < 0.6 \cdot 10^{-5}. \tag{5}$$

For exemple, if we cool the fuel D+T at 0.7 K and press 100,000 atm, we can reach thermonuclear fusion.

We can write criterion (2) throw density of fuel:
$$n = \frac{\rho}{\mu m_p}, \tag{6}$$
where ρ is density of fuel, kg/m^3; $\mu = m/m_p$ is molyar mass (for hydrogen H $\mu = 1$, for deiterium D $\mu = 2$, fo triteum T $\mu = 3$); $m_p = 1.67 \cdot 10^{-27}$ kg is mass of proton.

Substitute (6) to (2) we receive Criterium of Cool Fusion in form:
$$B = \frac{T}{\rho} < 0.25 \cdot 10^{-6}. \tag{7}$$

Methods for reaching the need low temperature.

Let us consider the possibility of current technology to reach the temperature and pressure requested for thermonuclear fusion.

The low temperature up 0.7K may be reached by pumping of helium vapor. The temperature low 0.3K up 0.001K is reached by magnetic refrigeration. The *nuclear* magnetic refrigeration allow to get temperature about 10^{-6}K. The mixing Helium-3 and Yelium-4 allows to get temperatures low 0.3K.

In laboratories, a record low temperature of 100 pK, or 1.0×10^{-10} K in 1999.

The current apparatus for achieving low temperatures has two stages. The first utilizes a helium dilution refrigerator to get to temperatures of millikelvins, then the next stage uses adiabatic nuclear demagnetisation to reach picokelvins.

There are many methods of getting low temperatures. For example, Dilution refrigerator:

A ^{3}He/^{4}He dilution refrigerator is a cryogenic device that provides continuous cooling to temperatures as low as 2 mK, with no moving parts in the low-temperature region. The cooling power is provided by the heat of mixing of the Helium-3 and Helium-4 isotopes. It is the only continuous refrigeration method for reaching temperatures below 0.3 K.

Methods for reaching the need high pressure.

In inercial fusion the scientists try to reach the high pressure by shock wive from laser vaporation. This method is very expensive and not suitable for us. One requests the gigantic instellation (1 ÷ 15B $), big energy (It has only 1÷1.5% efficiency) and works a shot time (10^{-8} s). Author offers to use cheap simple method described below (Fig.4). This method exploids the superhard allows widely used in industry.

The date (maximum pressure) of super hard allows are presented in Table 1.

Table 1. Vickers hardness some materials.

Material	Pressure in atm.	Material	Pressure in atm
Diamond	1150,000	B_4C	300,000
c-BC$_2$N	760,000	WB$_4$	300,000
c-BN	480,000	ReB$_2$	200,000
O$_5$B$_2$	370,000	Steel 40X	40,000

As you see from (5) we need pressure p_a = 100,000 atm for fuel temperature T = 0.7K.

Description and Innovations of Ultra-Cold Thermonuclear Reactor (fuel cupsule).

*Description and work **Version 1**.* The suggested thermonuclear fusion installation (more exactly: work capsule) is presented in Fig.4. The work capsule has the strong outer cover 1, explosive 2, pressure segments 3 (they convert low pressure of explosive 2 to high pressure of segment tip 5), fuel capsule 4, tip 5 of segment 3 from hardness material, canals for direct cooling of fuel capsule 6, elastic material between pressure segment 7.

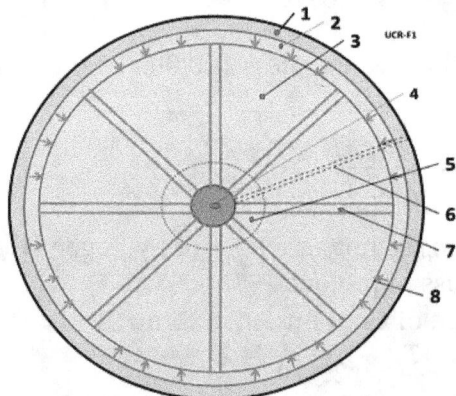

Fig.4. Ultra-Cold Thermonuclear Fusion Reactor (fuel capsule). **Version 1**. *Notations*: 1 - outer strong cover; 2 – layer of explosive; 3 – compress segment; 4 – fuel pellet; 5 – the tip from the super-hard alloy; 6 – canal for cooling fuel pellet by cooling liquid or gas; 7 - viscous grease (gasket from elastic material); 8 – pressure from explosive.

Version 1 (fuel capsule) works the next way. The fuel capsules keep in cryogenics vessel (for example liquid air). Before explosion the pellet gets an additional cooling through canal 6. After explosion the explosive layer 2, the explosive gas 8 presses to segments 3. The segments 3 increases this pressure in hundreds times and press by the hardness tips 5 the fuel pellet 4. After explosion the thermonuclear energy are used as it is described in [1] in thermonuclear reactor, or rocket engine, or weapon.

Description and work Version 2 (Fig.5). This version contains the outer cover 1, heat protection 2 (it may be vacuum); strong cover 3 (it can keep pressure from conventional explosive); fuze net 4; explosive 5; heat protection 6; thermonuclear fuel pellet 7; cooling canal 8.

Version 2 (fuel capsule) works the next way. The fuel capsules are kept into the cryogenics vessel (for example, in liquid helium). Before using, the pellet gets an additional cooling through canal 8. After explosion the explosive 5, the explosive gas presses to pellet 7. After thermonuclear explosion the thermonuclear energy are used as it is described in [1] in thermonuclear reactor, or MHD generator, or in rocket engine, or weapon.

The first version allows to get more high pellet pressure up the 1 million atmosphere and relatively high temperature up 2 K, but capsule has more size (diameter up 2 ÷ 4 sm), mass (5 ÷ 35 g) and needs in more complex design, having the pressure segments.

The second version needs in less temperature (up 0.6 K) because produces the lower pressure (up 70,000 atm). But one is simplest and has less size (diameter 0.7 ÷ 1.5 sm), and less mass (0.5 ÷ 8 g) (see estimation below).

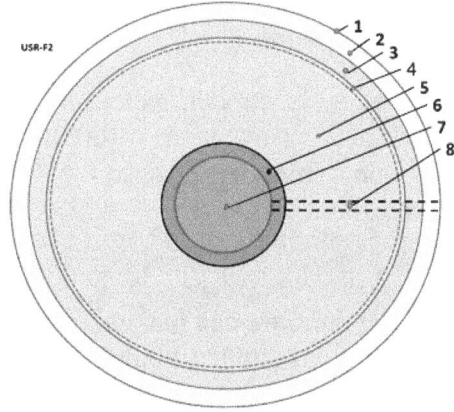

Fig.5. Ultra-Cold Thermonuclear Fusion Reactor (fuel capsule). **Version 2**. *Notations*: 1 - outer cover; 2 –heat protection (it may be vacuum); 3 – strong cover; 4 – fuse net; 5 – explosive; 6 – heat protection; 7 – fuel pellet; 8 – cooling canal.

Estimation. Let us estimate the suggested thermonuclear reactor. That is not optimal version. We demonstrate the method of estimation.

Version 1. Assume the fuel pellet has diameter 2 mm ($r = 1$ mm). Fuel is D+T. The fuel volume is

$$V = \frac{4}{3}\pi r^3 = 4.189 \times 1^3 \approx 4.2 \; mm^3 \qquad (8)$$

Fusion energy of couple nucleus D+T is $E_1 = 17.6$ MeV, density of frozen fuel D+T is $d = 0.2$ g/sm³ = 200 kg/m³, mass of fuel is $m = \rho V = 8.4 \cdot 10^{-6}$ kg. Number of fuel nucleolus and energy is:

$$n = \frac{M}{\mu m_p} = \frac{8.4 \cdot 10^{-6}}{2.5 \cdot 1.67 \cdot 10^{-27}} = 2.5 \cdot 10^{21},$$

$$E_2 = 0.5 n E_1 = 0.5 \cdot 2.5 \cdot 10^{21} \cdot 17.6 \cdot 10^6 \; eV = \qquad (9)$$
$$= 22 \cdot 10^{27} 1.6 \cdot 10^{-19} = 35 \cdot 10^8 \; J.$$

For efficiency coefficient $\eta = 0.3$ the received energy is

$$E = \eta E_2 = 0.3 \cdot 35 \cdot 10^8 \approx 10^9 \; J \qquad (10)$$

If installation produced one explosion in one second, the power is $P = 1$ Million kW. That is power average electric station.

If installation is used as a rocket engine and fuel capsule has mass $m = 40$ g = 0.04 kg, the speed of exhaust gas and thrust is

$$V = \left(\frac{2E}{m}\right)^{1/2} = \left(\frac{2 \cdot 10^9}{0.04}\right)^{0.5} = 225 \ km/s,$$

(11)

$$T = m \cdot \Delta V = 0.04 \cdot 2.25 \cdot 10^5 = 9 \cdot 10^3 \ N = 900 \ kgf$$

Conventional rocket has speed of exhaust gas about 3 km/s. Offer thermonuclear reactor has exhause speed in 75 times more. Increasing the frequency the fuel explosive, we can increase the rocket thrust. That means we can flight easy to any planets of Solar system.

If fuel capsule is used as weapon, its energy equals the 250 kg TNT (for specific energy of TNT $\approx 4.2 \cdot 10^6$ J). The initial pressure into pellet, when frozen fuel converted into gas is

$$p = n_0 kT, \quad \text{where} \quad n_0 = \frac{\rho}{\mu m_p} = \frac{200}{2.5 \cdot 1.67 \cdot 10^{-27}} = 4.8 \cdot 10^{28} \ \frac{1}{m^3}.$$

(12)

$$\text{For} \ T = 0.7 \ K, \quad p = 4.64 \cdot 10^5 \ \frac{N}{m^2} \approx 5 \ atm$$

Here $k = 1.38 \cdot 10^{-23}$ is Boltzmann constant, J/K; n_0 is number nucleus in 1 m³; $\rho = 200$ kg/m³ is density of frized (liquid) fuel in pellet. If compression is made in $T = $ const up $p = 100,000$ atm, the ratio of volume compression is $\varepsilon = 10^5/5 = 20,000$. Final radius of pellet from 1 mm decreases to $r = 1/\varepsilon^{1/3} = 1/27 = 0.037$ mm.

The full diameter of the fuel capsule will be about $1 \div 1.5$ cm.

Estimation of Version 2.

If we can produce the temperature lower $T = 0.6$K we can make the more simple fuel capcule (Fig.5). Conventional explosive be capable of pressure $p_a = 60,000 \div 80,000$ atm. For example, pressure of the explosive TNT having the specific energy $E_e = 4.2$ MJ/kg and density $\rho_e = 1654$ kg/m³ is:

$$p = E_c \rho_c = 4.2 \cdot 10^6 \times 1654 \approx 7 \cdot 10^9 \ \frac{N}{m^2} = 70,000 \ atm$$

(13)

That means criterion (5) can be applied and the fuel capsule may be made without additional segments 3 (Fig.4). Example: For $T = 0.5$K from (5) we get B-criterion

$$B = \frac{T^2}{p_a} = \frac{0.25}{7 \cdot 10^4} = 0.357 \cdot 10^{-5} < 0.6 \cdot 10^{-5}$$

(14)

The Version 2 has the pellet radius 0.5 mm. That means one produces in 8 times less power. But fuel capsule has less size (about 1 cm), less mass (about 3 g) and very simple design.

Other data and problems.

Compressing. In our consideration we assumed, the compressing the fuel pellet after explosion is isothermal process ($T = $ const). In reality one may be closed to adiabatic process (no adding and deleting heat from environment). For example, let us estimate the heading the pullet cooled up $T_2 = 0.01$K and pressed from $p_1 = 5$ atm up $p_2 = 70,000$ atm. The adiabatic process gives in end compressing the temperature

$$T_1 = T_2 \left(\frac{p_1}{p_2}\right)^{\frac{k-1}{k}} = 0.01 \left(\frac{7 \cdot 10^4}{5}\right)^{\frac{1.67-1}{1.67}} = 0.446 \ K$$

(15)

Here k is adiabatic rate. This value is from 1 up 1.67. One depends from structure of molecules and temperature. For isometric process $k = 1$, for air at room temperature one equals $k = 1.4$. We take the worst value $k = 1.67$.

If cover 6 (fig. 5) of the pellet contains the small granules having Helium-3 and Helium-4, they mixture in pressing and produce the mixture which has lower temperature than an initial components and not allows increasing temperature the fuel pellet.

The melding and boiling of Helium and fuel request a lot of energy. The ionization and dissociation of atom and molecules request the very big energy. That means one melding of Helium ($T = 0.95$K) stops the father increasing temperature.

We must use the explosive with low speed of burning; press speed must be less than the sound speed in fuel mixture in pellet. We must avoid the shock wave, use deeper cooling and protect the pellet from overheating, for example, by mixture of helium-3 and helium-4.

Below are some data which may be used for estimation.

Helium-4

Ionization energies	1st: 2372.3 kJ/mol 2nd: 5250.5 kJ/mol

Physical properties

Phase	gas
Melting point	0.95 K (−272.20 °C, −457.96 °F) (at 2.5 MPa)
Boiling point	4.222 K (−268.928 °C, −452.070 °F)
Density at stp (0 °C and 101.325 kPa)	0.1786 g/L
when liquid, at m.p.	0.145 g/cm³
when liquid, at b.p.	0.125 g/cm³
Triple point	2.177 K, 5.043 kPa
Critical point	5.1953 K, 0.22746 MPa
Heat of fusion	0.0138 kJ/mol
Heat of vaporization	0.0829 kJ/mol
Molar heat capacity	20.78 J/(mol·K)
Speed of sound	972 m/s
Thermal conductivity	0.1513 W/(m·K)

Hydrogen

Triple point	13.8033 K, 7.041 kPa
Critical point	32.938 K, 1.2858 MPa
Heat of fusion	(H_2) 0.117 kJ/mol
Heat of vaporization	(H_2) 0.904 kJ/mol
Molar heat capacity	(H_2) 28.836 J/(mol·K)

Ionization energies	1st: 1312.0 kJ/mol

Phase	gas
Melting point	13.99 K (−259.16 °C, −434.49 °F)
Boiling point	20.271 K (−252.879 °C, −423.182 °F)
Density at stp (0 °C and 101.325 kPa)	0.08988 g/L
when liquid, at m.p.	0.07 g/cm³ (solid: 0.0763 g·cm⁻³)
when liquid, at b.p.	0.07099 g/cm³

Magnetic ordering   diamagnetic

Helium-3.

Helium-3 boils at 3.19 K compared with helium-4 at 4.23 K, and its critical point is also lower at 3.35 K, compared with helium-4 at 5.2 K. Helium-3 has less than one-half of the density when it is at its boiling point: 59 gram per liter compared to the 125 gram per liter of helium-4—at a pressure of one atmosphere. Its latent heat of vaporization is also considerably lower at 0.026 kilojoules per mole compared with the 0.0829 kilojoules per mole of helium-4.

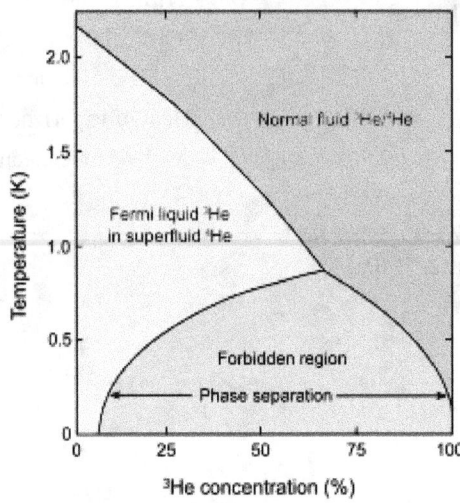

Fig.6. Phase diagram of liquid ^{3}He–^{4}He mixtures showing the phase separation.

Thickness of sphere cover. The Variant 2 useful if the sphere outer cover will keep the pressure in long time. According the Lawson criterion the received energy is proportional the time of reaction. If we can keep our pressure and temperature a long time, the probability thermonuclear reaction is increased. It is better, if the cover of fuel capsule must keep the internal pressure after conventional explosion.

Not difficult to get need equation for estimation the thickness and mass M the needed cover. The author received the next equations:

$$\overline{R} = \frac{R}{r} = \left(\frac{p}{\sigma} + 1\right)^{0.5}, \quad M = \frac{4}{3}\pi\gamma r^3\left[\left(\frac{R}{r}\right)^3 - 1\right], \quad (16)$$

where r is internal radios cover sphere, m; R is external radius of sphere, m; γ is density of cover, kg/m^3; p is pressure after conventional explosive, N/m^2; σ is safety tensile stress, N/m^2.

Property of some materials in Table 4.

Table 4. Property some material which can be used for cover.

Material	Ultimate tensile stress, MPa	Density, g/sm^3	Material	Ultimate tensile stress, MPa	Density, g/sm^3
Steel (AISI AII)	5205	7.45	Silicon (m-S1)	7000	2.33
Carbon fiber (Torcy +100G)	6370	1.80	Carbon nanotube	11,000÷63,000	0.037÷1.34
Zylon	5800	1.56	Grapheme	130,000	1

Example estimation of capsule cover for explosive press $p = 7 \cdot 10^9$ N/m^2 and tensile stress $\sigma = 2.5 \cdot 10^9$ N/m^2, $r = 2$ mm.

$$\overline{R} = \frac{R}{r} = \left(\frac{p}{\sigma}+1\right)^{0.5} = \left(\frac{7\cdot10^9}{2.5\cdot10^9}+1\right)^{0.5} 3.8^{0.5} \approx 2, \quad R = 4\, m\,,$$

$$M = \frac{4}{3}\pi\gamma r^3\left[\left(\frac{R}{r}\right)^3 - 1\right] = \frac{4}{3}3.14\cdot7450\cdot(2\cdot10^{-3})^3[2^3 - 1] = 1.75\cdot10^{-3}\, kg \approx 2\, g. \quad (16)$$

Consequently, the diameter cupsule-2 is about 8 mm.

Discussion

About sixty years ago, scientists conducted Research and Development of a thermonuclear reactor that promised then a true revolution in the energy. Existing thermonuclear reactors are very complex, expensive, large, and heavy. They cost many billions of US dollars and require many years for their design, construction and prototype testing.

For example, formation of the ITER Tokomak started in 2007 and the building costs are now over US $14 billion as of June 2015, some 3 times the original figure. The facility is expected to finish its construction phase in 2019 and will start commissioning the reactor that same year and initiate plasma experiments in 2020 with full deuterium-tritium fusion experiments starting in 2027. If ITER becomes operational, it will become the largest magnetic confinement plasma physics experiment in use, surpassing the Joint European Torus. The first commercial demonstration fusion power plant, named DEMO, is proposed to follow on from the ITER project.

The resulting design, now known as the National Ignition Facility, started construction at LLNL in 1997. NIF's main objective will be to operate as the flagship experimental device of the so-called nuclear stewardship program, supporting LLNLs traditional bomb-making role. Completed in March 2009, NIF has now conducted experiments using all 192 beams, including experiments that set new records for power delivery by a laser. The first credible attempts at ignition were initially scheduled for 2010 but ignition was not achieved as of September 30, 2012. As of October 7, 2013, the facility is understood to have achieved an important milestone towards commercialization of fusion, namely, for the first time a fuel capsule gave off more energy than was applied to it. This is still a long way from satisfying the Lawson criterion, but is a major step forward.

Many other magnetic reactors cannot stably achieve the nuclear ignition and the Lawson criterion. In future, they will have a lot of difficulties with acceptable cost of nuclear energy, with converting the nuclear energy to conventional energy, with small thermonuclear installation suitable for transportation or space exploration. Scientists promise an industrial application of thermonuclear energy after 10 – 15 years additional researches and new billions of US dollars in the future. But old methods not allow to reach it in nearest future.

In inertial confinement many scientists thought that short pressure (10^{-9} – 10^{-12} s), which they can reach by laser beam, compress the fuel capsule, but this short pressure only create the shock wave which produced the not large pressure and temperature in a limited range area in center of fuel capsule. The scientists try to reach it by increasing NIF, but plasma from initial vaporarization the cover of fuel capsule does not allow to delivery big energy. After laser beam, the fuel capsule is "naked" capsule. Capsule cannot to keep the high-energy particles of the nuclear ignition and loss them. Producing the power laser beam is very expensive and has very low efficiency (1 - 1.5%).

The author offers the new method the fusion of nucleuses. The old method try to reach very high speed of nucleus (very high temperature – in hundreds millions degree). The high kinetic energy of nucleus must overcome the repulsive force of nucleus. Author the blocks the repulsive force of nucleus by sphere Debye which allows to approach the nucleus distance when nucleus force produce the fusion. The very low temperature and high pressure decreases the Debye length for need value. The nucleus oscillations goes not depend from temperature and help the fusion.

The offered method possible allows to use reaction D+D (instead D+T) with cheap nuclear fuel D (Tritium is very expensive – about 20,000 USD for 1 g).

Conclusion

The author offers a new Method fusion in thermonuclear reaction and Installation for it. Author uses the well-known physical laws and shows the other opposed cheap way: very low temperatures $(0.01 \div 10K)$ and high pressure (some thousands or millions of atmospheres) allow to reach the same results in thermonuclear fusion. He uses not kinetic energy of nucleus again repulsive force of nucleus as in conventional methods. He uses the blocking the repulsive forces of nucleus by electrons (sphere Debye), very low temperature and high pressure. In current time to reach these temperature and pressure are easily than hundreds millions degrees by Magnetic or Inertial Confinement. New method the thermonuclear fusion very cheap and allows to use other thermonuclear fuel which are cheaper and produce the aleuronic reaction.

The offered reactor is small, cheap, may be used for cheap electricity, as engine for Earth transportation (train, truck, sea-going ships, aircraft), for space apparatus and for producing small and cheap and powerful weapons. Author offers the new Criterion for Ultra Cold Thermonuclear Fusion.

Useful data for estimation are in [1]-[3].

Acknowledgement

The author wishes to acknowledge R.B. Cathcart for correcting the author's English and offering useful advice and suggestions.

RÉFÉRENCES

(READER CAN FIND PART OF THESE ARTICLES IN WEBS: HTTPS://ARCHIVE.ORG/DETAILS/LIST5OFBOLONKINPUBLICATIONS, HTTP://BOLONKIN.NAROD.RU/P65.HTM, HTTP://ARXIV.ORG/FIND/ALL/1/AU:+BOLONKIN/0/1/0/ALL/0/1, HTTP://VIXRA.ORG).

1. AIP Physics Desk Reference, Third Edition, Springer.
2. Bolonkin A.A., **Ultra-Cold Thermonuclear Synthesis: Criterion of Cold Fusion.** 7 18 15. GSJornal: http://gsjournal.net/Science- Journals/%7B$cat_name%7D/View/6140 http://viXra.org/abs/1507.0158 , https://archive.org/details/ArticleColdFusionAfterRichard71815; www.IntellectualArchive.com , #1556; , https://www.academia.edu , https://www.researchgate.net/profile/Alexander_Bolonkin .
3. Bolonkin A.A., Cumulative Thermonuclear AB Reactor. 8 June 2015, http://viXra.org/abs/1507.0053
4. Wikipedia. Thermonuclear energy.

Published 18 June 2015

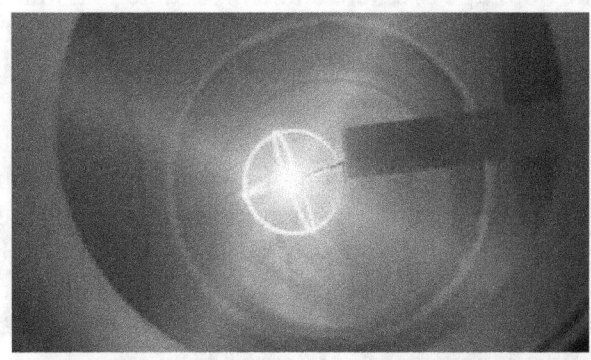

Chapter 12
Tritium Fusion: A Mistake for Thermonuclear Energy*

Abstract

For the past sixty years, scientists have spent approximately one hundred billion dollars in an attempt to develop tritium thermonuclear energy. They were unsuccessful. No stable thermonuclear reactions were achieved. Current plans are to design an expensive, but workable industrial installation. It will cost tens of billions of US dollars and will possibly only begin to produce electric energy 15 – 20 years from now. Even if the new designs were viable, they are economically unfeasible. Currently, Tritium is used for fusion ignition because the tritium-deuterium thermonuclear reaction (T+D) has the lowest ignition temperature (≈100 million degree) in contrast to deuterium thermonuclear reaction (D+D) which has a fusion ignition temperature 50 - 100 times hotter. This paper demonstrates that because tritium fuel is very expensive ($30,000/gram and more), the electricity generated by the tritium thermonuclear reactor will cost (≈ $1/kwh), at least 10 times more than conventional sources of energy (≈ $0.1/kwh, 2015). Even using Li-6, Li-7 blankets to breed tritium from fusion reactions cannot be a full solution, because, as we will show, they can only restore a maximum of 30% of the expensive tritium fuel. Hundreds of billions of dollars were spent in vain over the past sixty years for R&D of tritium fusion. It is the costliest mistake in the history of science! Research and Development (R&D) of huge, very expensive tritium fusion installations should be abandoned and in its stead, develop viable and economically feasible, inexpensive, small reactors that use deuterium fuel and high temperatures. That decreases the fuel cost by 30,000 times. Viable designs of small thermonuclear reactors have been offered by senior author in [8,9] where an analysis of the problems with the various configurations of the new small and cheap fusion reactors are detailed therein.

----
* This Chapter 12 is written together with Zarek Newman.
Keywords: *Cost of thermonuclear fuel, Cost of thermonuclear energy, Cost of thermonuclear reactor.*

INTRODUCTION

About sixty years ago, scientists conducted Research and Development of a thermonuclear reactor that promised then a true revolution in the energy industry and, especially, in aerospace. Using such reactor, aircraft could undertake flights of very long distance and extended periods of time significantly decreasing the cost of aerial transportation, freeing us from the reliance on ever-more expensive imported oil-based fuels.

Unfortunately, this task is not as easy, as scientists thought early on. Fusion reactions require a very large amount of energy to initiate in order to overcome the so-called Coulomb barrier or fusion barrier energy. The key to practical fusion power is to select a fuel that requires the minimum amount of energy to start, that is, the lowest barrier energy. The best fuel from this standpoint is a one-to-one mix of deuterium and tritium (D –T); both are heavy isotopes of hydrogen. The D - T mix has suitable low barrier energy. In order to create the required conditions, the fuel must be heated to tens of millions of degrees, and/or compressed to immense pressures. Tritium, however, is very expensive.

Brief Information about Thermonuclear Reactors

Fusion power is useful energy generated by nuclear fusion reactions. In this kind of reaction, two light atomic nuclei fuse together to form a heavier nucleus and release energy. In order for a reactor to be viable it must be able to reach *ignition* stage, that is, when the heating of the plasma by the products of the fusion reactions is sufficient to maintain the temperature of the plasma against all losses without external power input. The conditions needed for a nuclear fusion reactor to reach *ignition* stage is the "triple product" of density, confinement time, and plasma temperature T. In order to create the required conditions, the fuel must be heated to tens of millions of degrees, and/or compressed to immense pressures. The key to practical fusion power is to select a fuel that requires the minimum amount of energy to fuse, that is, the lowest barrier energy. The best-known fuel from this standpoint is a one-to-one mix of deuterium and tritium; both are heavy isotopes of hydrogen. The D-T mix has a low barrier. For the D-T reaction, the physical value is about $L = nT\tau > (10^{20} - 10^{21})$ in CI units, where T is temperature, [KeV], 1 eV = 1.16×10^4 K; n is matter density, $[1/m^3]$; τ is time, [s]. The thermonuclear reaction of $^2H + {}^3D$ realizes if $L > 10^{20}$ in CI (meter, kilogram, second) units. This number has not yet been achieved in any fusion reactor.

At present, D-T is used by two main methods of fusion: inertial confinement fusion (ICF) and magnetic confinement fusion (MCF)--for example, tokamak device.

In inertial confinement laser fusion (ICF), nuclear fusion reactions are initiated by heating and compressing a target. The target is a pellet that most often contains D –T (often only micro or milligrams). Intense focused laser or ion beams are used for compression of the pellets. The beams explosively detonate the outer material layers of the target pellet. That accelerates the underlying target layers inward, sending a shockwave into the center of each pellet mass. If the shockwave is powerful enough, and if high enough density at the center is achieved, some of the fuel will be heated enough to cause fusion reactions. In a target, which has been heated and compressed to the point of thermonuclear ignition, energy can then heat surrounding fuel to cause it to fuse as well, potentially releasing tremendous amounts of energy.

Magnetic confinement fusion (MCF). Since plasmas are very good electrical conductors, magnetic fields can also be configured to safely confine fusion fuel. A variety of magnetic configurations can be used, the basic distinction being between magnetic mirror confinement and toroidal confinement, the most popular designs being tokamaks and stellarators.

Short history of ICF thermonuclear fusion. Serious attempts at an ICF design was *Shiva*, a 20-armed neodymium laser system built at the Lawrence Livermore National Laboratory (LLNL) in California that started operation in 1978. Shiva was a "proof of concept" design, followed by the *NOVA* design with 10 times the power. Although net energy can be released even without ignition (the breakeven point), ignition is considered necessary for a *practical* power system. The resulting design, the National Ignition Facility (NIF), commenced construction at LLNL in the early 1990s, was six years behind schedule and over-budget by some $3.5 billion. Like earlier experiments, NIF failed to reach ignition and is, as of 2015, generating only about 1/3rd of the required energy levels needed to reach full fusion stage of operation.

Laser physicists in Europe have put forward plans to build a £500m facility, called HiPER, to study a new approach to laser fusion: a "fast ignition" laser facility would consist of a long-pulse laser with energy of 200 kJ to compress the fuel and a short-pulse laser with energy of 70 kJ to heat it. Basic data on a few of the current inertial laser installations:

1. NOVA uses laser NIF (USA), has 192 beams, impulse energy up 120 kJ. Can reach density of 20 g/cm^3, speed of cover is over 300 km/s. NIF has failed to reach ignition and is, as of 2013, generating about 1/3rd of the required energy levels. NIF cost is about $3.5B.
2. YiPER (EU) has impulse energy of 70 kJ.
2. OMEGA (USA) has impulse energy of 60 kJ.
3. Gekko-XII (Japan) has impulse energy of 20 kJ. Can reach density of 120 g/cm^3.
4. Febus (France) has impulse energy of 20 kJ.
5. Iskra-5 (Russia) has impulse energy of 30 kJ.

The largest current nuclear fusion experiment, JET, has resulted in fusion power production somewhat larger than the power put into the plasma, maintained for a few seconds.

The most well-known project of magnetic fusion is ITER. **ITER** (International Thermonuclear Experimental Reactor) is an international nuclear fusion research and engineering mega project, which will be the world's largest magnetic confinement plasma physics experiment. Construction of the ITER Tokamak complex started in 2013 and the building costs are now over US$14 billion as of June 2015. ITER began in 1985 as a Reagan–Gorbachev initiative and expected completion is in 2027. ITER reactor alone requires about one billion annually.

Similar projects. Other planned and proposed fusion reactors include DEMO, Wendelstein 7-X, NIF, HiPER, and MAST, as well as CFETR (China Fusion Engineering Test Reactor), a 200 MW tokamak.

Cost of Tritium Thermonuclear energy.

Cost. The lowest fuel ignition temperature for thermonuclear reaction is a mixture (ratio of weight 60%+40%) of tritium + deuterium (T+D). This temperature is tens of millions of degrees but it is still easier to attain than other possible fuels (for example D+D) which have ignition temperatures 50 - 100 times hotter than T+D fuel.

All present thermonuclear installations use tritium (T+D) fuel but they cannot reach the required temperature.

Tritium is very expensive: Currently it costs $30,000/gram [1]!

Deuterium is produced from seawater. It is cheap: Currently it costs about $1/gram.

Let us estimate the cost of energy produced from one milligram tritium (10^{-6} kg). As the cost of deuterium is negligible, it is insignificant for these computations. The estimation is very simple. We will give a detailed computation that will make it easy to understand.

Thermonuclear energy of two nuclei T+D is $E_1 = 17.6$ MeV $= 17.6 \cdot 10^6 \cdot 1.6 \cdot 10^{-19} = 2.8 \cdot 10^{-12}$ J

$$T + D \rightarrow {}^4He\ (3.5\ MeV) + n\ (12.1\ MeV). \tag{1}$$

One milligram of T contains this many nuclei:

$$N = \frac{M}{\mu m_p} = \frac{10^{-6}}{3 \cdot 1.6 \cdot 10^{-27}} = 2.1 \cdot 10^{20} \tag{2}$$

where $M = 10^{-6}$ kg is mass of one milligram; μ is number of nucleons in nucleus (in T, $\mu = 3$; in D, $\mu = 2$); $m_p = 1.6 \cdot 10^{-27}$ kg is the mass of one nucleon.

Helium energy ${}^4He$ (3.5 MeV) from (1) is easy to convert to heat. However, it is only 3.5/17.6 = 20% of the total nuclear energy. The neutron energy n (12.1 MeV) is difficult to harness, because the neutron has large penetration ability (tens of cm) and reacts with matter producing harmful radioactive isotopes.

The reaction probability is characterized by the cross section. Typical thermonuclear cross sections of main fuel particles are shown in fig.1.

The conventional heat engine has the efficiency coefficient of about $\eta = 0.3$. The total efficiency coefficient may be as low as $0.2 \cdot 0.3 = 0.06$. We will take the reactor efficiency $\eta = 0.2$. The total thermonuclear energy of one milligram of tritium is

$$E = E_1 N \eta = 2.8 \cdot 10^{-12} 2.1 \cdot 10^{20} 0.2 = 1.2 \cdot 10^8\ J \tag{3}$$

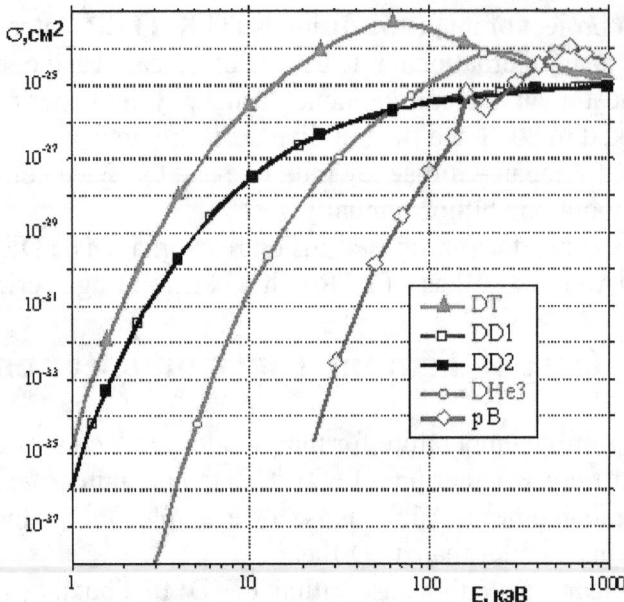

Fig.1. Thermonuclear cross section reaction of D+T, D+D (1), D+D (2), D+^{3}He, and p+B vs kinetic energy E [Kev] of the particles.

One kilowatt-hour has the energy $E_h = 10^3 \cdot 3600 = 3.6 \cdot 10^6$ W/h. Therefore, one milligram of tritium with the proper T+D ratio gives

$$C = \frac{E}{E_h} = \frac{1.2 \cdot 10^8}{3.6 \cdot 10^6} = 33.3 \text{ kwh}$$

(4)

Using current cost of $30,000/gram, one milligram of T costs $30 so we get 33.3 kwh of tritium thermonuclear energy. One kwh will cost

$$c = \frac{30}{33.3} = 0.9 \text{ dollars/kwh}$$

(5)

Currently, electric energy costs about 9 cents/kwh. That is the average price from conventional sources (gas, oil, water, wind, solar: 4÷14 cents/kwh, see [8] - [9]).

Value (5) is 10 times MORE than the current cost of energy. What did the best nuclear scientists who spent billions of dollars over tens of years accomplish? To get energy that is ten times more expensive than the present? And in addition, assuming that all of the above is free, you have to deal with the problems of radioactive waste and security. And even if we breed tritium with lithium blankets, we will show that the cost of tritium energy will still be **three-four** times more than the cost of current non-nuclear energy.

But is it possible that in the future the cost of tritium will decrease? Researchers predict the cost of tritium will increase. Currently tritium is produced in nuclear reactors as a by-product and its customers are mostly the thermonuclear research laboratories. It is very expensive. Now tritium costs $30,000/gram [1]! In the future, it is expected to cost $84,000÷130,000/g (400,000 ÷ 1,900,000 $/g, 2033) [2]. That increase in price inexorably raises the cost (5) of tritium energy by several times. It will not be acceptable as a viable power source. The only hope for a lower cost would be that a dedicated tritium production technology emerges that makes tritium at a cheaper cost than the present. The default, however, is that price will rise.

Note, we estimated only part of the full cost of tritium fusion energy, namely the cost of fuel. Gigantic thermonuclear installations, employing highly qualified staff, and additional required R&D further increase the cost of electricity produced by tritium.

Presently tritium is produced mainly from heavy water moderators in nuclear reactor. Processing several thousand tons of heavy water to extract only a few kg of tritium. Scientists working with current tritium reactors argue that Tritium fusion facilities will be able to produce more tritium than it consumes by means of lithium breeder blankets. Thus, the cost of tritium will decrease.

Upon closer scrutiny, however, this argument fails. We will subsequently consider neutron loss. But even assuming negligible loss, tritium reactors produces high energy neutrons. Capturing these neutrons requires a very thick lithium blanket. Also, the consumption of nuclear fuel is very small. Both of these factors makes the concentration of tritium in lithium very small on the order of some kg Thisonly increases the cost of its extraction and overall raises the price of tritium.
We will consider breeder blankets in more detail below.

Detailed consideration of tritium production in ICF

Possible candidates for fusion fuel include deuterium (D, ^{2}H) and tritium (T, ^{3}H) as well as helium-3 (^{3}He). These are not the only candidates, many other elements can also be fused together, but the larger electrical charge of their nuclei requires a much higher temperature for ignition. Only the fusion of the lightest elements is seriously considered as a future energy source. Although the energy density of fusion fuel is even higher than that of fission fuel, and fusion reactions sustained for a few minutes have been achieved, utilizing fusion fuel as a net energy source remains only a theoretical possibility.
The easiest nuclear reaction which requires the lowest energy, is (1). This reaction is common in research, industrial and military applications, usually as a convenient source of neutrons. Deuterium is a naturally occurring isotope of hydrogen and is commonly available. The large mass ratio of the hydrogen isotopes makes their separation easy compared to the difficult uranium enrichment process. Tritium is a natural isotope of hydrogen, but because it has a short half-life of 12.32 years, it is hard to find, store, produce, and is expensive. Consequently, the deuterium-tritium fuel cycle requires the breeding of tritium from lithium using one or two of the reactions (6) (we will show that the second reaction is uncommon):

$$n + {}^6Li \rightarrow {}^4He\ (1.92\ MeV) + T\ (2.58\ MeV)\ ,\quad n\ (>2.5\ MeV) + {}^7Li = {}^4He + T + n'. \quad (6)$$

The reactant neutron is supplied by the D-T fusion reaction shown above (1), and the one that has the greatest yield of energy. The reaction with ^{6}Li is exothermic, providing a small energy gain for the reactor. The reaction with ^{7}Li is endothermic and though it does not consume the neutron. In order to have any type of gain in neutron production at least some ^{7}Li reactions must take place to replace the neutrons lost to absorption by other elements. Most reactor designs use the naturally occurring mix of lithium isotopes.

Several drawbacks are commonly attributed to D-T fusion power:

1. It produces substantial amounts of neutrons that result in the neutron activation of the reactor materials.
2. Only about 20% of the fusion energy yield appears in the form of charged particles with the remainder carried off by neutrons, which limits the extent to which direct energy conversion techniques might be applied.

3. It requires the handling of the radioisotope tritium. Similar to hydrogen, tritium is difficult to contain and may leak from reactors in some quantity. Some estimates suggest that this would represent a fairly large environmental release of radioactivity.

The neutron flux expected in a commercial D-T fusion reactor is about 100 times that of current fission power reactors. This poses great problems for material design. To illustrate this point, after a series of D-T tests at JET, the vacuum vessel was sufficiently radioactive that remote handling was required for the year following the tests.

Production and demand of tritium.

According to the 1996 report from the Institute for Energy and Environmental Research, only 225 kg (496 lb.) of tritium has been produced in the United States since 1955. At the time of the report, only about 75 kg (165 lb.) remained due to its continual decay into helium-3.

Special heavy water reactors at the Savannah River Site produced tritium for American nuclear weapons until their closures in 1988. With the Strategic Arms Reduction Treaty (START) after the end of the Cold War, the existing supplies were sufficient for the smaller stockpile of nuclear weapons for some time.

The production of tritium was resumed with lithium irradiation rods at the reactors of the commercial Watts Bar Nuclear Generating Station in 2003–2005 followed by extraction of tritium from the rods at the new Tritium Extraction Facility at the Savannah River Site beginning in November 2006.

Canada has 21 heavy water reactors (CANDU reactors) that produce significant amounts of tritium (2.5-3.5 kg) for civilian applications and is the only source of non-military tritium.Tritium is produced in **heavy water-moderated reactors** whenever a **deuterium** nucleus captures a neutron. This reaction has a quite small absorption **cross section**, making **heavy water** a good **neutron moderator**, and relatively little tritium is produced. Even so, cleaning tritium from the moderator may be desirable after several years to reduce the risk of its escaping into the environment. The company Ontario Power Generation formerly known as "Ontario Hydro" in Darlington commissioned a **Tritium Removal Facility (TRF)** for the isolation of the isotope from the heavy water moderators. This facility chemically extracts tritium from the moderator water of all of Ontario Power Generation's CANDU reactors, using a two-stage process. Stage 1 is a *vapor phase catalytic extraction (VPCE) process* which extracts the tritium in vapor form. Stage 2 is a *cryogenic distillation process* which then distills the tritium at low temperatures and immobilizes it.They can process up to 2,500 tons of heavy water a year, and separate out about 2.5 kg (5.5 lb.) of tritium, with a purity greater than 98%, making it available for other uses. The CANDU are mostly scheduled to retire around the year 2025.

Large amounts of tritium are required for experiments and testing of thermonuclear power facilities. For example, to run ITER will require a minimum of about 3 kg of tritium. The start of the DEMO will need 4 -10 kg. Hypothetical tritium reactor would consume **56** kg of tritium to produce 1 GW of electricity per year, while global stocks of tritium for 2003 were a total of 18 kg.

Estimation of the tritium production by T+D ICR reactors.

Some scientists think: T+D Reactors are capable of producing more tritium than they consume. They show thereactions:

n + ^{6}Li → ^{4}He (1.92 MeV) + T (2.58 MeV), n (>2.5 MeV) + ^{7}Li = ^{4}He + T + n'.   (6)'

Cross sections of thesereactions are in fig.2.

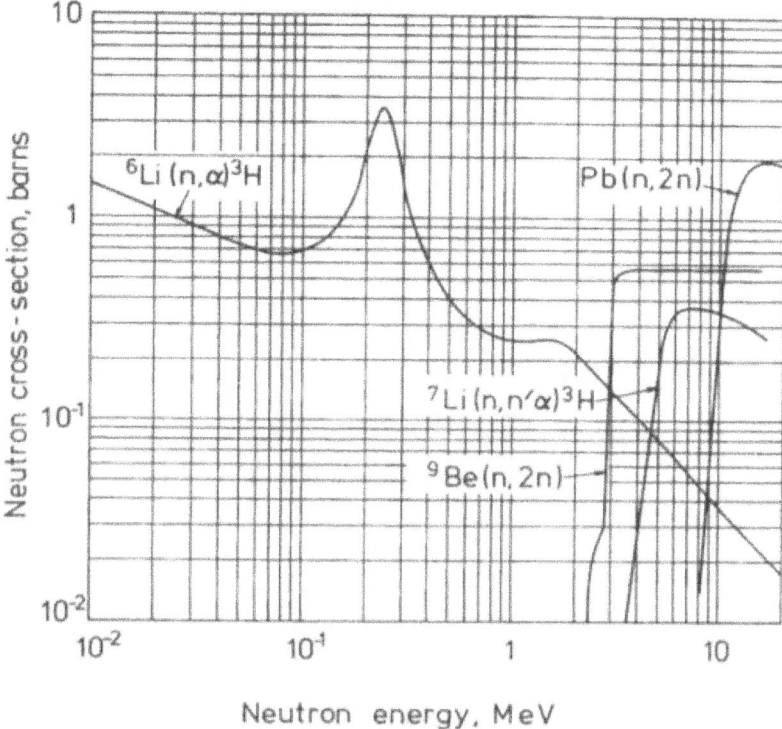

Neutron energy, MeV

Fig.2. Cross section for reaction ^{6}Li(n,α)^{3}H, ^{7}Li(n,nα)^{3}H, ^{9}Be(n,2n), Pb(n,2n). Other important cross sections include elastic and inelastic scattering cross sections for Li, Be, Pb, needed for the slowing down (moderation) of the 14 MeV primary neutrons and neutron absorption cross sections for structural material (Figure from M. Sawan).

Let us consider this suggestion in more detail.
Tritium production by T+D reactor may be represented by the equation:

$$\eta = \sum_{i=1}^{i=6} \eta_i .$$

(7)

Where η is the relative remainder of the total neutrons after loss (which dictates the amount of tritium able to be produced); η_i is the remainder in local stages (regions).

But before we estimate the losses in each of the stages we will show the impossibility of achieving η > 1 by showing the unlikeliness of the second reaction in (6). Let us consider the ideal case where there is no loss except from collisions of neutrons with other particles. The average neutron loses most of its energy in scattering (elastic collisions which converts its kinetic energy to heat) BEFORE making any useful inelastic absorption collisions (the second equation in (6)).
 Number of collisions the moving neutron has with the surrounding motionless particles for a given loss of energy is (see Kikoin [10], p. 924):

$$\xi = 1 + \frac{(A-1)^2}{2A} \ln \frac{A-1}{A+1},   N = \frac{1}{\xi} Ln \frac{E_2}{E_1},$$

(8)

Where N is number of collisions; A is nuclear number of motionless particles; E_1 is the initial energy of neutron, MeV; E_2 is the final energy of the neutron, MeV.

For ^{7}Li the value $A = 7$, maximum $E_2 = 14.1$ MeV, minimum $E_1 = 2.5$ MeV. So $N = 6.65$.

Let us estimate the probability of an absorption collision with Li-7. For E = 5 MeV the cross section of neutron scattering by Li-7 and Li-6 is about $\sigma \approx 1$ barn (1 barn = 10^{-24} cm^2) [10] p.904. For Li-7 absorption $\sigma = 0.15$ barn. Consequently, free paths$l = 1/n\sigma$ before a scattering collision is 19 cm, before an absorptioncollision with the 50% Li-7 the free paths is 292 cm. Before colliding with Li-7 the neutron has $N_s = 292/19 \approx 15$ scattering collisions with Li before an absorption collision with Li-7 and has already lost enough kinetic energy that it is below the required minimum of 2.5 MeV. That means that the probability of a reaction with Li-7 is close to zero (1/15) and $\eta < 1$. We will show that it is closer to$\eta \approx 0.31$.This shows how the breeder reactors possessing$\eta > 1$is not possible even though all of them produce additional neutrons from the first reaction in (6). As for the neutron n' produced in the second equation of (6), it is possible to show that its energy will be $E = 1.45$ MeV and cannot take part in a reaction with Li-7.

Let us now give a detailed estimate of the losses in the$\eta_i$main stages(fig.3):

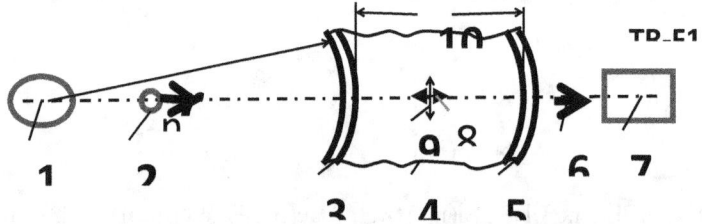

Fig.3. Loss of tritium in breeder production in T+D nuclear ICF reactor. *Notations*: 1 –nuclear reaction zone; 2 – neutron produced from reaction (1); 3 – strong spherical wall that can containthe very high nuclear pressure and temperature; 4 – layer (zone, area, mixture) of the Li-6 and Li-7 "blanket" (used as retardant, tritium manufacturer, heat transfer agent); 5 – outer spherical wall (possible neutron reflector); 6 - moving mixture 4 totritium extraction factory (once per 2 – 3 years); 7 – tritium extraction factory (plant); 8 – slow neutrons moving in radial direction; 9 - slow neutrons moving perpendicular to the radius; 10 – thickness δ of mixture 4.

1) η_1 is the percentage of fuel that is actually fused during the thermonuclear reaction. Reaction speed is linearly dependent on density of the fuel components. Reaction speed decreases when fuel density decreases and practically stops when the plasma has considerably expanded. The combustion time of ICF reactor is very small (about 10^{-7} s). All of the tritium cannot undergo combustionduring this short time and some tritium is lost by incomplete combustion. We take $\eta_1 \approx$ 0.7 (70% combustion).

2) $\eta_2$is the fraction of neutrons that managed to get past sturdy wall 3. This strong, thick wall must contain the high pressure and temperature of the small nuclear explosion. The wall absorbs part of the neutrons and producesradioactive isotopes. These harmful isotopes force the replacement of the reactor every 1 – 2 years. The fraction of neutrons not absorbed by wall 3 is $\eta_2 \approx 0.8$.

3) 4 (fig.3) is the area of the Li-6 and Li-7 blanket. The blanket serves a few purposes. It is used as the retardant, the tritium manufacturer and the heat transfer agent. As an example, we consider the mixture50% Li-6 and 50% Li-7 and reactions (6). The second reaction (with Li-7) in (6)requires a neutron with minimum kineticenergy of 2.5 MeV. That means thatif the remainder of the neutron's kineticenergy after its dissipation collisions is lower than 2.5 MeV, this reaction cannot be initiated and the energy is used to heat the lithium.

We use thefollowing data for our estimation:

Li-6. If neutron has energy E=10 MeV, Li-6 cross section area is $\sigma = 2.5 \cdot 10^{-2}$ barns (1 barn = $10^{-24} cm^2$) (Fig. 2), and mileage (free path) $l = 1/n\sigma = 750$ cm (here mixture density $n = 5.33 \cdot 10^{22}$ $1/cm^3$). This mileage is very big and the layer 10 must be very large. That means, we must use the mix in with the Li-6 an efficient retardant or have a very thick blanket. If $E \approx 0$, the Li-6 cross section is huge with$\sigma \approx 940$ barns, mileage of neutron$l = 0.02$ cm which makes it an efficient retardant.

Li-7. If neutron has energy E=10 MeV, Li-7 cross section area is $\sigma = 3.5.10^{-1}$ barns and mileage $l = 73$ cm. If $E \approx 0$, then$\sigma \approx 0$ barns because the reaction cannot occur since the neutron lacks the required energy.

Ratio of remaining neutrons is $\eta_3 \approx 1$ with no gain.
Cost of lithium is 270 \$/kg (2015).

4) Let us estimate the neutrons leak through walls 3, 5. Assume that only wall 5 has the neutron reflector with albedo $a = 0.5$ (wall 3 cannot have reflector), for neutron 2 from 6 directions we get the coefficient of neutron leak from zone 4:

$$\eta_4 = 1 - \left[\frac{1}{6} + \frac{1}{6}(1-a) \right] = 0.75 \ . \tag{9}$$

The cost of a beryllium reflector is 4480 \$/kg.

5) The loss of tritium in decay. The lithium blanket is located in the reactor for a minimum of two years. Tritium half-life is 12.3 years. That means the average remainder coefficient is about η_5 =1 - 1/12.3 = 0.92.

6) The loss of tritium in extraction. Tritium extraction is a specialized technology and a very expensive manufacturing process. But even at best we cannot hope to make a full tritium extraction. A realistic extraction coefficient is about $\eta_6 = 0.9$.

The result, we get the relative tritium (after more than two years) according to (7):

$$\eta = \eta_1\eta_2\eta_3\eta_4 \cdot \eta_5\eta_6 = 0.7 \cdot 0.8 \cdot 1 \cdot 0.75 \cdot 0.92 \cdot 0.9 \approx 0.35 < 1. \tag{10}$$

That means we can only restore 1/3of initial amount of tritium and its cost will be about current price.

In 2011, the US Department of Energy (DOE) after expending billions of dollars for R&D of nuclear energy financed the JASON organization to estimatewhether tritium reproduction of η= 1 is feasible. In report [3] (without authors??) JASON wrote that the correct installation (reactor) can haveη = 1.04 ÷ 1.10, i.e. reactor may produce 4 ÷ 10% more tritium then it consumes during 2 ÷ 3 years of operation. In ten years, it would produce an excess in the range of $1.04^{10/2}$= 1.21 to $1.1^{10/2}$ = 1.61, i.e. JASON hopes to get 21% to 61% excess tritium after 10 years of operation. They proposed to buy tritium for 10,000 ÷ 30,000 \$/gram and sell it for 100,000 \$/gram (?) and concluded that the income of the electric station from tritium production will be 10 times more than the income from electricity. Currently, in order to get1 gram of tritium we must spend 10 times more energy than what we can get from nuclear reaction (1)(T+D → ^{4}He+n).

The JASON estimation is wrong because they do not account for many losses in (10). The detailed analysis shows the probability of reaction ^{7}Li + n + 2.5 MeV → T + 2α + n' (7) is very small (<1/15).

JASON also suggested that adding trace amounts of^{9}Be or Pb to the lithium can act as neutron multipliers increasing the overall tritium production. This cannot work, however, because their reactions request high negative additional energy of neutron (3 – 10 MeV):

$$n + {}^9Be + 3 \text{ MeV} \rightarrow 2\alpha + n' + n'', \quad n + Pb + 10 \text{ MeV} \rightarrow Pb + n' + n''. \qquad (11)$$

The energy required for these reactions makes them rare occurrences. Neutrons lose energy very quickly in scattering collisions and are very soon under the 3 MeV thresholds. The probability of these reactions taking place while the required negative energy is above these values (3 MeV, 10 MeV) is very small. (0.01 – 0.02).

If a fusion station would be designed specially to produce tritium it will lose more energy than is produced since $\eta < 1$. Furthermore, the extraction of the tritium from the lithium "blanket" requires a specialized factory which consumes more energy than is produced by the obtained tritium. Scientists have been experimenting with tritium since 1960, but cannot get excess tritium ($\eta > 1$) by reactions (6).

It may be said that the efficient electric station is abad tritium production plant. The good tritium production plant is an inefficient electric station.

Summary: In reality, the cycles (1), (6) produce much less tritium than it consumes ($\eta < 1$).

Alternative methods of fusion

Every year about one billion dollars is expended in building ITER, a gigantic installation which will produce very expensive energy tens of times more expensive then what we have now. It is expected to begin to generate electricity in another 10 – 15 years after expenditure of additional tens of billions of dollars. Comparable situations exist in current tritium installations in other countries.

We must stop the profligate funding of these expensive tritium thermonuclear installations. We must not, however, stop R&D of thermonuclear energy. Our attention and funding must be focused on new ideas and designs using small cheap thermonuclear installations and cheaper fuel. Some designs of these small thermonuclear reactions are presented in [4-9].

At present in order to reach the high temperature needed for fusion, scientists use expensive laser pressure (ICF) or heating by an electric current in one direction (MCF). The author [8] offers using direct fuel heating by means of electric field and opposed fuel jets. That allows to get very high fuel temperature (up to a billion degrees). In articles [4]-[7] the author offers cumulative explosion to attain high-pressure and electric heating necessary for fusion.

Brief information about the current cost of thermonuclear fuel is presented below:

- *Tritium.* Only certain specially designed nuclear reactors can produce it. Presently Tritium cost is 30,000 \$/g [1]. In the future, the expected cost will be from 84,000÷130,000 \$/g (up to 400,000÷1,900,000 \$/g)[2].
- *Deuterium.* Seawater contains deuterium. Mean abundance in ocean water (from VSMOW) 155.76 ± 0.1 ppm (a ratio of 1 part per approximately 6420 parts), that is, about 0.015% of the atoms in a sample (by number, not weight). The World produces tens of thousands of tons in year. Cost 1 \$/g.

- *Helium-3*. Very rare isotope currently cheaper than it would be because of natural gas production. Helium-4 contains $1.3*10^{-6}/1$ of the Helium-3. Cost is 30,000 \$/g now.
- *Lithium 6 -7*. Natural mixture (Li-7 92%, Li-6 8%) costs 270 \$/kg.
- *Boron*. Cost 11,140 \$/kg.
- *Beryllium*. Cost 4480 \$/kg.
- *Uranium-238* Naturally contains 0.7% of Uranium-235. Natural uranium costs 90÷250 \$/kg.
- *Plutonium-239*. Costs 5600 \$/g.

As you can see the thermonuclear fuel D+D is the cheapest (by 30,000 times!). Moreover, the reaction D+D produces less and lower energy neutrons. However, D+T has the lowers temperature for thermonuclear reaction/low reactivity.

The required temperature for most of the thermonuclear fuels is around 100 times more than for T+D. That is why it is a popular choice for ignition experiments.

How did it happen that scientific community did not take into account the estimated cost of tritium energy?

Perhaps discussions about the future cost of thermonuclear energy was discouraged via articles not published. Or maybe it was simply assumed that the cost of the thermonuclear energy will be cheap. Perhaps it was assumed that in the future fossil fuel will become prohibitively expensive. Or maybe the assumption is that while expensive today, there would be in the future cheaper ways to produce tritium.

How much cheaper must tritium be to compete with coal? Currently by EIA figures 1 kwh takes about 1.8 cents of coal at \$40/ton, To be competitivewith coal Tritium would need to be around 40-45 times cheaper or ~\$750 a gram. *Using current cost of \$30,000/gram, one milligram of T costs \$30 and can produce 33.3 kwh of tritium thermonuclear energy. One kwh will cost*

$$c = \frac{30}{33.3} = 0.9 \text{ dollars/kwh}$$

Even if the cost of Tritium could be cheaper, current approaches are still left with the problem of thermonuclear ignition and the problems which appear in MCF (non-stable of plasma) and laser ICF (uneven compression). The present solution to these problems require gigantic, very expensive installations which provide stable jobs for scientists for many years to come.

But perhaps the approach for thermonuclear ignitions (MCF, laser ICF) was based upon a wrong assumption. Ignition by pressure (very strong magnetic field in MCF or rocket evaporation in laser ICF) have low efficiency and are very expensive methods. The strong magnetic field requires superconductivity and very low temperature. The laser pressure requires powerful lasers with low efficiency. The temperature of the liner increases from shock waves and in time the liner will need replacement along with many other components. To reach the desired temperature and pressure is a very difficult challenge with current technology.

Among the authors new ideas to achieve fusion, are new methods to achieve high temperature. This is a more efficient strategy of increasing the nuclear reactivity. In the temperature range from 10 to 100 million degrees, increasing the temperature by 10 times increases the thermonuclear reactivity by thousands of times. Increasing the pressure (density of nucleus) by ten times increases the thermonuclear reactivity only by ten times.

Temperature significantly increases the probability of thermonuclear reaction and produces fuel that can be used for other reactors. We can use inexpensive fuel to produce small neutrons, large protons, expensive elements, including tritium which can be a fuel for thermonuclear reactors. civil and military industry.

Some of new fusion ideas previously proposed (for example, ultra-cold fusion [6]), are very flexible in the nuclear fuels they can use and are not reliant solely on tritium.

Discussion

Existing thermonuclear reactors are very complex, expensive, large, and heavy. They cost many billions of US dollars and require many years for their design, construction and prototype testing. They cannot stably achieve nuclear ignition and the Lawson criterion. In the future, they will have great difficulty justifying the high cost of nuclear energy, the additional cost of converting the nuclear energy to conventional energy, and greater difficulty in designing a small thermonuclear installation suitable for transportation or space exploration. While scientists optimistically promise an industrial application of thermonuclear energy (for T+D) after 10 – 15 years which hinge upon additional research and even morefunding of billions of US dollars in the future, in the near future these old methods will not have any industrial applications nor any feasible transport engine.

Consistent failure to achieve a desired result often requires a paradigm shift, looking at the same thing from a different perspective. The pressure, time and temperature required for any particular fuel to fuse is known as the Lawson criterion L (for T+D). Lawson criterion relates to plasma production temperature, plasma density and time. The thermonuclear reaction is realized when L is more than a certain magnitude. To achieve this, two main methods of nuclear fusion have been employed: inertial confinement fusion (ICF) and magnetic confinement fusion (MCF). In inertial confinement, many scientists thought that short pressure ($10^{-9} - 10^{-12}$ s), which they can achieve by laser beam wouldsufficiently compress the fuel capsule, but this short pressure only creates a shock wave which produces insufficient high temperature and pressure to the target area in center of fuel capsule. Scientists tried to reach ignition by increasing laser NIF, but plasma from initial vaporization of the cover of fuel capsule does not deliver sufficient energy. After laser beam, the fuel capsule is essentially a "naked" capsule. Capsule cannot retain the high-energy particles for the duration of the nuclear ignition and loses them. Producing the required quality of laser beam is very expensive and has very low efficiency (1 - 3%).

The main disadvantage of all current reactors is a gigantic cost of installation and using the very expensive T fuel. As it is shown in given research the cost of tritium thermonuclear energy will be at minimum ten times more than current conventional energy. It renders meaningless all current tritium researches and installations.

The pressure strategy cannot be used for thermonuclear reaction in its classical form. The produced pressure and temperature by laser ICF and magnetic MCF are not enough for tritium thermonuclear reaction.

The paradigm that is self-limiting seems to be checkmated in an unsolvable dilemma: Because inertial confinement fusion (ICF) and magnetic confinement fusion (MCF) are the two methods employed does NOT mean that other methods would be just as ineffectual. There are other methods which are published and are all prior art and there must be other methods which have yet not been thought up, but it is these creative solutions that deserve funding.

Alternative methods to trigger fusion have been published by senior author since 1986, methods that can be adaptable for spacecraft propulsion and electricity generation. The simplest and mostperspective method to attain usable fusion energy is by means of a high voltage (70 – 100kV) condenser 100kJ and special fuel capsules containing 0.1mg fuel. [9]. The condensers require no special material and can be made from aluminum foil and film. Author is ready to consult with any interested electrical engineer who would like to verify. The important innovations are method for compressing the fuel gas into a fuel cartridge at room temperature and an electric impulse for heating the fuel up to thermonuclear temperatures.

A different butmore complex approach [6] is a new method for achieving thermonuclear reaction using very low temperatures (0.01 ÷ 10K) and high pressure (some thousands or millions of atmospheres). In this method, instead of using the kinetic energy of nucleus against repulsive force of nucleus, (as in all conventional methods under R&D), he uses the blocking of the repulsive forces of nucleus by electrons (the Debye sphere), very low temperature and high pressure. Using today's technology, it is easier to reach these temperatures and pressures than the hundreds of millions of degrees required by Magnetic or Inertial Confinement. The new method for thermonuclear fusion is very cheap and allows the use of other thermonuclear fuels which are cheaper and can produce the aneutronic reaction. The offered fusion reactor is small in bulk, cheap to construct and operate, may be used for the copious production of very cheap electricity, can be used as an engine for Earth-biosphere transportation (train, truck, sea-going ships, aircraft), for outer space apparatus propulsion and for producing small, cheap and powerful deadly explosive weapons. In brief, the author has offered a comprehensive new Criterion for Ultra Cold Thermonuclear Fusion!

In another innovation by main author [5,7] is the use of rocket electric explosive for acceleration of very small amounts of fuel to very high speeds (from 0 km/s up to 1000 km/s and more), that increases the kinetic energy (temperature) of the fuel by hundreds of times and allows the use of other (not tritium) fuel. Author noted that the mass of fuel is very small allowing to reach the high temperature, speed and pressure required for fusion.

The current ICF uses frozen fuel at close to absolute zero. That is not acceptable for practice. Author also suggested the transport nuclear engine and nuclear rocket.

These methods make possible the use ofD+D reaction (instead D+T) with cheap nuclear fuel D (Tritium is very expensive – about 30,000 USD per 1g, deuterium costs 1 $/g). These methods also allowthe use of compressed fuel-gas at room temperature obviating the requirement for super cooled fuel as is necessary in ICF fusion.

Conclusion

The estimation of tritium reproduction [3] is wrong. Tritium reproduction is a long, very expensive process and yields a maximum of only about 30% from initial. And even if there was an extremely cheap extraction process the price of T+D nuclear electricity would still be 3 – 4 timesmore expensive than it is currently.

Because tritium fuel is very expensive ($30,000/gram and more) the energy produced by a tritium thermonuclear reactor will cost ($\approx$ $1/kwh) which is at least 10 times more expensive than existing sources of energy ($\approx$ $0.1/kwh, 2015), the expenditure of hundreds of billions of dollars and sixty years for R&D of the tritium fusion were spent in vain. It is the costliest mistake in science history. The authors propose abandoning (freezing) R&D of huge very expensive tritium fusion installations and R&D instead cheap small reactors using deuterium fuel and high temperatures which decreases the fuel cost by 30,000 times.

Acknowledgment
We thank and acknowledge Joseph Friedlander's contribution to this article.

References

[1] https://www.google.com/#q=cost+of+tritium+per+gram

[2] Wittenberg, L. J., "Comparison of Tritium Production Reactors", Fusion Tech., 19, ... Tritium costs in the range $84,000 to $130,000 per gram, depending on ... www.fusion.ucla.edu/ITER.../Tritium%20Supply%20Considerations.ppt

[3] JASON: Tritium (2011). Report JSR-11-345. The Mitre Corporation, McLean, Virginia https://fas.org/irp/agency/dod/jason/

[4] Bolonkin,A.A., "Inexpensive Mini Thermonuclear Reactor". International Journal of Advanced Engineering Applications, Vol.1, Iss.6, pp.62-77 (2012). http://viXra.org/abs/1305.0046 , http://archive.org/details/InexpensiveMiniThermonuclearReactor,

[5] Bolonkin A.A. , Cumulative Thermonuclear AB-Reactor.. Vixra 7/ 8/2015,
http://viXra.org/abs/1507.0053
https://archive.org/details/ArticleCumulativeReactorFinalAfterCathAndOlga7716

[6] Bolonkin A.A., Ultra-Cold Thermonuclear Synthesis: Criterion of Cold Fusion. 7 18 2015.
http://viXra.org/abs/1507.0158, GSJornal:
http://gsjournal.net/Science- Journals/%7B$cat_name%7D/View/6140 .

[7] Bolonkin A.A., Cumulative and Impulse Mini Thermonuclear Reactors. 3 30 16,
http://viXra.org/abs/1605.0309, https://archive.org/download/ImpulseMiniThermonuclearReactors

[8]Bolonkin A.A.,Small, Non-Expensive Electric Impulse Thermonuclear Reactorwith colliding jets.
7 11 16, Preprint http://viXra.org/abs/1611.0276 ,

[9] Bolonkin A.A., Electric Cumulative Thermonuclear Reactors (17 July 2016). Under publication.

[10] Ed. Kikoin I.K., Tables of Physics Values, Moscow, AtomPublicHouse, 1976 (in Russian).

[11] Wikipedia. Cost of electricity by source,
https://en.wikipedia.org/wiki/Cost_of_electricity_by_source

[12] Wikipedia. Electricity pricing.https://en.wikipedia.org/wiki/Electricity_pricing
11 September 2016.

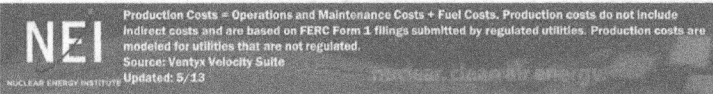

U.S. Electricity Production Costs
1995-2012, *In 2012 cents per kilowatt-hour*

2012
Coal - 3.27
Gas - 3.40
Nuclear - 2.40
Petroleum - 22.48

28.0
24.0
20.0
16.0
12.0
8.0
4.0
0.0

1995 1996 1997 1998 1999 2000 2001 2002 2003 2004 2005 2006 2007 2008 2009 2010 2011 2012

NEI
NUCLEAR ENERGY INSTITUTE

Production Costs = Operations and Maintenance Costs + Fuel Costs. Production costs do not include indirect costs and are based on FERC Form 1 filings submitted by regulated utilities. Production costs are modeled for utilities that are not regulated.
Source: Ventyx Velocity Suite
Updated: 5/13

Chapter 13
Transparent Fuel Capsule for Fusion Reactor

Abstract

For more 60 years scientists havewanted to reach confined, stable thermonuclear reaction state. They are using two main methods: ICF – Inertial Confinement Fusion and MCF – Magnetic Confinement Fusion. In ICF they have tried to heat a frozen thermonuclear fuel by highly compressing the reactive force of the fuel's vaporized cover and tohold (confine) it by inertial forces of the fuel used. In MCF they heatrarefied plasma by electric current and hold it a relative long time by enshrouding magnetic field. In ICF, only 10-20% the laser energy is used for compression and significantly less for further fuel heating.

The author is offering a significantly new design the fuel pellet (capsule) for laser ICF reactor which allows using about 90% the laser energy for pellet heating and compression work. The second advantage of the author's innovative suggested method is significantly increasing (by a hundred-fold) the time of nuclear reaction (reactivity) as well as the possibility to use the compressed gas fuel at room temperature,instead of the frozen fuel held at absolute Kelvin zero. The suggested pellet (capsule) design requires few collimated light beams (maximum 6, not 192 as with NIF) because it is using offered multi-reflect capsule(pellet). That greatly simplifies the design laser system.Possible of getting conditions will be enough for using the D-D nuclear fuel, which is monetarily less costly by 30,000 times than T-D fuel.

Keywords: *Inertial Confinement Fusion, Thermonuclear reactor, inertial thermonuclear reactor, fuel pellet of thermonuclear reactor, transparent fuel capsule for fusion reactor.*

INTRODUCTION
BRIEF INFORMATION ABOUT THERMONUCLEAR REACTORS

Fusion power is useful energy generated by nuclear fusion reactions. In this kind of reaction, two light atomic nuclei fuse together to form a heavier nucleus, releasing energy that may be confined and harnessed for various human purposes. In order for a reactor to be viable it must be able to reach *ignition* stage, that is, when the heating of the plasma by the products of the fusion reactions is sufficient to maintain the temperature of the plasma against all losses without external power input. The conditions needed for a nuclear fusion reactor to reach *ignition* stage are the "triple product" of density, confinement time, and plasma temperature T. In order to create the required conditions, the fuel must be heated to temperatures beyond mere ordinary workshop plasmas, and/or compressed to immense pressures. The key to practical fusion power is to select a fuel that requires the minimum amount of energy to fuse, that is, the lowest barrier energy. The best-known fuel from this standpoint is a one-to-one mix of deuterium and tritium; both are heavy isotopes of hydrogen. The T-D mix has a low barrier. For the T-D reaction, the physical value is about $L = nT\tau > (10^{20} - 10^{21})$ in CI units, where T is temperature, [KeV], 1 eV = 1.16×10^4 K; n is matter density, [$1/m^3$]; τ is time, [s]. The thermonuclear reaction of $^3H + {}^2D$ realizes if $L > 10^{20}$ in CI (meter, kilogram, second) units. This number has not yet been achieved in any fusion reactor.

At present, T-D is used by two main methods of fusion: inertial confinement fusion (ICF) and magnetic confinement fusion (MCF)--for example, tokomak device.

In inertial confinement laser fusion (ICF), nuclear fusion reactions are initiated by compressing and heating a target. The target is a pellet that most often contains T –D (often only micro or milligrams). Intense focused laser or ion beams are used for compression of the pellets. The beams explosively detonate the outer superficial material layers of the target pellet. That accelerates the underlying target layers inward, sending a shockwave into the center of each pellet mass. If the shockwave is powerful enough, and if high enough density at the center is achieved, some of the fuel will be heated enough to cause fusion reactions. In a target, which has been heated and compressed to the point of thermonuclear ignition, energy can then heat surrounding fuel to cause it to fuse as well, potentially releasing tremendous amounts of energy.

Magnetic confinement fusion (MCF). Since plasmas are very good electrical conductors, magnetic fields can also be configured to safely confine fusion fuel. A variety of magnetic configurations can be used, the basic distinction being between magnetic mirror confinement and toroidal confinement, the most popular designs being tokomaks and stellarators.

Short history of ICF thermonuclear fusion. Serious attempts at an ICF design was *Shiva*, a 20-armed neodymium laser system built at the Lawrence Livermore National Laboratory (LLNL) in California that started operation in 1978. Shiva was a "proof of concept" design, followed by the *NOVA* design with 10 times the power. Although net energy can be released even without ignition (the breakeven point), ignition is considered necessary for a *practical* power system. The resulting design, the National Ignition Facility (NIF), commenced construction at LLNL in the early 1990s, was six years behind schedule and over-budget by some $3.5 billion. Like earlier experiments, NIF failed to reach ignition and was, as of 2015, reportedly generating only about 1/3rd of the required energy levels needed to reach full fusion stage of operation.

Laser physicists in Europe have put forward plans to build a £500m facility, called HiPER, to study a new approach to laser fusion: a "fast ignition" laser facility would consist of a long-pulse laser with energy of 200 kJ to compress the fuel and a short-pulse laser with energy of 70 kJ to heat it. Basic data on a few of the current inertial laser installations:

1. NOVA uses laser NIF (USA), has 192 beams, impulse energy up 120 kJ. Can reach density of 20 g/cm³, speed of cover is over 300 km/s. NIF has failed to reach ignition and is, as of 2013, generating about 1/3rd of the required energy levels. NIF cost is about $3.5B.
2. HiPER (EU) has impulse energy of 70 kJ.
2. OMEGA (USA) has impulse energy of 60 kJ.
3. Gekko-XII (Japan) has impulse energy of 20 kJ. Can reach density of 120 g/cm³.
4. Febus (France) has impulse energy of 20 kJ.
5. Iskra-5 (Russia) has impulse energy of 30 kJ.

The largest current nuclear fusion experiment, JET, has resulted in fusion power production some larger than the power put into the plasma, maintained for a few seconds.

The most well-known project of magnetic fusion is ITER. **ITER** (International Thermonuclear Experimental Reactor) is an international nuclear fusion research and engineering mega project, which will be the world's largest magnetic confinement plasma physics experiment. Construction of the ITER Tokomak complex started in 2013 and the reported building costs exceeded US$14 billion by June 2015. ITER began in 1985 as a Reagan–Gorbachev initiative and expected completion is in 2027. ITER reactor alone requires about one billion USD annually to operate.

Similar projects. Other planned and proposed fusion reactors include DEMO, Wendelstein 7-X, NIF, HiPER, and MAST, as well as CFETR (China Fusion Engineering Test Reactor), a 200 MW tokamak.

Innovations in offered fuel capsule

Before consideration the offered transparent capsule, let us show the work of the current capsule (Fig.1a). Current capsule contains pellet 1 having the frizzed T-D ($\approx$ 2 mm)and cover plastic ablators 2,($\approx$ 5 mm) typically polystyrene (CH). Many power laser beams 3 (up 192) irradiate the capsule from all sides. In result the material (ablator) is heated by the absorbed laser energy and evaporates or sublimates (4). If process acts a very short time (ns), one produces the shock-wave (5). The shock-wave cumulative the high pressure in a center of fuel pellet 1. If pressure and temperature of fuel is enough the nuclear reaction occurs. If reactivity is enough, the fuel is ignited.

This method has many difficult technical problems not easily overcome successfully. One requests many identical light beams acting simultaneously for uniform radiation bathing. The shock wave acts very short time and does not produce enough heat. Only 10-20% of the laser energy is actually used for pressurization of the fuel pellet. The pellet 1 after sublimation is "naked" and cannot steadily hold the pressure for enough time for any useful nuclear reaction result.

The author herein offers two versions of his more efficient capsule design (1b, 1c).
Both pellets have a compressed (up 200 ÷ 600 atm) gas fuel (or frozen fuel).
In first version (fig.1b) the cover 6 is made from a good transparence material not heated by focused laser beam. The frequency of laser beam or fuel (or in the fuel additive) is taken such that fuel well absorbs the imposed laser radiation. The result is that the fuel absorbs 90% (or more) of the laser energy and has enough heating for sustainedignition all available fusion reaction fuel (not only small region in center pellet as in conventional method).
The massive transparencecover from heavy nuclei reflects the light nuclei of fuel (and reaction products) and significantly increases the duration time of a desired reaction.

It is difficult to select absorption frequency or absorption additive matter, the capsule is covering the strong mirror 8 (having the small holes for laser beams). It is better if mirror has the high reflectivity Fresnel cells. The mirror multiple reflects the laser beams back to the pellet and good increases the pressure to fuel.

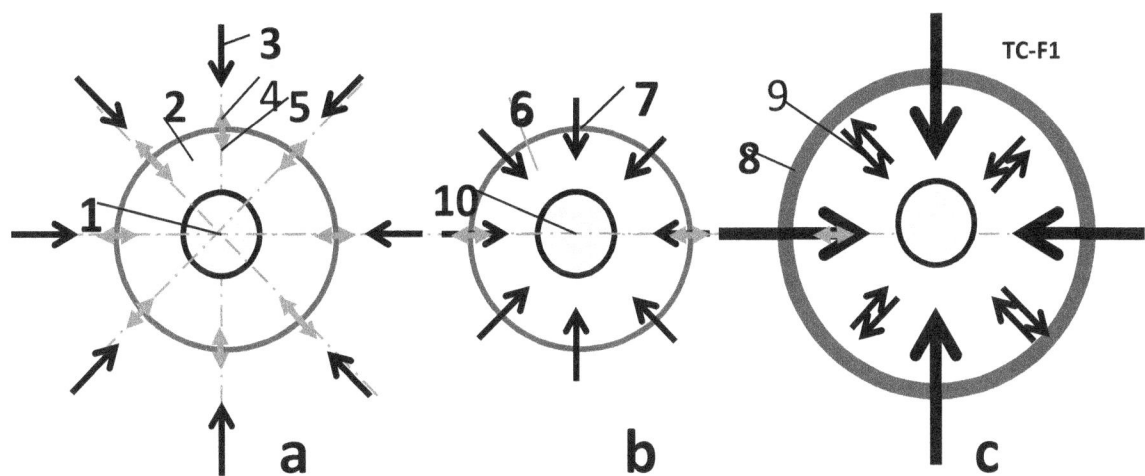

Fig.1. The current ICF capsule for laser driver (a); the offered transparency ICF capsule (b); the offered transparency-reflective capsule for laser drive. *Notations*: 1 – pellet having the thermonuclear fuel; 2 – cover which is vaporized by laser beam and produced the shock wave; 3 – laser beam; 4 – cover gas; 5 – repulsive force (pressure); 6 - high transparence cover; 7 – laser beam through the transparency cover; 8 – high efficiency light (laser beam) reflector(possible Fresnel cells); 9 – issue and reflected laser beam; 10 – fuel pellet (absorption additives are possible in version "b" and "c").

Advantages of the offered design

The herein offered method and design has workability advantages, especially in comparison the conventional fuel capsule:

1. Cconsiderably increasesthe necessary high temperature (in 10 – 20 times and more). In conventional capsule only 10 ÷ 20% the beam energy converts to pressure. The part of the residual energy converts to energy the shock wave, the part of remaining wave energy converts again in pressure at centerpellet, the part of last energy produces the high temperature at small central volume at a very short time. As result, very few nucleuses have the nuclear reaction and cannot to heat to needed temperature the rest fuel (cannot ignite the nuclear reaction).

In suggested capsule, no studies pressure cover, producing shock-wave, again compressing the fuel by shock wave, producing the temperature by shock wave. The 90% and more the laser energy become directly converted into the fuel temperature. The temperature influences to nuclear reactivity significantly more than pressure.

2. Greatly increase the reaction time (up hundred and more times). Ithappens because the mass of the cover 6 (Fig.1b) resists an expansion the hot fuel gas. The mass of the cover may be in hundreds times more than mass of fuel (mass of fuel is 0.01 – 0.1 mg). Besides the nuclear number of cover is ten times more than fuel and reaction fragments (cover has $N = 50 – 200$, fuel and reaction fragment has $N = 2 – 6$). This means the cover nucleus will be reflect the light hot fuel and fragment nucleus. The increasing reaction time greatly produces the full combustion of fuel and improves the efficiency coefficient.

3. The offered design my uses the compressed gas fuel pellets (up 800 atmospheres). That is more comfortable then the very coldcapsules (about absolute zero Kelvin) used standardly at the present time.

4. For testing the offered design may be applied any current ICF laser system, having the laser beam more 20 kJ.

5. It is possible for herein offered design the laser beam energy 50 kJ will be enough for ignition fuel D+D. In this case we solve the very important problem – final cost of the nuclear energy produced. At present, cost of T+D nuclear energy in 10 times more than cost the energy of commercial electricity generation infrastructure facilities using natural gas and oil as fuels. Change fuel T+D by D+D decreases the fuel cost in 30,000 times the cost of the nuclear energy [14](see estimation below).

Estimations and Computations
Brief Information about Plasmas and Nuclear Reactions

Below aresome equation useful for computation:

2. The Deep of Penetration of outer radiation into plasma is

$$d_p = \frac{c}{\omega_{pe}} = 5.31 \cdot 10^5 n_e^{-1/2} . \text{ [cm]} \tag{1}$$

For plasma density $n_e = 10^{22}$ 1/cm^3 $d_p = 5.31 \times 10^{-6}$ cm.

3. The Gas (Plasma) Dynamic Pressure, p_k, is

$$p_k = nk(T_e + T_i) \quad \text{if} \quad T_e = T_k \quad \text{then} \quad p_k = 2nkT , \tag{2}$$

where $k = 1.38 \times 10^{-23}$ is Boltzmann constant; T_e is temperature of electrons, ^oK; T_i is temperature of ions,^oK. These temperatures may be different; n is plasma density, 1/m^3; p_k is plasma pressure, N/m^2.

4. The gas (plasma) ion pressure, p, is

$$p = \frac{2}{3}nkT , \tag{3}$$

Here n is plasma density in 1/m^3.

5. The magnetic p_m and electrostatic pressure, p_s, are

$$p_m = \frac{B^2}{2\mu_0}, \quad p_s = \frac{1}{2}\varepsilon_0 E_s^2 , \tag{4}$$

where B is electromagnetic induction, tesla; $\mu_0 = 4\pi \times 10^{-7}$ electromagnetic constant; $\varepsilon_0 = 8.85 \times 10^{-12}$, F/m,is electrostatic constant; E_S is electrostatic intensity, V/m.

6. Ion thermal velocity is

$$v_{Ti} = \left(\frac{kT_i}{m_i}\right)^{1/2} = 9.79 \times 10^5 \mu^{-1/2} T_i^{1/2} \quad \text{cm/s}, \tag{5}$$

where $\mu = m_i/m_p$, m_i is mass of ion, kg; $m_p = 1.67 \times 10^{-27}$ is mass of proton, kg.

7. Transverse Spitzer plasma resistivity

$$\eta_\perp = 1.03 \times 10^{-2} Z \ln \Lambda T^{-3/2}, \quad \Omega \text{ cm} \quad \text{or} \quad \rho \approx \frac{0.1 Z}{T^{3/2}} \quad \Omega \text{ cm}, \tag{6}$$

where $\ln \Lambda = 5 \div 15 \approx 10$ is Coulomb logarithm, Z is charge state.

8. Reaction rates $\langle \sigma v \rangle$ (in cm^3 s^{-1}) averaged over Maxwellian distributions for low energy (T<25 keV) may be represent by

$$(\overline{\sigma v})_{DD} = 2.33 \times 10^{-14} T^{-2/3} \exp(-18.76 T^{-1/3}) \quad \text{cm}^3\text{s}^{-1}, \tag{7}$$

$$(\overline{\sigma v})_{DT} = 3.68 \times 10^{-12} T^{-2/3} \exp(-19.94 T^{-1/3}) \quad \text{cm}^3\text{s}^{-1},$$

where T is measured in keV.

9. The power density released in the form of charged particles is

$$P_{DD} = 3.3 \times 10^{-13} n_D^2 (\overline{\sigma v})_{DD}, \quad \text{W cm}^{-3}$$

$$P_{DT} = 5.6 \times 10^{-13} n_D n_T (\overline{\sigma v})_{DT}, \quad \text{W cm}^{-3} \tag{8}$$

$$P_{DHe^3} = 2.9 \times 10^{-12} n_D n_{He^3} (\overline{\sigma v})_{DHe^3}, \quad \text{W cm}^{-3}$$

Here in P_{DD} equation it is included D+T reaction.

10. Reaction rates are presented in Table 1 below:

Table 1. Reaction rates $\langle \sigma v \rangle$ (in cm^{-3} s^{-1}) averaged over Maxwellian distributions

Temperature, keV	D+D, (2a + 2b)	D+T, (1)	D+^{3}He, (3)	T+T, (4)	T+^{3}He, (6a-c)
1.0	1.5×10^{-22}	5.5×10^{-21}	10^{-26}	3.3×10^{-22}	10^{-28}
2.0	5.4×10^{-21}	2.6×10^{-19}	1.4×10^{-23}	7.1×10^{-21}	10^{-25}
5.0	1.8×10^{-19}	1.3×10^{-17}	6.7×10^{-21}	1.4×10^{-19}	2.1×10^{-22}
10.0	1.2×10^{-18}	1.1×10^{-16}	2.3×10^{-19}	7.2×10^{-19}	1.2×10^{-20}
20.0	5.2×10^{-18}	4.2×10^{-16}	3.8×10^{-18}	2.5×10^{-18}	2.6×10^{-19}
50.0	2.1×10^{-17}	8.7×10^{-16}	5.4×10^{-17}	8.7×10^{-18}	5.3×10^{-18}
100.0	4.5×10^{-17}	8.5×10^{-16}	1.6×10^{-16}	1.9×10^{-17}	7.7×10^{-17}
200.0	8.8×10^{-17}	6.3×10^{-16}	2.4×10^{-16}	4.2×10^{-17}	9.2×10^{-17}
500.0	1.8×10^{-16}	3.7×10^{-16}	2.3×10^{-16}	8.4×10^{-17}	2.9×10^{-16}
1000.0	2.2×10^{-16}	2.7×10^{-16}	1.8×10^{-16}	8.0×10^{-17}	5.2×10^{-16}

Source: AIP, Desk Reference, Third Edition, p.644.

Theory, computation and estimation of nuclear reactors and comparison one with current laser ICF.

Estimation of Laser method (ICF).

For comparison the laser and offer methods, we estimate the current ICF laser method.

Typical laser installation for ICF has the power 5 MJ and deliver to pellet about 20÷50 kJ energy. The pullet has the 1 – 10 mg liquid (frozen) fuel T+D (density 200 kg/m³), diameter of the spherical fuel pullet about1- 2 mm, diameter of an evaporative coating 4 – 10 mm.

Let us take the delivered energy $E = 50$ kJ, volume of the coating $v = 50$ mm³,Specific weight of coating $\gamma = 400$ kg/m³ (molar weight $\mu = 10$).

For these data and instant delivery of laser energy the maximum pressure in cover is

$$p = \frac{E}{v} = \frac{5 \times 10^4}{50 \cdot 10^{-9}} = 10^{12} \frac{N}{m^2} = 10^7 atm \tag{9}$$

But we don't know what part this pressure transfer to the fuel pellet.

Number of nuclear in 1 m³ of covering is

$$n = \frac{\gamma}{\mu m_p} = \frac{0.4 \cdot 10^3}{10 \cdot 1.67 \cdot 10^{-27}} = 2.4 \cdot 10^{28} \quad [m^{-3}] \tag{10}$$

Here m_p=1.67·10⁻²⁷ is mass of nucleon (proton) [kg].

Temperature of evaporating cover is

$$T = \frac{p}{nk} = \frac{10^{13}}{2.4 \cdot 10^{28} 1.38 \cdot 10^{-23}} = 3 \cdot 10^7 \quad [K] \tag{11}$$

Here $k = 1.38 \times 10^{-23}$ Boltzmann constant, J/K.

Speed of evaporated covering is

$$V = \left(\frac{8kT}{\pi \mu m_p}\right)^{0,5} = \left(\frac{8 \cdot 1.38 \cdot 10^{-23} 3 \cdot 10^7}{3.14 \cdot 10 \cdot 1.67 \cdot 10^{-27}}\right)^{0.5} = 2.51 \cdot 10^5 \; m/s = 251 \; km/s \tag{12}$$

Time of evaporating for thickness of covering $l = 2 \cdot 10^{-3}$ m is

$$t = \frac{l}{V} = \frac{2 \cdot 10^{-3}}{2.51 \cdot 10^5} = 8 \cdot 10^{-9} \quad s \tag{13}$$

Let us to consider now the process into pellet.
The density of T+D fuel particles is

$$n_f = \frac{\gamma}{\mu m_p} = \frac{200}{2.5 \cdot 1.67 \cdot 10^{-27}} = 4.8 \cdot 10^{28} \; \frac{1}{m^3} \tag{14}$$

where $\mu = 2.5$ is average molar mass of fuel T+D.

The frozen (liquid) fuel, after converting in gas, has a temperature of about $T = 20$ K. The pressure average speed V_n of particles after conversion of the fuel into gas (plasma) and sound speed V_f to fuel gas at temperature 20K are:

$$p_f = n_f kT = 4.8 \cdot 10^{28} \times 1.38 \cdot 10^{-23} \times 20 = 1.325 \cdot 10^7 N/m^2 = 132.5 \quad atm,$$

$$V_n = \left(\frac{8kT}{\pi \mu m_p}\right)^{1/2} = \left(\frac{8 \cdot 1.38 \cdot 10^{-23} \cdot 20}{3.14 \cdot 2.5 \cdot 1.67 \cdot 10^{-27}}\right)^{1/2} = 410 \; \frac{m}{s}, \tag{15}$$

$$V_f = \left(\frac{p_f}{\rho_f}\right)^{1/2} == \left(\frac{1.325 \cdot 10^7}{200}\right)^{1/2} = 257 \quad m/s.$$

The laser beam is very short – some picoseconds(10^{-12} s) in duration. That way the laser pressure can produced only shock-wave. The pellet cover protects the pellet from laser heating.

Additional fuel pressure in *center* of pellet from two opposing sound wave bump-up is

$$p_s = \rho_f (2V_f)^2 / 2 = 200 \cdot (2 \cdot 250)^2 / 2 = 25 \cdot 10^6 \; N/m^2 = 250 \; atm \tag{16}$$

Fuel temperature in *center* of small mass pellet where two opposing sound (shock) wave bump-up happens is

$$T = \frac{\pi \mu m_p (V_n + V_f)^2}{8k} = \frac{3.14 \cdot 2.5 \cdot 1.67 \cdot 10^{-27}(410+250)^2}{8 \cdot 1.38 \cdot 10^{-23}} = 51.7\,K$$

(17)

In reality, the full pressure and temperature in center of capsule is much more, but not enough for the full nuclear reaction. The author herein computes ONLY the sound wave. Any shock-wave becomes fast at short distance than the sound wave. However, in our case this computation is very complex.

Current inertial reactors have the maximal rate of fuel compressing in center of pellet about

$$\xi \approx 600$$

Criterion of ignition (for radius of pullet $R_0 = 0.02$ cm and solid or liquid fuel $\rho_0 = 0.2$ g/cm³) is

$$\rho R = \rho_o R_o \xi^{2/3} = 0.2 \cdot 0.02 \cdot (600)^{2/3} = 0.28 < 1$$

(18)

where ρ in g/cm³, R in cm. This value is insufficient ($0.28 < 1$).

You can imagine – with just a small effort and we will fulfill the criterion of ignition! Look your attention in very low temperature of fuel (17). For this temperature, the criterion may be wrong, or area of the ignition located into center of pullet may be very small, that energy is very few for ignition of all fuel?

Estimation of some parameters the nuclear reactor and pellet.

Below is not mega-project. Instead, below, are the estimations of the typical parameters of nuclear reactors.

1. Suitable thermonuclear reactions.

The corresponding reactions are

D + T →⁴He (3.5MeV) + n (14.1MeV);
D + D →T (1.01MeV) + p (3.02MeV) 50%,   (19)
D + D →³He(0.82MeV) + n (2.45MeV) 50% .

The deuterium cannot be used in the laser reactor because one requests in 100 times more ignition criterion then T + D. But D+D may be used in AB reactors with an additional heating by electric field [3 = 14].

The ³He is received in deuterium reaction may be used in next reactions:

D + ³He →⁴He (3.6MeV) + p (14.7MeV);
³He + ³He →⁴He +2p (12.9MeV).   (20)

They produce only high-energy protons, which can be directly converted in electric energy. Last reactions do not produce radio isotopic matters (no neutrons). But ³He is very expensive (30,000 $/g) as T.

Reaction D + D has the other distinct advantages:

1. One produces the protons which energy theoreticallycan be converted directly to electric energy.
2. One produces the tritium, which is expensive and may be used for thermonuclear reaction.
3. One produces less and low energy neutrons, which create radioactive matters.

The other important advantage is using the pellets with compression gas fuel. Let us take a micro-balloon (pellet) having fuel gas with $p_o > 200$ atm., radius 0.05 cm., temperature 300K. The mass fuel will be less 1 mg. The thickness of pellet having pressure 500 atmospheres is about 0.05÷0.1 mm.

Compressed gas micro-balloon (pellet) is more comfortable for working because it is unnecessary to store the fuel at lower (frozen)temperature (<10 K).

Next, we will consider mainly the fuel mass of $M = 0.1$ µg (10^{-7} kg), fuels T + D, D + D and the pellet volume$v_1 = 2$ mm³and $v_2 = 4.19$ mm³ (diameter of pellet is $d = 2$ mm).

2. Number of nucleus N in given volume $v = 2$ mm³ and density n of fuel until the nuclear reaction
   is:

Density. For $T = 300^\circ C$, Boltzmann constant $k = 1.38 \cdot 10^{-23} J/K$, we have :

$$T + D: \quad N = \frac{M}{\mu m_p} = \frac{10^{-7}}{2.5 \cdot 1.67 \cdot 10^{-27}} = 2.4 \cdot 10^{19}, \quad n = \frac{N}{v} = \frac{2.4 \cdot 10^{19}}{2 \cdot 10^{-9}} = 1.2 \cdot 10^{28} \frac{1}{m^3},$$

$$D + D: \quad N = \frac{M}{\mu m_p} = \frac{10^{-7}}{2 \cdot 1.67 \cdot 10^{-27}} = 3 \cdot 10^{19}, \quad n = \frac{N}{v} = \frac{3 \cdot 10^{19}}{2 \cdot 10^{-9}} = 1.5 \cdot 10^{28} \frac{1}{m^3}, \quad (21)$$

Pressure $p = nkT$. For $T = 300^\circ C$, Boltzmann constant $k = 1.38 \cdot 10^{-23} J/K$, we have :

$$T + D: \quad p = 1.2 \cdot 10^{28} 1.38 \cdot 10^{-23} \cdot 300 = 4.97 \cdot 10^7 \ N/m^2 = 497 \ atm,$$

$$D + D: \quad p = 1.5 \cdot 10^{28} 1.38 \cdot 10^{-23} \cdot 300 = 6.21 \cdot 10^7 \ N/m^2 = 621 \ atm, \qquad (22)$$

For volume $v = 4$ mm^3 the pressure will be in two time less (for T+D: $p = 237$ atm, for D+D $p = 296$ atm). Temperature of gas fuel before the laser heating is $T = 300$ºC.

3. *Temperature and pressure of gas fuel in pellet after heating by laser beam having energy E = 50 kJ, coefficient efficiency η = 1 is:*

Temperature:

$$T = \frac{2E}{3kN}; \quad T + D: \quad T = \frac{2 \cdot 5 \cdot 10^4}{3 \cdot 1.38 \cdot 10^{-23} 2.4 \cdot 10^{19}} = 100 \cdot 10^6 K = 8.6 \ keV,$$

$$D + D: \quad T = \frac{2 \cdot 5 \cdot 10^4}{3 \cdot 1.38 \cdot 10^{-23} 3 \cdot 10^{19}} = 80 \cdot 10^6 K = 6.9 \ keV.$$

$$(23)$$

Pressure is (for volume of pellet $v = 2 \cdot 10^{-9}$ m^3):

$$p = \frac{E}{v} = \frac{5 \cdot 10^4}{2 \cdot 10^{-9}} = 2.5 \cdot 10^{13} \frac{N}{m^2} = 2.5 \cdot 10^8 \ atm$$

$$(24)$$

4. . *Temperature and pressure of gas fuel in pellet after full Thermonuclear reaction, coefficient efficiency η = 1 is* (for volume of pellet $v = 2 \cdot 10^{-9}$ m^3, without neutron energy):

$T + D: \quad E = 0.5 \cdot N \cdot E_1 = 0.5 \cdot 2.4 \cdot 10^{19} \cdot 3.5 \cdot 10^6 = 4.2 \cdot 10^{25} \ MeV = 6.72 \cdot 10^6 \ J,$

$$p = \frac{\eta E}{v} = \frac{1 \cdot 6.72 \cdot 10^6}{2 \cdot 10^{-9}} = 3.36 \cdot 10^{15} \frac{N}{m^2} = 3.36 \cdot 10^{10} \ atm, \qquad (25)$$

$$T = \frac{p}{nk} = \frac{3,36 \cdot 10^{15}}{1.2 \cdot 10^{28} \cdot 1.38 \cdot 10^{-23}} = 2 \cdot 10^{10} K.$$

$D + D: \quad E = 0.5 \cdot N \cdot E_1 = 0.5 \cdot 3 \cdot 10^{19} \cdot 2.42 \cdot 10^6 = 3.63 \cdot 10^{25} \ MeV = 5.81 \cdot 10^6 \ J,$

$$p = \frac{\eta E}{v} = \frac{1 \cdot 5.81 \cdot 10^6}{2 \cdot 10^{-9}} = 2.9 \cdot 10^{15} \frac{N}{m^2} = 2.9 \cdot 10^{10} \ atm,$$

$$T = \frac{p}{nk} = \frac{2.9 \cdot 10^{13}}{1.5 \cdot 10^{28} \cdot 1.38 \cdot 10^{-23}} = 1.4 \cdot 10^{10} K.$$

$$(26)$$

5. *The plasma confinement time.*

From the equations of uniformly accelerated motion, we can derive an equation for the assessment of the plasma conformation time:

$$E = \frac{mV^2}{2}, \quad r \approx V_a t = \frac{Vt}{2}, \quad t = \sqrt{2} r \left(\frac{m}{E}\right)^{1/2}, \qquad (27)$$

where E is kinetic energy, J; m is mass of fuel + pellet cover, kg; V_a is average speed, m/s; V is final speed of fuel product, m/s; r is radius of pellet, m; t is conformation time, s.

Estimation are (for $r = 1 mm = 10^{-3}$ m):
For pellet **without** cover, $m = 10^{-4}$ g $= 10^{-7}$ kg, heating only the laser beam $E = 5.10^4$ J):

$$t = \sqrt{2} \cdot 10^{-3} \left(\frac{10^{-7}}{5 \cdot 10^4} \right)^{1/2} = 2 \cdot 10^{-9} \ s \ . \tag{28}$$

For pellet **with** cover, $m = 10^{-3}$ kg, heating only the laser beam:

$$t = \sqrt{2} \cdot 10^{-3} \left(\frac{10^{-3}}{5 \cdot 10^4} \right)^{1/2} = 2 \cdot 10^{-7} \ s \ . \tag{29}$$

For pellet without cover, $m = 10^{-7}$ kg, heating by nuclear reaction T+D $E = 6.72.10^6$ J.:

$$t = \sqrt{2} \cdot 10^{-3} \left(\frac{10^{-7}}{6.72 \cdot 10^6} \right)^{1/2} = 2.1 \cdot 10^{-10} \ s \ . \tag{30}$$

For pellet with cover, $m = 10^{-3}$ kg, heating by nuclear reaction T+D $E = 6.72.10^6$ J.:

$$t = \sqrt{2} \cdot 10^{-3} \left(\frac{10^{-3}}{6.72 \cdot 10^6} \right)^{1/2} = 2.1 \cdot 10^{-8} \ s \tag{31}$$

As you see the transparent cover increases the conformation time by a hundred times!
For reaction D+D having E = 5.71 MJ the conformation time is closed.
Duration of laser impulse is some very few pico-seconds (10^{-12}).

6. Radiation absorption by the pellet transparency cover.

Radiation absorption by the pellet transparency cover is very important for offered method. If cover hot has enough transparency and the cover begin to active absorb (blocked) the laser radiation, the cover will evaporate and we return to current old method, which is not efficiency as a practice shows.

Author offers to use for cover the very efficiency material used in fiber optic for the transferring the light signal (fig.2).

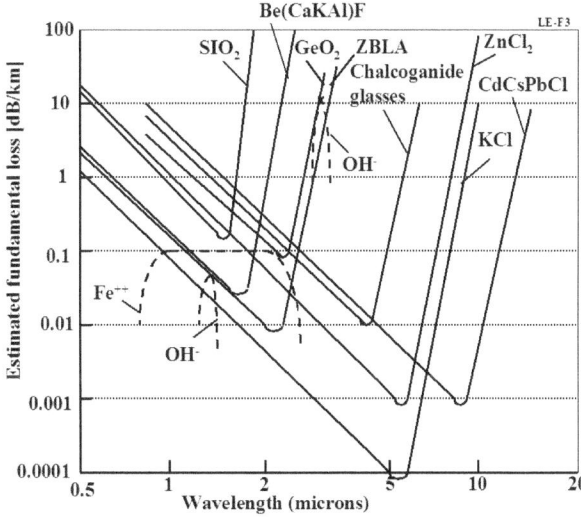

Fig. 2.Estimation of basic attenuation of some possible very low-loss materials [2], p.376.

The conventional optical matter widely produced currently in industry has an attenuation coefficient equal to $a = 2$ dB/km.

Let us estimate the loss of beam energy in the transparence cover. Coefficient a loss for thickness of cover $\delta = 2$ mm is

$$\gamma = 2.2 \cdot 10^{-4} a \delta = 2.2 \cdot 10^{-4} \cdot 2 \cdot 2 \cdot 10^{-3} = 8.8 \cdot 10^{-7} \ . \tag{32}$$

For the laser beam energy equals$E = 5 \cdot 10^4$ J, the loss energy is

$$E_l = \gamma E = 8.8 \cdot 10^{-7} 5 \cdot 10^4 = 4.4 \cdot 10^{-2} \ J \qquad (33)$$

Assume the pellet cover is from a quartz having the heat capability 880 J/kg·K and permissible temperature 300°C. For heating of cover to this temperature is necessary 7.7 J. That is significantly more $4.4 \cdot 10^{-2}$ J. That means no problem with heating of pellet cover (no loss the energy into good transparency cover).

7. Heating, absorption, reflection and transparency the laser beam by fuel.
Heating, absorption, reflection and transparency of the fuel by the laser beam is important problem in offered method. The fuel must absorb most of the beam energy. If fuel is plasma, then the absorption$x$ [cm] the radiation by gas fuel may be estimated by next equation:

$$x = \frac{1}{n\sigma}, \qquad (34)$$

where$n$ is density of fuel, 1/cm^3; σ is cross section area of photon into plasma, cm^2.
The crosssection of photons is presented in fig.3 [18].As you see the elements having nuclear mass 2 ÷ 6 have $\sigma = 100 \div 10,000$ cm^2. The pellet having the density $n = 1.2 \cdot 10^{22}$1/cm^3 and $\sigma = 2 \cdot 10^3$barns has

$$x = \frac{1}{1.2 \cdot 10^{22} 2 \cdot 10^3 10^{-24}} = 0.042 \ cm \qquad (35)$$

That is enough for transmitted beam energy to the receiving fuel.

If fuel pellet will be reflects or passes radiation, you must use the pellet design fig.1c, having the reflector over pellet cover.

The beam pressure is

$$p = \frac{(1+\rho)E}{cst} = \frac{(1+0) \cdot 5 \cdot 10^4}{3 \cdot 10^8 12.6 \cdot 10^{-9} \cdot 3 \cdot 10^{-12}} = 4.4 \cdot 10^{15} \ \frac{N}{m^2} = 4.4 \cdot 10^{10} \ atm \qquad (36)$$

Here ρ is coefficient reflectivity, < 1; E is energy of beam, J; $c = 3 \cdot 10^8$ m/s is light speed; s is surface area of spherical pellet ($d = 2$ mm); t is beam time , sec.

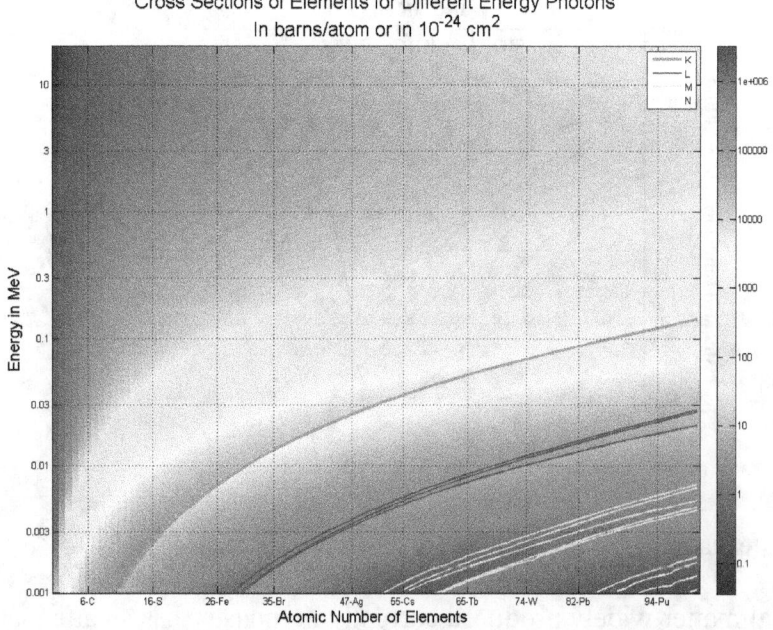

Fig.3. Cross Section of Elements for Different Energy Photon.

The beam pressure is gigantic. In case multi reflectivity the pressure will increases in some times. But time is very short and this pressure can only create the shock wave.

For increasing the beam energy absorption we can add in fuel the matter which active absorbs the need light, for example, polystyrene (OH), or Fe (see fig.2). NIF used the wave $\lambda = 0.351 \cdot 10^{-6}$, $1.053 \cdot 10^{-6}$ m.

8. Lawson's criterion.

Lawson's criterion of nuclear reaction is

$$L = nT\tau \text{ f } 10^{20} \div 10^{21},$$

(37)

where n is density of plasma, $1/m^3$; T is temperature, keV; τ is time, sec.

After heating the pellet $v = 2$ mm^3 by laser beam $E = 50$ kJ the fuel T+D we have $n = 1.2 \cdot 10^{28}$ 1/m^3, temperature $T = 10^8$K $= 10^8 \cdot 0.86 \cdot 10^{-4}$ eV $= 8.6$ keV, $\tau = 2 \cdot 10^{-7}$ s, (see above). Substitute these values into (37) we have:

For fuel T + D $L = 1.2 \cdot 10^{28} \cdot 8.6 \cdot 2 \cdot 10^{-7} = 2.06 \cdot 10^{22}$ f $(10^{20} \div 10^{21})$.

For fuel D+D after heating the pellet $v = 2$ mm^3 by laser beam $E = 50$ kJ we have $n = 1.5 \cdot 10^{28}$ 1/m^3, temperature $T = 0.8 \cdot 10^8$K $= 0.8 \cdot 10^8 \cdot 0.86 \cdot 10^{-4}$ eV $= 6.9$ keV, $\tau = 2 \cdot 10^{-7}$ s. Substitute this values into (37) we have:

For fuel D + D $L = 1.5 \cdot 10^{28} \cdot 6.9 \cdot 2 \cdot 10^{-7} = 2.06 \cdot 10^{22}$ f $(10^{20} \div 10^{21})$.

Lawson criterion is fulfilling.

For conventional pellet WITHOUT transparency cover having fuel T+D, $v = 2$ mm^3, $n = 1.2 \cdot 10^{28}$ 1/m^3, temperature $T = 0.15 \cdot 10^8$K $= 10^8 \cdot 0.15 \cdot 0.86 \cdot 10^{-4}$ eV $= 1.29$keV (here 0.15 is part of beam energy used for heating the fuel by current method), $\tau = 2 \cdot 10^{-9}$ s, (see above), we have:

For fuel T + D $L = 1.2 \cdot 10^{28} \cdot 1.29 \cdot 2 \cdot 10^{-9} = 3.1 \cdot 10^{19}$ p $(10^{20} \div 10^{21})$.

Lawson criterion is **not** fulfilling.

9. The mean free path of the nuclear reaction particles.

For transferringenergy,the mean free path of the nuclear reaction particles must be into the fuel i.e. less then the radius of the fuel pellet. Otherwise, the particles leave the fuel without transfer the full its energy to nuclear fuel. Thatway all neutrons having the big free path (great penetrating power) leavethe fuel without heating the fuel. They heat only the reactor body and produce the radioactive isotopes in body matter. The heating the fuel can only the charged particles.

Let us to estimate the mean free path the charge particles the reaction (19). Author has only the mean free path the α-particles (⁴He)having mass $m = 4$ and charge $Z=2$ [15] p.955, Fig.44.8 versus energy for air ($m = 14$) and pressure 1 atm. He recalculated them to other particles mass ($m = 1, 3$) other pressure ($p = 300$ atm), other charges ($Z = 1$) and other fuel mass ($m = 2, 2.5$). The results of estimation are shown in Table 2 immediately following:

Table 2. Estimation of the mean free path of the nuclear reaction particles

Particles. Free path as α, (energy, MeV)	Free pathas α, mm, in air at 1 atm	Free path as α,mm, in air at 300atm	Correct. coeffic. on mass particle	Free path as α, mm, correct. in mass	Correct. coeffic. of charge particle 2/Z	Free path as α, mm, correct. in Z	Correct. coeffic. of invar. mass (m air/ m fuel)	Final Free path ,mm, in fuel at 300atm
⁴He (3.5 MeV)	20	0.067	1	0.067	1 (Z=2)	0.067	14/2.5	0.375
T=³H (1 MeV)	5	0.017	3/4	0.0127	2 (Z=1)	0.0254	14/2	0.178
³He (0.8 MeV)	4	0.013	3/4	0.0097	1 (Z=2)	0.0097	14/2	0.068
p (3 MeV)	18	0.067	1/4	0.0167	2 (Z=1)	0.0334	14/2	0.234

The free pass of reaction particles limits the minimal radius (mass) of fuel. The result of computation shows the direct efficiency using the electric energy of the charged particles is very difficult because the most part of their energy will absorb the fuel for pellet heating.

10. Cost of the thermonuclear fuel. (2016)

Deuterium. The sea water contains deuterium about $1.55 \cdot 10^{-4}$ %. The World produces about tens thousand tons in year. Cost 1 \$/g.

Tritium. The special nuclear reactors can produced it. Now the cost is 30,000 \$/g. In future an expected cost will be from 100K÷200K \$/g.

Helium-3. Very rare isotope. The Helium-4 contains $1.3 \cdot 10^{-6}/1$ of the Helium-3. Cost is 30K \$/g. One project offers to extract it on Moon and delivery to Earth.

Lithium 6 -7. Nature mixture cost 270 \$/kg.

Barium. Cost 11140 \$/kg.

Uranium-238 contains 0.7% of Uranium-235. It cost 90÷250 \$/kg.

Plutonium-239. Cost 5600 \$/g.

As you see the thermonuclear fuel D+D is the cheapest, but T+D has the lowest temperature for thermonuclear reaction. All the current experimental thermonuclear installations are using the T+D.

Look your attention, the offered method allows to get very high thermonuclear temperature. We take $U = 15 \div 50$ kV [3-14], but no limit take $U = 100, 200, 500$ kV. The 200 kV produce the temperature $T = 200 \cdot 10^3 \cdot 1.18 \cdot 10^4 = 2.36 \cdot 10^9$ K (two billions!). As you see in fig. 4 and estimations over [12], that significantly increase the probability of thermonuclear reaction and produce a fuel for the other reactor. We canuse the cheap fuel produced few neutrons, many protons, expensive elements, which can be to made a fuel for thermonuclear reactors.

Discussion

About sixty years ago, scientists conducted Research and Development of a thermonuclear reactor that promised then a true revolution in the energy industry and, especially, in humankind's aerospace activities. Using such reactor, aircraft could undertake flights of very long distance and for extended periods and that, of course, decreases a significant cost of aerial transportation, allowing the saving of ever-more expensive imported oil-based fuels. The pressure, time and temperature required for any particular fuel to fuse is known as the Lawson criterion L. Lawson criterion relates to plasma production plasma density, temperature and time.

The thermonuclear reaction is realized when L is more certain magnitude. There are two main methods of nuclear fusion: inertial confinement fusion (ICF) and magnetic confinement fusion (MCF).

Existing thermonuclear reactors are very complex, expensive, large, and heavy. They cost many billions of US dollars and require many years for their design, construction and prototype testing. They cannot stably achieve the nuclear ignition and the Lawson criterion. In future, they will have many difficulties with acceptable cost of nuclear energy, with converting the nuclear energy to conventional energy, with small thermonuclear installation suitable for transportation or space exploration. Scientists promise an industrial application of thermonuclear energy after 10 – 15 years additional researches and new billions of US dollars in the future. However, old methods do not allow us to reach an industrial or transport engine in nearest future.

In inertial confinement many scientists thought that short pressure ($10^{-9} - 10^{-12}$ s), which they can reach by laser beam, compress the fuel capsule, but this short pressure only create the shock wave which produced the not large pressure and temperature in a limited range area in center of fuel capsule. The scientists try to reach it by increasing NIF, but plasma from initial vaporization the cover of fuel capsule does not allow to delivery big energy. After laser beam, the fuel capsule is "naked" capsule. Capsule cannot to keep the high-energy particles of the nuclear ignition and loss them. Producing the power laser beam is very expensive and has very low efficiency (1 - 3.5%).

The offered method EIF (Electric Impulse Fusion) does not have these disadvantages [3-14]. One uses the primary high pressed gas fuel ampoules and directly heats them to need high temperature by special electric impulse in special cartridge. The shell of capsule protects the fuel by the heavy elements ($\mu = 200$) having high number of nucleons A and charges Z. They reflect the light protons, D, T, repels high-energy reacted particles (D, T, ^{3}He, ^{4}He, p) back to fuel and significantly increasing the pressure and conformation time.

The laser ICF, MCF ideas cannot be used for thermonuclear reaction in its classical form. Produced temperature and pressure by laser ICF and magnetic MCF are not enough for thermonuclear reaction. In given article the author offers the new design of ICF pellet. That design allows more full using the beam energy for heating the pellet and to reach the need temperature (up 100 MeV) and using the primary compressing the gas fuel (up 700 atm) in special ampoules and increase the time of reaction in 100 times by heaver pullet cover. That increases the intensity of nuclear reaction (and temperature) in hundreds times.

The important innovations are the compressed the fuel gas into fuel cartridge at room temperature and using laser beam for *direct* heating of fuel up the thermonuclear temperatures. The current ICF uses the frozen fuel about absolute zero. That is not acceptable for practice. Author also suggested the transport nuclear engine and nuclear rocket [3-14].

The method possible allows to use reaction D+D (instead T+D) with cheap nuclear fuel D (Tritium is very expensive – about 30,000 USD per 1 g, deuterium costs 1 \$/g). One also allows using the compressed fuel-gas at room temperature. We can use the nuclear reactions, which do not produce many neutrons and gamma radiation. They are dangerous for people.

Conclusion

The author offers a new design the fuel pellet for the laser impulse thermonuclear reactors, which increases the temperature of a primary compressed nuclear fuel in hundreds of times, reaches the ignition and full thermonuclear reaction. New design of the nuclear pellet offered by its originator contains several innovations and inventions.

Main of them is using a high transparence cover of pellet, which allows to use the laser beam for direct high efficiency heating of fuel by laser beam to high temperature the hundredmilliondegrees.The second innovation is using the heavytransparent pullet cover used for increasing the compressing fuel state (reaction time).Important innovation is compressed gas fuel at room temperature and using efficiency cover reflector (fig.3c) for increasing the temperature, pressure fuel and nuclear reactivity.

The offered method is cheap and easy for testing. For its testing may be used the most currentICF installations. Closed ideas are in [3]-[14].

Acknowledgement

The author wishes to acknowledge R.B. Cathcart for correcting the author's English and offering useful advices and suggestions.

REFERENCES
(READER CAN FIND PART OF THESE ARTICLES IN WEBS:
HTTPS://ARCHIVE.ORG/DETAILS/LIST5OFBOLONKINPUBLICATIONS,HTTP://BOLONKIN.NAROD.RU/P65.HTM, HTTP://ARXIV.ORG/FIND/ALL/1/AU:+BOLONKIN/0/1/0/ALL/0/1, HTTP://VIXRA.ORG).

[1]Bolonkin, A.A., Light Pressure Engine. Patent (Auther Certificate) #1183421, USSR (priority on 5 January, 1983), 1985 (in Russian).
[2] Bolonkin A.A., Non-Rocket Space Launch and Flight.Elsevier, 2005. Or Journal of British Interplanetary Society, Vol. 57, No.11/12, 2004; pp. 379-390, p.376.
https://archive.org/details/Non-rocketSpaceLaunchAndFlightv.3

[3] Bolonkin,A.A., "Inexpensive Mini Thermonuclear Reactor". International Journal of Advanced Engineering Applications, Vol.1, Iss.6, pp.62-77 (2012). http://viXra.org/abs/1305.0046 http://archive.org/details/InexpensiveMiniThermonuclearReactor,

[4] Bolonkin A.A. , Cumulative Thermonuclear AB-Reactor.. Vixra 7/ 8/2015, http://viXra.org/abs/1507.0053 , https://archive.org/details/ArticleCumulativeReactorFinalAfterCathAndOlga7716

[5] Bolonkin A.A., Ultra-Cold Thermonuclear Synthesis: Criterion of Cold Fusion. 7 18 2015. http://viXra.org/abs/1507.0158, GSJournal: http://gsjournal.net/Science-Journals/%7B$cat_name%7D/View/6140 .

[6] Bolonkin A.A., Cumulative and Impulse Mini Thermonuclear Reactors. 3 30 16, http://viXra.org/abs/1605.0309, https://archive.org/download/ImpulseMiniThermonuclearReactors ,

[7] Bolonkin A.A.,Electric Cumulative Thermonuclear Reactors. https://archive.org/download/abolonkin_gmail_201610, http://vixra.org/abs/1610.0208 .

[8] Bolonkin, A.A., "Non Rocket Space Launch and Flight". Elsevier, 2005. 488 pgs. ISBN-13: 978-0-08044-731-5, ISBN-10: 0-080-44731-7 . http://vixra.org/abs/1504.0011 v4, Journal of BritishInterplanetary Society, Vol. 57,No.11/12, 2004; pp. 379-390. https://archive.org/details/Non-rocketSpaceLaunchAndFlightv.3.

[9] Bolonkin, A.A., "New Concepts, Ideas, Innovations in Aerospace, Technology and the Human Sciences", NOVA, 2006, 510 pgs. ISBN-13: 978-1-60021-787-6. http://viXra.org/abs/1309.0193,

[10] Bolonkin, A.A., Femtotechnologies and Revolutionary Projects. Lambert, USA, 2011. 538 p. 16 Mb. ISBN:978-3-8473-0839-0, http://viXra.org/abs/1309.0191.

[11] Bolonkin, A.A., Innovations and New Technologies (v.2).Lulu, 2014. 465 pgs. 10.5 Mb, ISBN: 978-1-312-62280-7. https://archive.org/details/Book5InnovationsAndNewTechnologiesv2102014/

[12] Bolonkin, A.A.,Small, Non-Expensive Electric Impulse Thermonuclear Reactor with Colliding Jets. 7 11 16, Preprint http://viXra.org/abs/1611.0276 .

[13] Bolonkin,A.A., "Inexpensive Mini Thermonuclear Reactor". International Journal of Advanced Engineering Applications, Vol.1, Iss.6, pp.62-77 (2012). http://viXra.org/abs/1305.0046 http://archive.org/details/InexpensiveMiniThermonuclearReactor.

[14] Bolonkin, A.A., Neumann Z., Cost of a nuclear fuel energy. In press.

[15] Kikoin I.K., Tables of Physical Values, Moscow, Atomizdat, 1975 (Russian).

[16] Koshkin N.I., Shirkevich M.G., Handbook of Elementary Physics, Moscow, Nauka, 1982 (Russian).

[17] AIP, Physics Desk Reference, 3-rd Edition. Springer, AIP PRESS.

[18] Wikipedia. Absorption of radiation. https://en.wikipedia.org/wiki/Absorption_cross_section

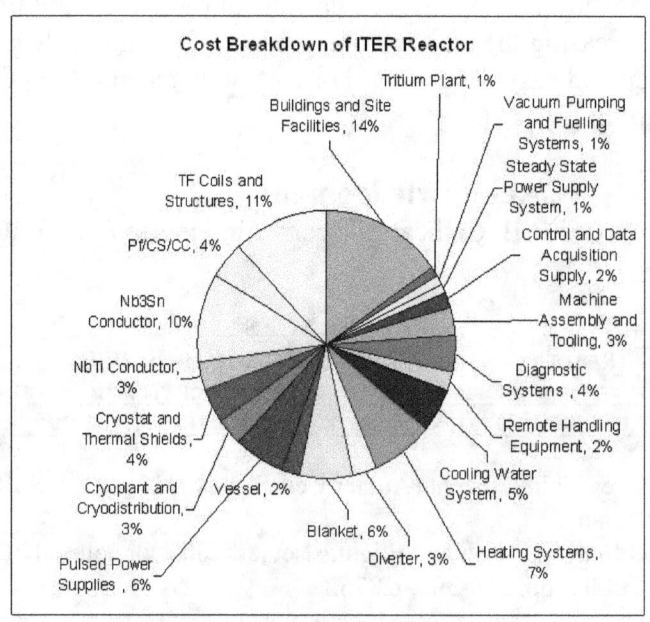

Applications

1. Thermonuclear Reactions.

Nuclear energy.

The energy released in a nuclear reaction can appear mainly in one of three ways:

- kinetic energy of the product particles;
- emission of very high energy photons, called gamma rays;
- some energy may remain in the nucleus, as a metastable energy level.

When the product nucleus is metastable, this is indicated by placing an asterisk ("*") next to its atomic number. This energy is eventually released through nuclear decay.

Kinetic energy may be released during the course of a reaction (exothermic reaction) or kinetic energy may have to be supplied for the reaction to take place (endothermic reaction). This can be calculated by reference to a table 1 of very accurate particle rest masses.

Table 1. Relative stable isotope mass. Note: 3 Li 6 is $_3^6 Li$ and so on.

H 1	1.007 825 032 23(9)	2 He 3	3.016 029 3201(25)
D 2	2.014 101 778 12(12)	4	4.002 603 254 13(6)
T 3	3.016 049 2779(24)		
Li 6	6.015 122 8874(16)	4 Be 9	9.012 183 065(82)
3 7	7.016 003 4366(45)		
5 B 10	10.012 936 95(41)	6 C 12	12.0000000(00)
11	11.009 305 36(45)	13	13.003 354 835 07(23)
		14	14.003 241 9884(40)
7 N 14	14.003 074 004 43(20)	8 O 16	15.994 914 619 57(17)
15	15.000 108 898 88(64)	17	16.999 131 756 50(69)
		18	17.999 159 612 86(76)

Source: http://physics.nist.gov/physRefData/compositions/index.html,

Example: For reaction

$$^6Li + {}^2H \rightarrow 2\,^4He$$

is follows: according to the reference tables 1 the ₃Li-6 nucleus has a relative atomic mass of 6.015 atomic mass units (abbreviated u), the deuterium ²H has 2.014 u, and the helium-4 nucleus has 4.0026 u. Thus:

- total rest mass on left side = 6.015 + 2.014 = 8.029 u;
- total rest mass on right side = 2 × 4.0026 = 8.0052 u;
- missing rest mass = 8.029 – 8.0052 = 0.0238 atomic mass units.

In a nuclear reaction, the total (relativistic) energy is conserved. The "missing" rest mass must therefore reappear as kinetic energy released in the reaction; its source is the nuclear binding energy. Using Einstein's mass-energy equivalence formula $E = mc^2$, the amount of energy released can be determined. We first need the energy equivalent of one atomic mass unit: $1\ u\ c^2 = 931.49$ MeV.

Hence, the energy released is 0.0238 × 931 MeV = 22.2 MeV.

Expressed differently: the mass is reduced by 0.3%, corresponding to 0.3% of 90 PJ/kg is 270 TJ/kg.

If result nuclear reaction has only two matter components, the energy distributed inversely between them.

$$E_1 = E \frac{\mu_2}{\mu_1 + \mu_2}, \quad E_2 = E \frac{\mu_1}{\mu_1 + \mu_2}, \quad E_1 + E_2 = E,$$

where E_1, E_2, E are energy the first, second and common components respectively of the left energy side; μ_1, μ_2 are molar mass of component in the right energy side.

Exemple.

For reaction

$$^2H + {}^3H \rightarrow {}^4He + {}^1n + 17.6 \text{ MeV}$$

we get

$$E_1 = E \frac{\mu_2}{\mu_1 + \mu_2} = 17.6 \frac{1}{4+1} = 3.5, \quad E_2 = E \frac{\mu_1}{\mu_1 + \mu_2} = 17.6 \frac{4}{4+1} = 14.1,$$

$$^2H + {}^3H \rightarrow {}^4He \ (3.5 \text{ MeV}) + {}^1n \ (14.1 \ (\text{MeV})).$$

Graph of Avarage binding energy per nucleon MeV vs Number of nucleons in ucleus.

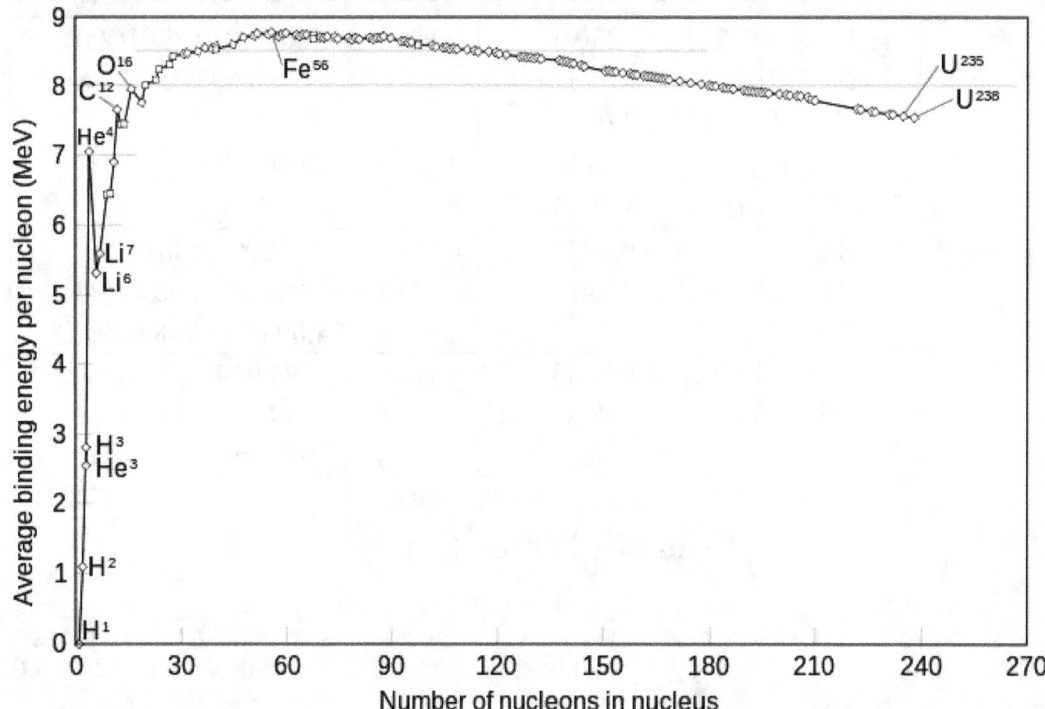

This graph may be used for quick aproximatly estimation the reaction energy. For example: Estimate energy reaction D+T → α+n, or:

$$^2H + {}^3H \rightarrow {}^4He + {}^1n .$$

The sum of mass and charges in left and right mast be same. From grapf we have:

In left side we have energy $E_1 = 2 \cdot 1.1 + 3 \cdot 2.9 = 10.9$ MeV; in right side we have $E_2 = 4 \cdot 7 = 28$ MeV. The difference is $E_2 - E_1 = 28 - 10.9 = 17.1$ MeV. We spent about $0.3 \div 0.7$ MeV for electrostatic barrier.

$$E_c = \frac{k Z_1 Z_2 e^2}{r_0},$$

where $k = 9 \cdot 10^9$, Z is charge state of 1 and 2-nd particles respectively; $e = 1.6 \cdot 10^{-19}$ is charge of electron; $r_0 \approx 2 \cdot 10^{-15}$ m is distance between charged particles. E_c is $\approx 0.3 \div 0.7$ MeV. If we add $E_c \approx 0.5$ to 17.1, we get kinetic energy of particles after reaction approximately $E = 17.1 + 0.5 \approx 17.6$ MeV.

Table of reaction energy (Kikoin, p.942)

In computation you must to take account the energy of the short time isotopes $\approx 10^{-15}$ sec. (Kikoin, p. 943)

$$^5He \rightarrow {}^4He + n + 0.958 , \quad {}^5Li \rightarrow {}^4He + {}^1H + 1.965 , \quad {}^8Be \rightarrow 2\,{}^4He + 0.094$$

and energy in decay ^6He ($T_{1/2} = 0.802$ sec)

$$^6He \rightarrow {}^6Li + \beta^- + v^- + 3.510,$$

where v^- is anti-neutrino. Check stability and half-life elements in right.

Below is Tables 2 some theoretical thermonuclear reactions (Notations: β^- - electron, β^+ - positron, v - neutrino, γ – gamma rays).

Table 2. (Kikoin, p. 943. Mistakes are corrected)

1. $^1H + {}^1H \rightarrow {}^2H + \beta^+ + v + 0.420$
2. $^2H + {}^1H \rightarrow {}^3He + \gamma + 5.494$
3. $^2H + {}^2H \rightarrow {}^3H + {}^1H + 4.033 \quad 50\%$
    $\rightarrow {}^3He + n + 3.270 \quad 50\%$
4. $^3H + {}^1H \rightarrow {}^4He + \gamma + 19.814$
5. $^3H + {}^2H \rightarrow {}^4He + n + 17.590$
    $\rightarrow {}^5He + \gamma + 16.632$
6. $^3H + {}^3H \rightarrow {}^4He + 2n + 11.332$
    $\rightarrow {}^5He + n + 10.374$
7. $^3He + {}^2H \rightarrow {}^4He + {}^1H + 18.354$
    $\rightarrow {}^5Li + \gamma + 16.388$
8. $^3He + {}^3H \rightarrow {}^4He + {}^2H + 14.321 \quad 43\%$
    $\rightarrow {}^4He + {}^1H + n + 12.096 \quad 51\%$
    $\rightarrow {}^5He + {}^1H + 11.138$
    $\rightarrow {}^5Li + n + 10.131$
    $\rightarrow {}^6Li + \gamma + 15.793$
9. $^3He + {}^3He \rightarrow {}^4He + 2\,{}^1H + 12.860$
    $\rightarrow {}^5Li + {}^1H + 10.895$
10. $^4He + {}^2H \rightarrow {}^6Li + \gamma + 1.472$
11. $^4He + {}^3H \rightarrow {}^7Li + \gamma + 2.467$
12. $^4He + {}^3He \rightarrow {}^7Be + \gamma + 1.587$
13. $^6Li + {}^1H \rightarrow {}^4He + {}^3He + 4.021$
    $\rightarrow {}^7Be + \gamma + 5.608$
14. $^6Li + {}^2H \rightarrow {}^4He + {}^3H + {}^1H + 2.561$
    $\rightarrow {}^4He + {}^3He + n + 1.797$
    $\rightarrow 2\,{}^4He + 22.375$
    $\rightarrow {}^5He + {}^3He + 0.839$
    $\rightarrow {}^5Li + {}^3H + 5.028$
    $\rightarrow {}^7Li + {}^1H + 5.028$
    $\rightarrow {}^7Be + n + 3.384$
    $\rightarrow {}^8Be + \gamma + 22.280$
15. $^6Li + {}^3H \rightarrow 2\,{}^4He + n + 16.117$
    $\rightarrow {}^5He + {}^4He + 15.160$
    $\rightarrow {}^7Li + {}^2H + 0.995$
    $\rightarrow {}^8Li + {}^1H + 0.803$
    $\rightarrow {}^8Be + n + 16.023$
16. $^6Li + {}^3He \rightarrow {}^5Li + {}^4He + 14.916$
    $\rightarrow {}^8Be + {}^1H + 16.787$
17. $^6Li + {}^4He \rightarrow {}^{10}B + \gamma + 4.461$
18. $^6Li + {}^6Le \rightarrow {}^7Li + {}^4He + {}^1H + 3.556$
    $\rightarrow {}^7Be + {}^4He + n + 1.912$

$\rightarrow {}^8Be + {}^4He + 20.808$
$\rightarrow {}^9Be + {}^3He + 1.895$
$\rightarrow {}^9B + {}^3H + 0.808$
$\rightarrow {}^{10}B + {}^2H + 2.989$
$\rightarrow {}^{11}B + {}^1H + 12.220$
$\rightarrow {}^{11}C + n + 1.895$

19. $^7Li + {}^1H \rightarrow 2\,{}^4He + 17.347$
    $\rightarrow {}^8Be + \gamma + 17.252$
20. $^7Li + {}^2H \rightarrow 2\,{}^4He + n + 15.122$
    $\rightarrow {}^5He + {}^4He + 14.164$
    $\rightarrow {}^8Be + n + 15.028$
    $\rightarrow {}^9Be + \gamma + 16.693$
21. $^7Li + {}^3H \rightarrow {}^5He + {}^4He + n + 7.907$
    $\rightarrow {}^6He + {}^4He + 9.834$
    $\rightarrow {}^8Be + 2n + 8.770$
    $\rightarrow {}^9Be + n + 10.436$
22. $^7Li + {}^3He \rightarrow {}^6Li + {}^4He + 13,325$
    $\rightarrow {}^8Be + {}^2H + 11.759$
    $\rightarrow {}^9Be + {}^1H + 11.199$
    $\rightarrow {}^9Be + n + 9.349$
    $\rightarrow {}^{10}Be + \gamma + 17.786$
23. $^7Li + {}^4He \rightarrow {}^{11}B + \gamma + 8.664$
24. $^7Li + {}^6Li \rightarrow {}^9Be + {}^4He + 15.220$
    $\rightarrow {}^{10}B + {}^3H + 1.994$
    $\rightarrow {}^{11}B + {}^2H + 7.192$
    $\rightarrow {}^{12}B + {}^1H + 8.337$
    $\rightarrow {}^{11}C + 2n + 2.204$
    $\rightarrow {}^{12}C + n + 20.924$
25. $^7Li + {}^7Li \rightarrow {}^8Be + {}^6He + 7.272$
    $\rightarrow {}^{10}Be + {}^4He + 14.783$
    $\rightarrow {}^{11}B + {}^3H + 6.197$
    $\rightarrow {}^{12}B + {}^2H + 3.309$
    $\rightarrow {}^{13}B + {}^1H + 5.964$
    $\rightarrow {}^{12}C + 2n + 13.672$
    $\rightarrow {}^{13}C + n + 18.619$
26. $^9Be + {}^1H \rightarrow {}^6Li + {}^4He + 2.126$
    $\rightarrow {}^8Be + {}^2H + 0.559$
    $\rightarrow {}^{10}B + \gamma + 6.587$
27. $^9Be + {}^2H \rightarrow 2\,{}^4He + {}^3He + 4.687$
    $\rightarrow {}^7Li + {}^4He + 7,154$
    $\rightarrow {}^8Be + {}^3H + 4.592$
    $\rightarrow {}^{10}Be + {}^1H + 4.590$
    $\rightarrow {}^{10}B + n + 4.363$
    $\rightarrow {}^{11}B + \gamma + 6.197$

28. $^9Be + ^3H \rightarrow ^8Li + ^4He + 2.930$
$\rightarrow ^{11}B + n + 9.561$
$\rightarrow ^{12}B + \gamma + 10.463$

29. $^9Be + ^3He \rightarrow ^8Be + ^4He + 18.913$
$\rightarrow ^{10}B + ^2H + 1.094$
$\rightarrow ^{11}B + ^1H + 10.325$
$\rightarrow ^{11}C + n + 7.562$

30. $^9Be + ^4He \rightarrow ^{12}C + n + 5.704$
$\rightarrow ^{13}C + \gamma + 10.650$

31. $^9Be + ^6Li \rightarrow ^8Be + ^7Li + 5.587$
$\rightarrow ^{10}B + ^5He + 1.933$
$\rightarrow ^{11}B + ^4He + 14.347$

33. $^{10}B + ^1H \rightarrow ^7Be + ^4He + 1.148$
$\rightarrow ^{11}C + \gamma + 8.693$

34. $^{10}B + ^2H \rightarrow 3^4He + 17,914$
$\rightarrow ^8Be + ^4H + 17.819$
$\rightarrow ^{11}B + ^1H + 9,231$
$\rightarrow ^{11}C + n + 6,468$
$\rightarrow ^{12}C + \gamma + 25,182$

35. $^{10}B + ^3H \rightarrow ^9Be + ^4He + 13,227$
$\rightarrow ^{11}B + ^2H + 5,199$
$\rightarrow ^{12}B + ^1H + 6,343$
$\rightarrow ^{12}C + n + 18,941$

36. $^{10}B + ^3He \rightarrow 3 ^4He + ^1H + 12,420$
$\rightarrow ^8Be + ^4He + ^1H + 12.326$
$\rightarrow ^9B + ^4He + 12,140$
$\rightarrow ^{11}C + ^2H + 3.199$
$\rightarrow ^{12}C + ^1H + 19.695$
$\rightarrow ^{12}N + n + 1.5$

37. $^{10}B + ^4He \rightarrow ^{12}C + ^2H + 1.341$
$\rightarrow ^{13}C + ^1H + 4.063$
$\rightarrow ^{13}N + n + 1.060$
$\rightarrow ^{12}N + ^2H + 3.199$

38. $^{10}B + ^6Li \rightarrow ^{11}C + ^5He + 4.038$
$\rightarrow ^{12}C + ^4He + 23.716$
$\rightarrow ^{13}C + ^3He + 8.085$
$\rightarrow ^{13}N + ^3H + 5.845$
$\rightarrow ^{14}N + ^2H + 10.141$
$\rightarrow ^{15}N + ^1H + 18.751$
$\rightarrow ^{14}O + 2n + 1.990$
$\rightarrow ^{15}O + n + 15.209$

39. $^{10}B + ^7Li \rightarrow ^{12}C + ^4He + n + 16.463$
$\rightarrow ^{13}C + ^4He + 21.410$
$\rightarrow ^{14}N + ^3H + 9.146$

39. $^{10}B + ^7Li$ (continue) $\rightarrow ^{15}N + ^2H + 13.723$

$\rightarrow ^{13}C + ^2H + 9.178$
$\rightarrow ^{14}C + ^1H + 15.130$
$\rightarrow ^{13}N + 2n + 3.951$
$\rightarrow ^{14}C + n + 14.504$

32. $^9Be + ^7Li \rightarrow ^8Be + ^8Li + 0.367$
$\rightarrow ^{12}B + ^4He + 10.463$
$\rightarrow ^{13}C + ^3H + 8.183$
$\rightarrow ^{14}C + ^2H + 10.102$
$\rightarrow ^{15}C + ^1H + 9.095$
$\rightarrow ^{14}N + 2n + 7.251$
$\rightarrow ^{15}N + n + 18.086$

39. $^{10}B + ^7Li$ (continue) $\rightarrow ^{16}N + ^1H + 13.985$
$\rightarrow ^{15}N + 2n + 7.957$

40. $^{11}B + ^1H \rightarrow ^8Be + ^4He + 8.588$
$\rightarrow ^{12}C + \gamma + 15.957$

41. $^{11}B + ^2H \rightarrow ^8Be + ^4He + n + 6.363$
$\rightarrow ^9Be + ^4He + 8.028$
$\rightarrow ^{12}B + ^1H + 1.144$
$\rightarrow ^{12}C + n + 6.363$
$\rightarrow ^{13}C + \gamma + 18.679$

42. $^{11}B + ^3H \rightarrow ^{10}Be + ^4He + 8.586$
$\rightarrow ^{13}C + n + 12.422$

43. $^{11}B + ^3He \rightarrow ^8Be + ^6Li + 4.566$
$\rightarrow ^{10}B + ^4He + 9.122$
$\rightarrow ^{12}C + ^2H + 10.463$
$\rightarrow ^{13}C + ^1H + 13.185$
$\rightarrow ^{13}N + n + 10.182$
$\rightarrow ^{14}N + \gamma + 20.735$

44. $^{11}B + ^4He \rightarrow ^{14}C + ^1H + 0.784$
$\rightarrow ^{14}N + n + 0.157$

45. $^{11}B + ^6Li \rightarrow ^{12}C + ^4He + n + 12.260$
$\rightarrow ^{13}C + ^4He + 17.207$
$\rightarrow ^{14}N + ^3H + 4.942$
$\rightarrow ^{15}N + ^2H + 9,520$
$\rightarrow ^{16}N + ^1H + 9.520$
$\rightarrow ^{15}O + 2n + 3.753$

46. $^{11}B + ^7Li \rightarrow ^{13}C + ^4He + n + 9.954$
$\rightarrow ^{15}C + ^3He + 18.131$
$\rightarrow ^{15}N + ^3H + 8.525$
$\rightarrow ^{16}N + ^2H + 4.758$
$\rightarrow ^{17}N + ^1H + 8.415$
$\rightarrow ^{16}O + 2n + 12.162$

Table 3. High nuclear cross section *aneutronic* reactions[2]

Isotopes	Reaction
Deuterium–helium-3	$^2D + {}^3He \rightarrow {}^4He + {}^1p + 18.3$ MeV
Deuterium–lithium-6	$^2D + {}^6Li \rightarrow 2\,{}^4He + 22.4$ MeV
Proton–lithium-6	$^1p + {}^6Li \rightarrow {}^4He + {}^3He + 4.0$ MeV
Helium-3–lithium-6	$^3He + {}^6Li \rightarrow 2\,{}^4He + {}^1p + 16.9$ MeV
Helium-3-helium-3	$^3He + {}^3He \rightarrow {}^4He + 2\,{}^1p + 12.86$ MeV
Proton–lithium-7	$^1p + {}^7Li \rightarrow 2\,{}^4He + 17.2$ MeV
Proton–boron	$^1p + {}^{11}B \rightarrow 3\,{}^4He + 8.7$ MeV
Proton–nitrogen	$^1p + {}^{15}N \rightarrow {}^{12}C + {}^4He + 5.0$ MeV

Table 3a. Reactions by slow neutrones produced the charges particles (Kikoin, p.888)

Reactions (Energy in MeV)	Cross section, barn
$^3He + n \rightarrow {}^3H + p + 0.764$	5400
$^6Li + n \rightarrow {}^3H + \alpha + 4.785$	945
$^7Be + n \rightarrow {}^7Li + p + 1.65$	51000
$^{10}B + n \rightarrow {}^7Li + \alpha + 2.791$	3837
$^{14}N + n \rightarrow {}^{14}C + p + 0.626$	1.75
$^{17}O + n \rightarrow {}^{14}C + \alpha + 1.72$	0.5

Stable isotopes (Kikoin, p.825)

Stable isotopes are: ^1H, ^2H, ^3He, ^4He, ^6Li, ^7Li, ^9Be, ^{10}B, ^{11}B, ^{12}C, ^{13}C, ^{14}N, ^{15}N .

http://hyperphysics.phy-astr.gsu.edu/hbase/Nuclear/nucrea.html#c3

Some Nuclear Reactions. Table 3.

Reaction	Measured Q (MeV)	Reaction	Measured Q (MeV)
$^2H(n,\gamma)^3H$	6.257 +/- 0.004	$^9Be(p,\alpha)^6Li$	2.132 +/- 0.006
$^2H(d,p)^3H$	4.032 +/- 0.004	$^{10}B(n,\alpha)^7Li$	2.793 +/- 0.003
$^6Li(p,\alpha)^3H$	4.016 +/- 0.005	$^{10}B(p,\alpha)^7Be$	1.148 +/- 0.003
$^6Li(d,p)^7Li$	5.020 +/- 0.006	$^{12}C(n,\gamma)^{13}C$	4.948 +/- 0.004
$^7Li\,(p,n)^7Be$	-1.645 +/- 0.001	$^{13}C(p,n)^{13}N$	-3.003 +/- 0.002
$^7Li\,(p,\alpha)^4He$	17.337 +/- 0.007	$*^{14}N(p,n)^{14}C$	-0.627 +/- 0.001
$^9Be(n,\gamma)^{10}Be$	6.810 +/- 0.006	$^{14}N(n,\gamma)^{15}N$	10.833 +/- 0.007
$^9Be\,(\gamma,n)^8Be$	-1.666 +/- 0.002	$^{18}O(p,n)^{18}F$	-2.453 +/- 0.002
$^9Be\,(d,p)^{10}Be$	4.585 +/- 0.005	$^{19}F(p,\alpha)^{16}O$	8.124 +/- 0.007

Index

* The nuclear reaction in the atmosphere which produces carbon-14 for radiocarbon dating.

Data from C. W. Li, W. Whaling, W. A. Fowler, and C. C. Lauritson, Physical Review 83:512 (1951)

HyperPhysics***** Nuclear

Go
Back

Number of collising the moving neutron in the moveless surrounding particles for given loss of energy is (see Kikoin [10], p. 924) :

$$\xi = 1 + \frac{(A-1)^2}{2A} \ln \frac{A-1}{A+1}, \quad N = \frac{1}{\xi} Ln \frac{E_2}{E_1}, \tag{9}$$

where N is number of collisings; A is nuclear number of moveless particle; E_1 is the initial energy of neutron, MeV; E_2 is final energy of neutron, MeV.

For ^{7}Li the value $A = 7$, maxinun $E_2 = 14.1$ MeV, minimum $E_1 = 2.5$ MeV. The $N = 6.65$.

Properties of some retarders in Tables 4 below (Kikoin, p.925):

Table 4. Number of callusing for retarding
from 2 MeV to warm energy

Retarder	ξ	N
^{1}H	1.00	18
^{2}D	0.725	25
^{4}He	0.425	43
^{7}Li	0.268	67
^{9}Be	0.209	86
^{12}C	0.158	114
^{16}O	0.120	150
A is big	2/(A+2/3)	9A+6
H_2O	0.948	
D_2O	0.570	
BeO	0.173	

The time of neutron retendering estimate by equation:

$$t = \frac{2}{\xi} \overline{\lambda}_{scat} \left(\frac{1}{v_f} - \frac{1}{v_i} \right).$$

Here t is regarding time, sec; $\overline{\lambda}_{scat}$ - is the free path of neutrons in relation to dispersal, cm; v_f, v_i - are final and initial speed of neutrone respectively, cm/sec. Result of computation in Table 5.

Table 5. The regarding time of neutron from
energy $E = 2$ MeV to $E = 0.025$ eV.

Regarded	$\overline{\lambda}_{scat}$ · cm	t, 10^{-5} sec
Water	1.1	1.0
Heavy water	2.6	4.6
Beryllium	1.6	6.7
Beryllium oxide	1.5	7.8
Graphite	2.6	15.0

Cross section areas of some thermonuclear reactions

Cross section appears in diapason < 200 keV may be estimated by formula of George Gamov

$$\sigma(E) = (A/E)exp(-B/E^{0.5}) ,$$

where σ is cross-section of reaction; E is energy. A, B are the test coefficients.

Rough estimation the B coefficient is

$$BE^{-0.5} \approx 2\pi Z_1 Z_2 e^2/(hv)$$

Here Z_1, Z_2 are charges of colliding particles; e is charge of electron; h is Plank constant.

Below are the **test** cross-sections [Kikon, pp.947 - 950]. If no the reaction product, the σ is for all products.

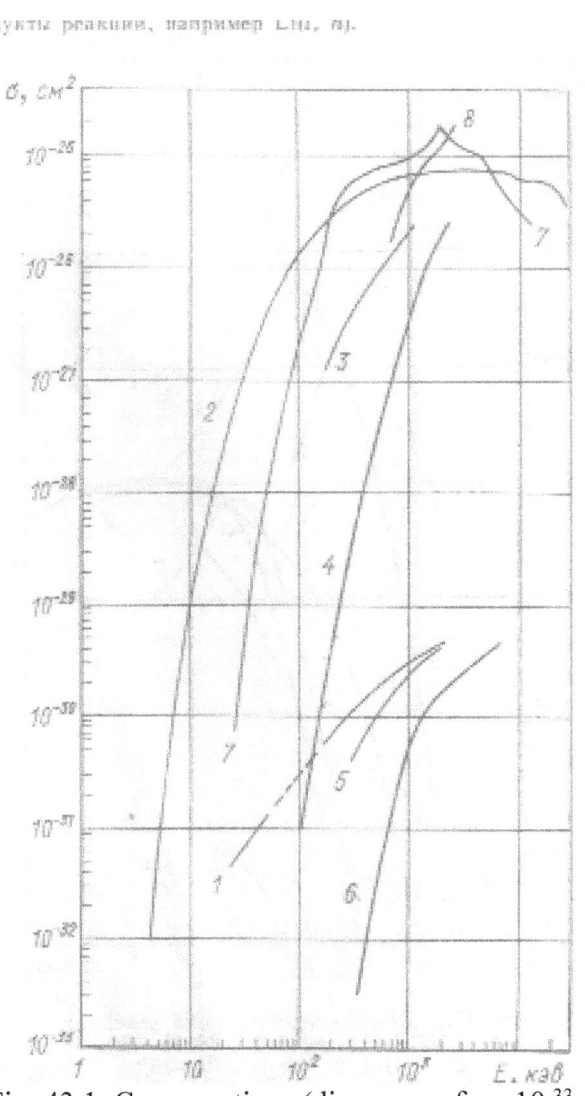

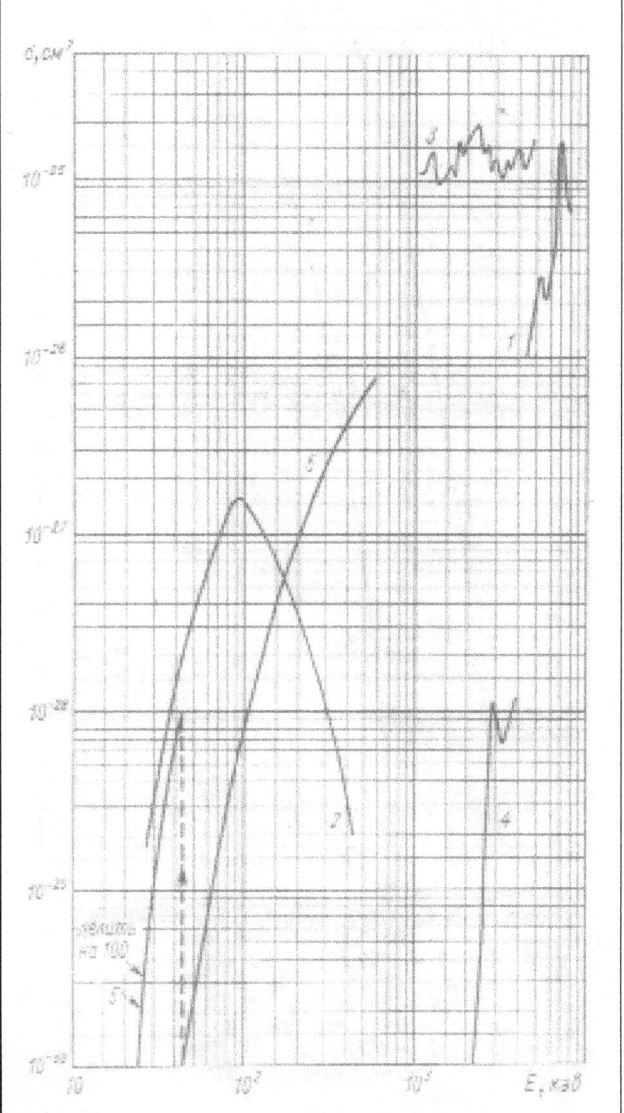

Fig. 43.1. Cross-sections (diapason σ from 10^{-33} to 10^{-25} cm^2) vs. Energy (E 1 ÷ 10^4 keV).
1 – ^{2}H(p,γ)^{3}He; 2 – ^{2}H(d,p)^{3}He; 3 – ^{3}He(t, d)^{4}He;
4 – ^{3}He(^{3}He,p)...; 5 – ^{3}H(ά, γ)^{7}Li; 6 – ^{3}He(ά,γ)^{7}Be;
7 – ^{6}Li(p,ά)^{3}He; 8 – ^{9}Li(t,ά)... .

Fig. 43.7. Cross-sections (diapason σ from 10^{-30} to 10^{-25} cm^2) vs. Energy (E 1 ÷ 10^4 keV).
1 – ^{7}Li(ά,n)^{10}B; 2 – T(d,γ)^{5}He; 3 – ^{12}C(d,n)^{13}N;
4 – ^{18}O(α,γ)^{22}Ne; 5 – ^{7}Li(p,α)^{4}He;

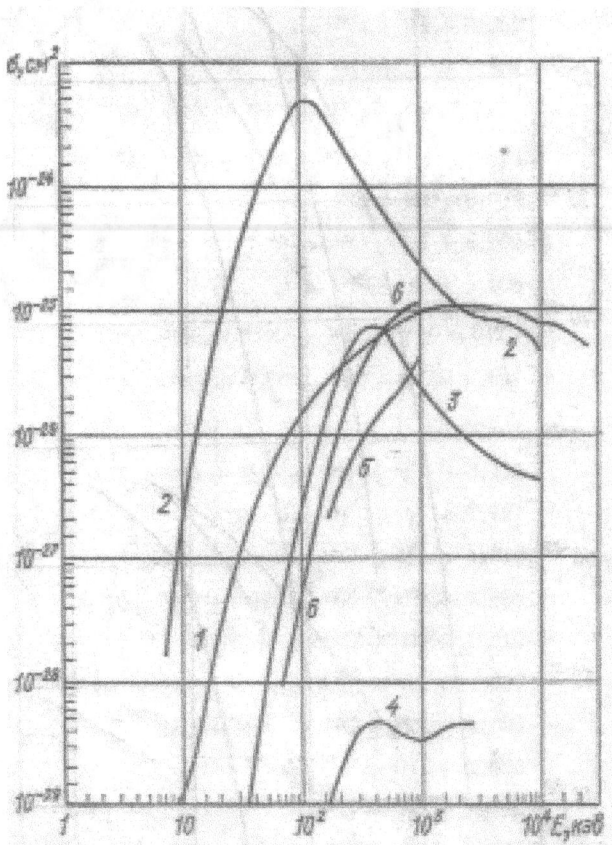

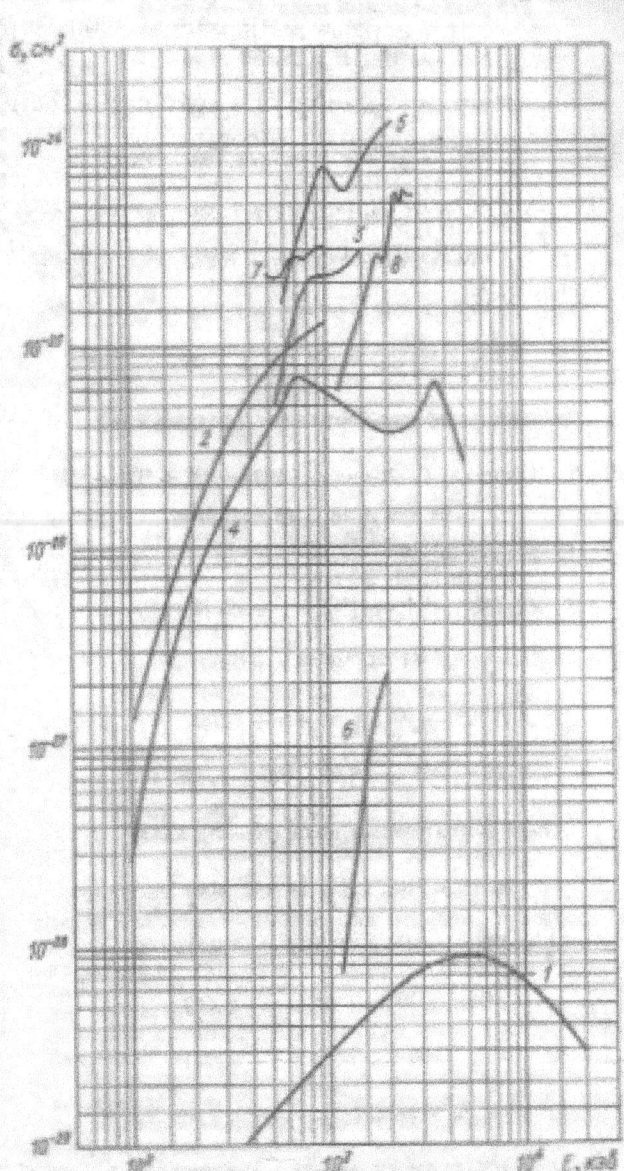

Fig. 43.2. Cross-sections (diapason σ from 10^{-29} to 10^{-25} cm^2) vs. Energy (E $1 \div 10^4$ keV).
1 – ^{2}H(d,n)^{3}He; 2 – ^{3}H(d,n)^{4}He; 3 – ^{3}He(d,p)^{4}He;
4 – ^{3}He(d, γ)^{5}Li; 5 – ^{3}He(t, pn)^{4}He; 6 – ^{6}Li(d,p)^{7}Li;

Fig. 43.3. Cross-sections (diapason σ from 10^{-30} to 10^{-24} cm^2) vs. Energy (E $10^2 \div 10^4$ keV).
1 – ^{3}H(p,γ)^{4}He; 2 –^{6}Li(d,n)^{7}Be; 3 – ^{6}Li(d,t)...;
4 – ^{6}Li(d,α)^{4}He; 5 – ^{7}Li(t,n)…;
6 - ^{7}Li(^{7}Li,p)^{13}B + ^{7}Li(^{7}Li,d)^{12}B;
7 – ^{9}Be(p,d)^{8}Be; 8 - ^{9}Be(t,n)…

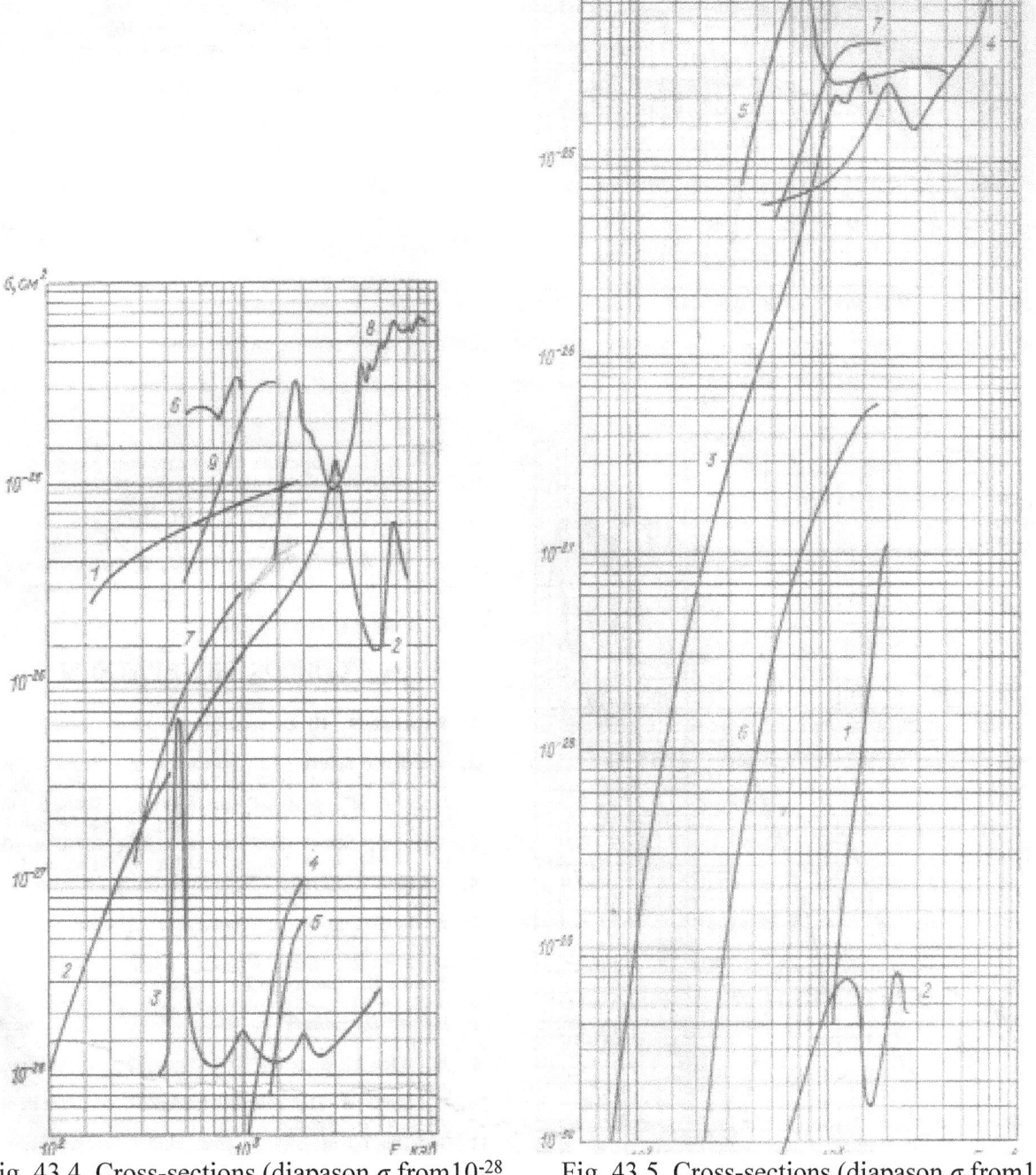

Fig. 43.4. Cross-sections (diapason σ from 10^{-28} to 10^{-25} cm²) vs. Energy (E $10^2 \div 10^4$ keV).
1 – ^{3}H(t,n)...; 2 – ^{6}Li(p,d)^{4}He; 3 – ^{7}Li(p,γ)^{8}Be;
4 – ^{6}Li(^{7}Li,p)^{12}B; 5 – ^{7}Li(^{7}Li,^{8}Be)^{6}He;
6 – ^{9}Be(p,α)^{6}Li ; 7 - ^{9}Be(d,p)^{10}Be ; 8 -^{9}Be(α,n)^{12}C ;
9 - ^{10}B(d,n)…

Fig. 43.5. Cross-sections (diapason σ from 10^{-30} to 10^{-24} cm²) vs. Energy (E $10^2 \div 10^4$ keV).
1– ^{9}Be(^{7}Li,^{8}Li)^{8}Be; 2 –^{10}B(p,γ)^{11}C;
3–^{10}B(p,α)^{7}Be; 4 – ^{10}B(α,n)^{13}N; 5–^{11}B(p,α)^{8}Be;
6 – ^{11}B(d,p)^{12}B; 7 – ^{11}B(d,n)…;

172

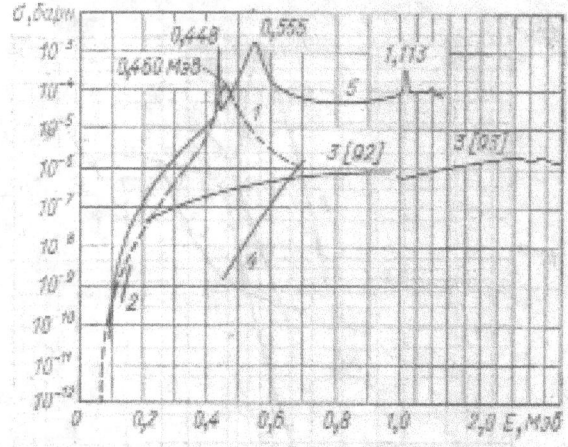

Fig. 43.6. Cross-sections in barn (diapason σ from 10^{-12} to 10^{-1} barn) vs. Energy (E 0 ÷ 2 MeV).
1 – ^{12}C(p,γ)^{13}N; 2 – ^{14}N(p, γ)^{15}O; 3 – ^{7}Be(p,γ)^{8}B; 4– ^{13}C(α,n)^{16}O; 5 –^{13}C(p,γ)^{14}N.

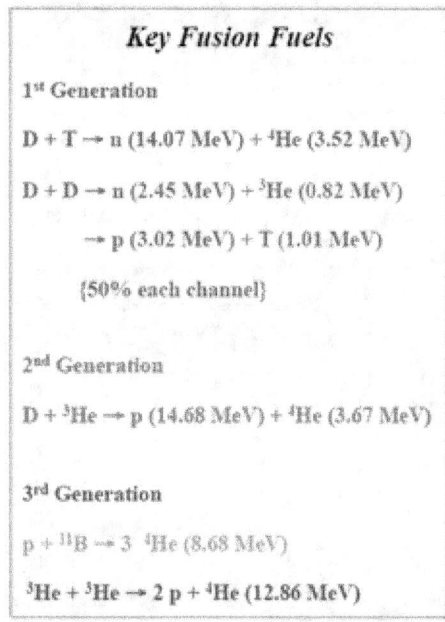

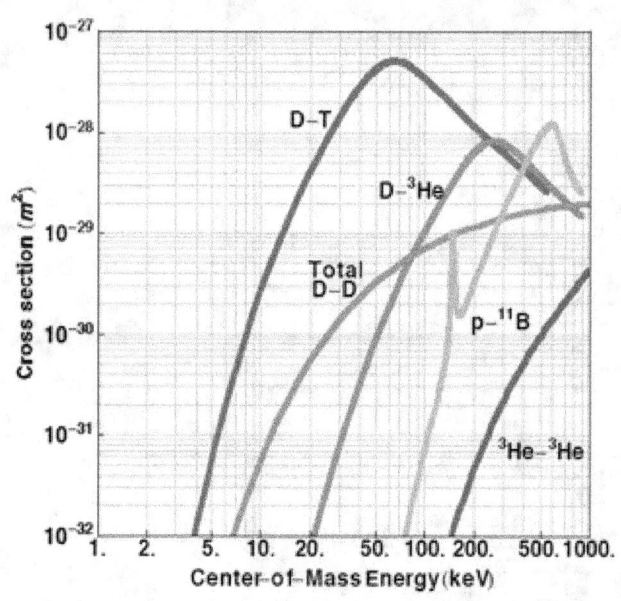

https://www.quora.com/What-are-the-pros-and-cons-of-controlled-nuclear-fusion-with-inertial-confinement-like-NIF-National-Ignition-Facility-versus-magnetic-confinement-like-ITER-or-TOKAMAK

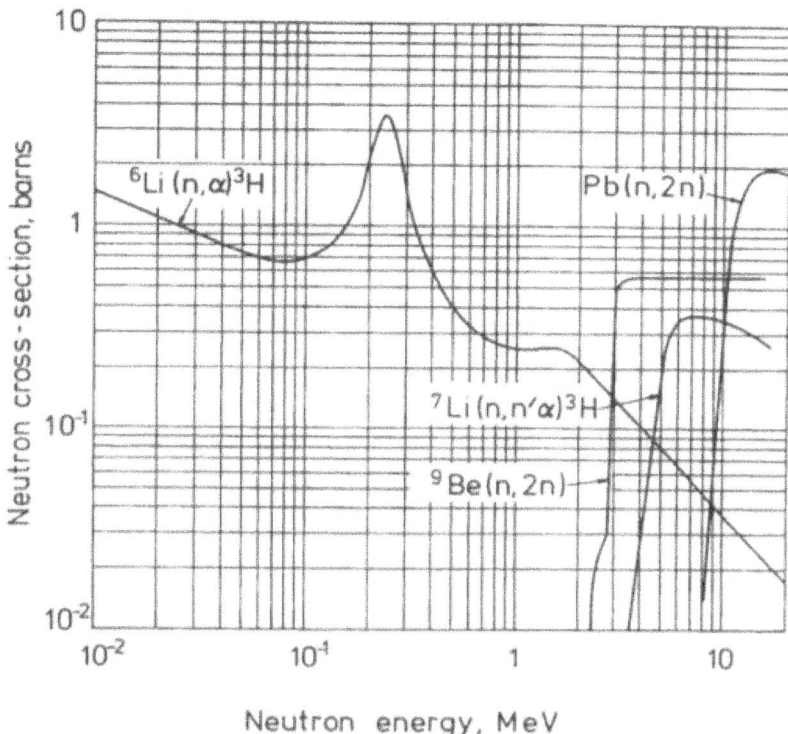

Cross section for reaction $^6Li(n,\alpha)^3H$, $^7Li(n,n\alpha)^3H$, $^9Be(n,2n)$, $Pb(n,2n)$. Other important sections include elastic and inelastic scattering cross sections for Li, Be, Pb, needed the slowing down (moderation) of the 14 MeV primery neutrons and neutron absorption cross sections for structural material (Figure from M. Sawan).
http://www.kayelaby.npl.co.uk/atomic_and_nuclear_physics/4_7/4_7_4.html

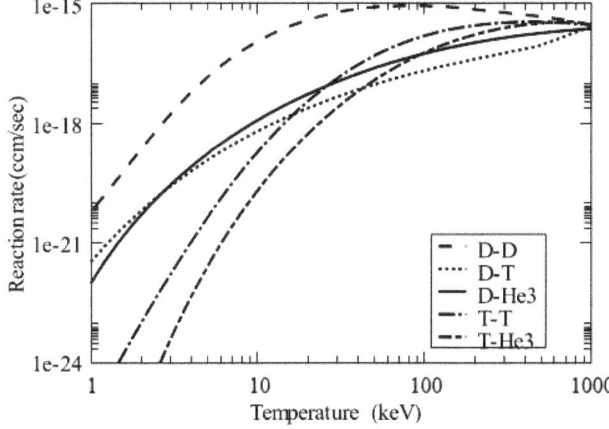

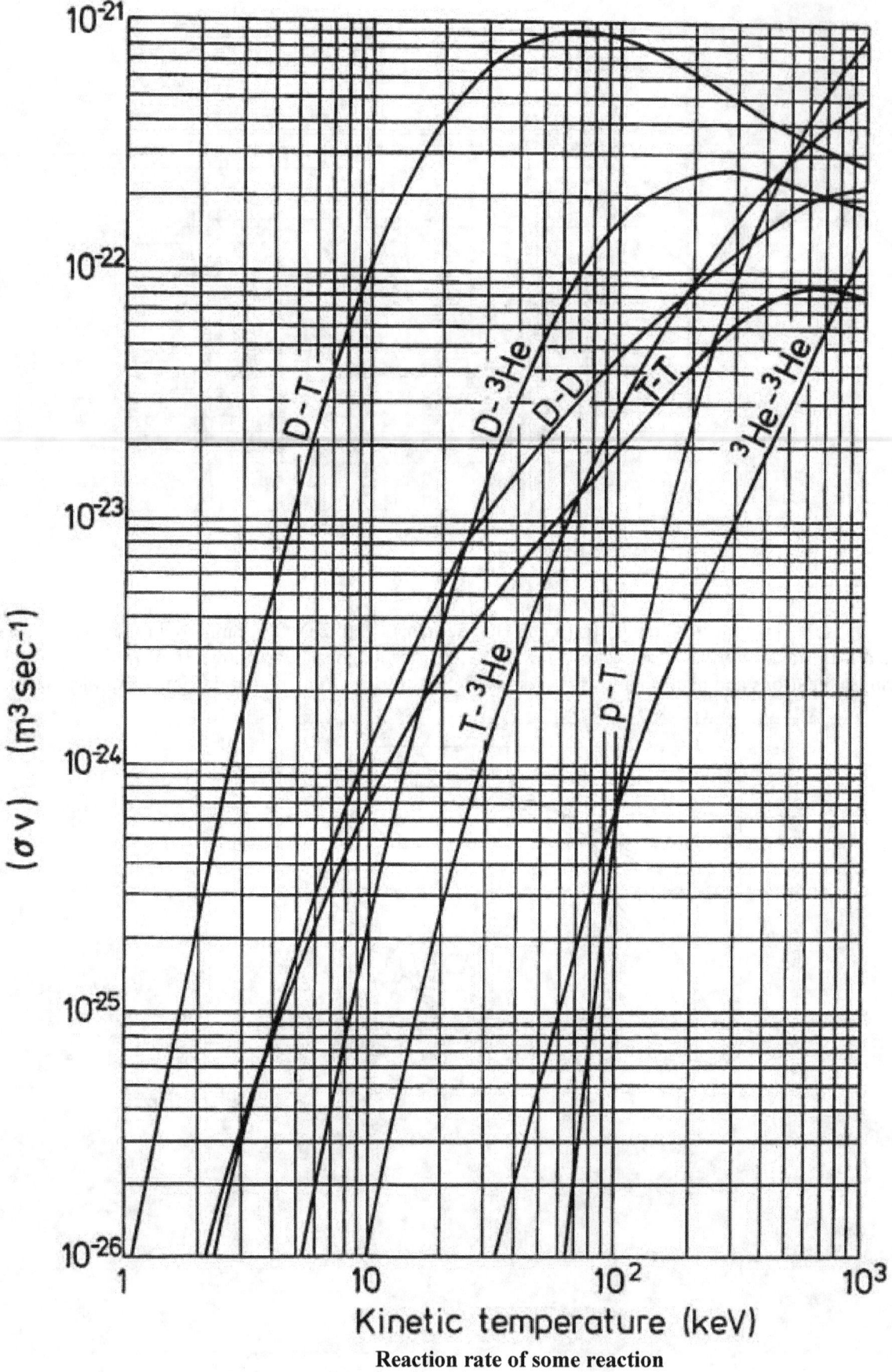

Reaction rate of some reaction

Cross section for reactions absorption (solid line) and scattering (points) neutrons.

Thermonuclear Reaction from Nuclear Encyclopedia 7 8 18

TABLE 32.1 Most Important Fusion Reactions (This is in book)

D + T → He4 (3.5 MeV) + n (14.1 MeV)

D + D → T(1.01 MeV) + p (3.02 MeV) 50%

He3 (0.82 MeV) + n (2.45 MeV) 50%

D +He3 → He4 (3.6 MeV) + p (14.7 MeV)

T + T → He4 + 2n + 11.3 MeV

He3 +T → He4 + p + n + 12.1 MeV 51%

   → He4 (4.8 MeV) + D (9.5 MeV) 43%

   → He5 (2.4 MeV) + p (11.9 MeV) 6%

p +Li6 → He4 (1.7 MeV) +He3 (2.3 MeV)

p +Li7 → 2He4 + 17.3 MeV 20%

   → Be7+n−1.6 MeV 80%

D +Li6 → 2He4 + 22.4 MeV

p +B11 → 3He4 + 8.7 MeV

n +Li6 → He4 (2.1 MeV) + T (2.7 MeV)

n +Li7 → He4 + T + n−2.5 MeV

Cross section of neutrons

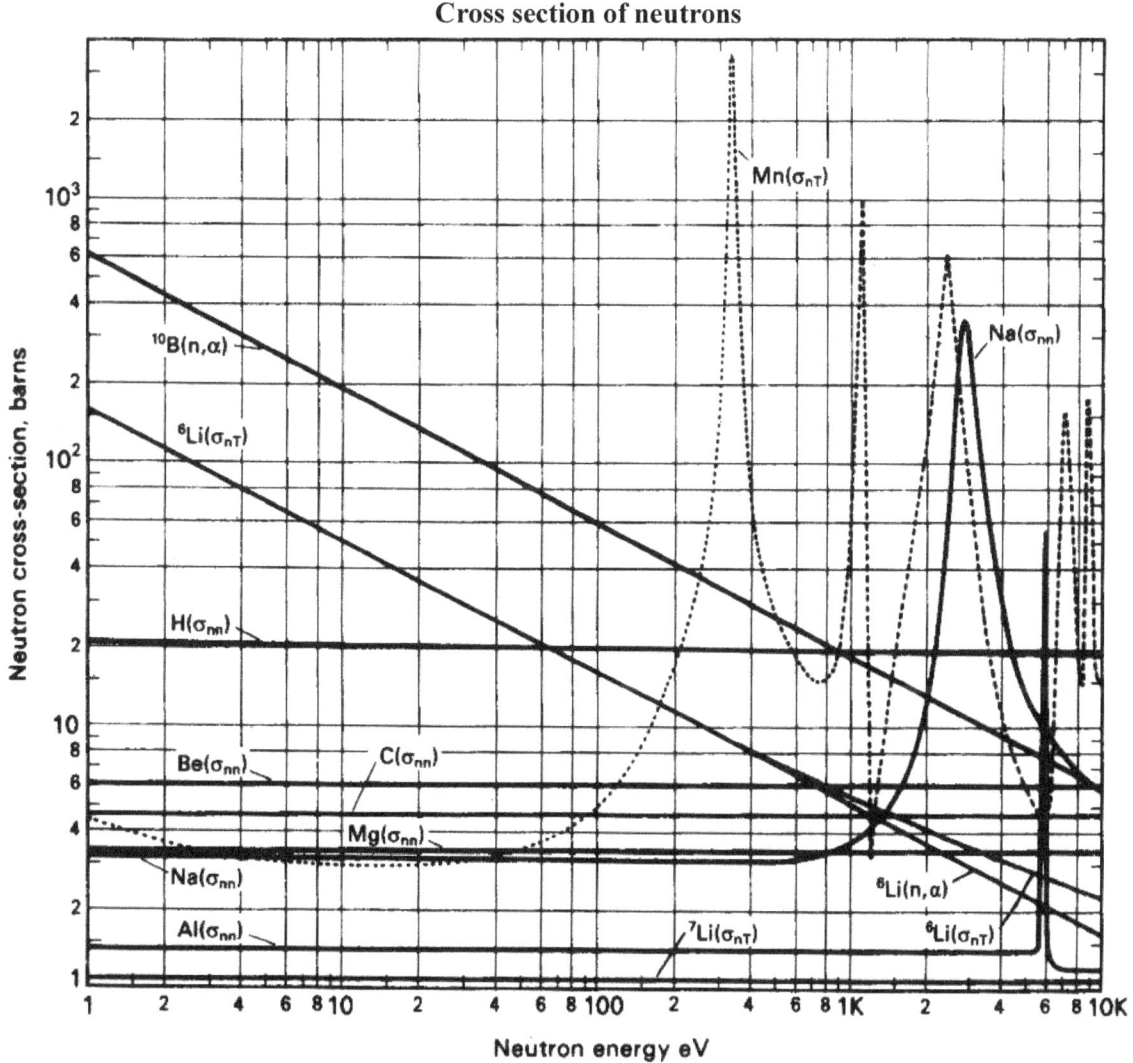

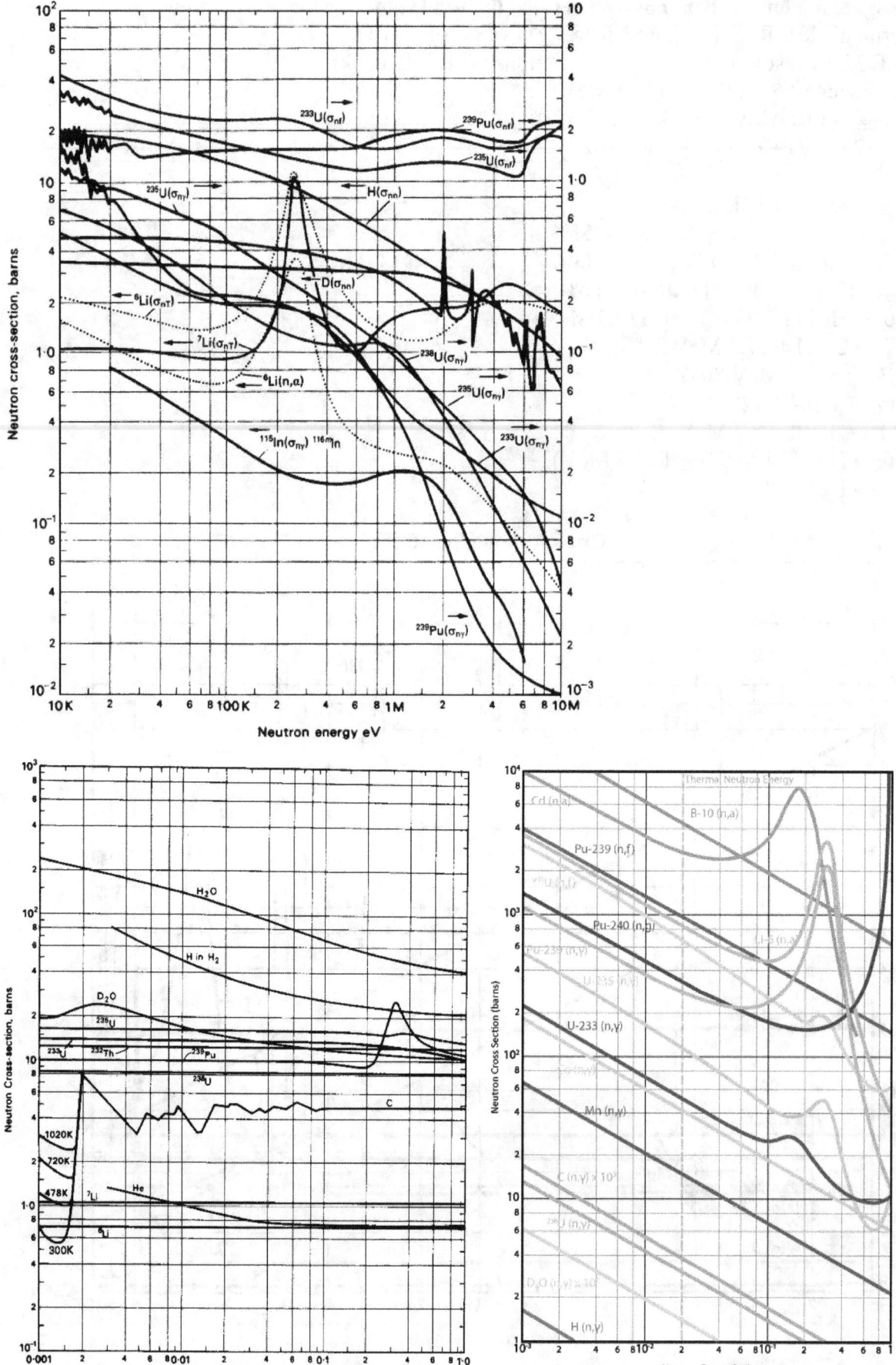

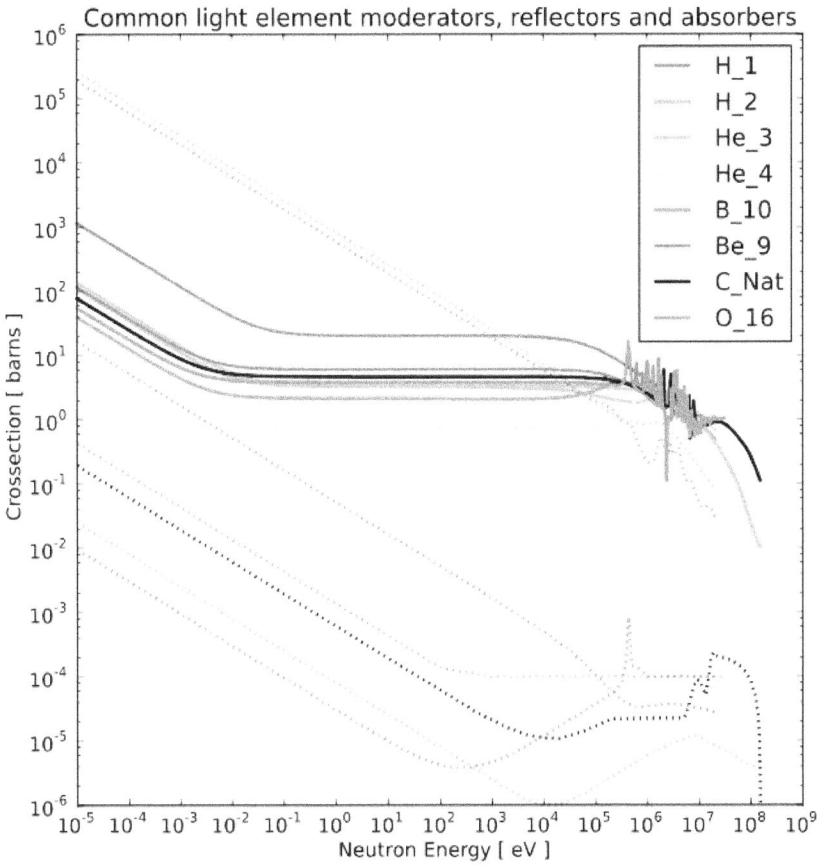

Common light element moderators, reflectors and absorbers

https://upload.wikimedia.org/wikipedia/commons/8/80/Isotopes_and_half-life.svg

==

Cross section warm neutrons:

https://environmentalchemistry.com/yogi/periodic/crosssection.html

==

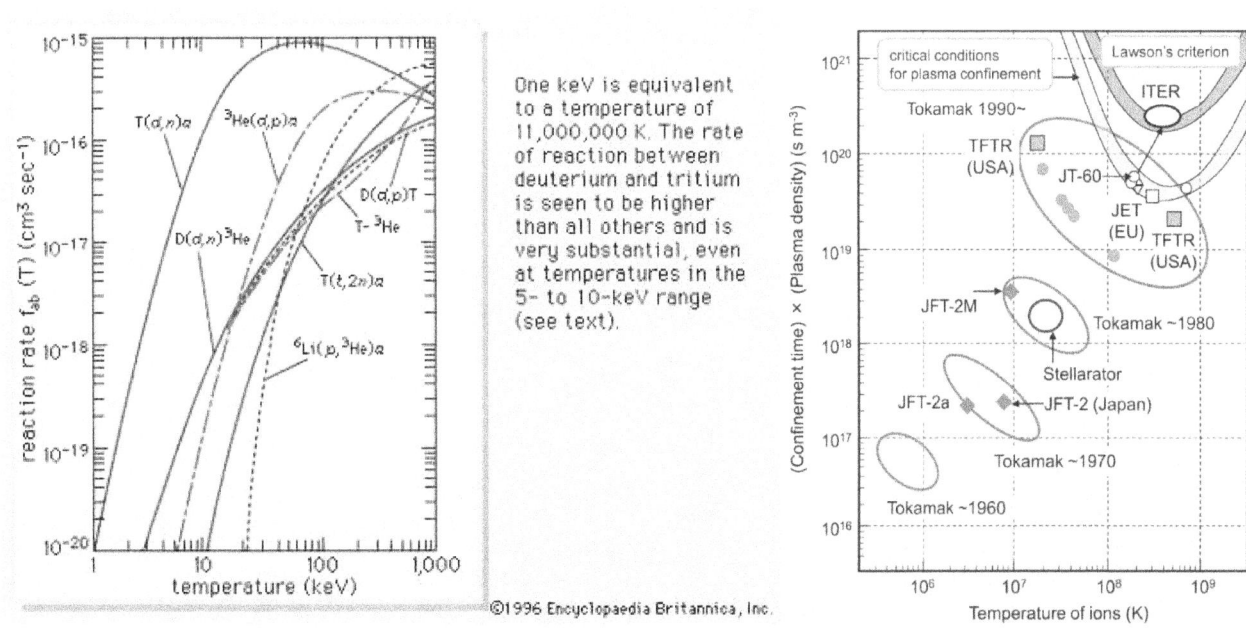

One keV is equivalent to a temperature of 11,000,000 K. The rate of reaction between deuterium and tritium is seen to be higher than all others and is very substantial, even at temperatures in the 5- to 10-keV range (see text).

©1996 Encyclopaedia Britannica, Inc.

3. Decay mode

https://upload.wikimedia.org/wikipedia/commons/7/79/NuclideMap_stitched_small_preview.png

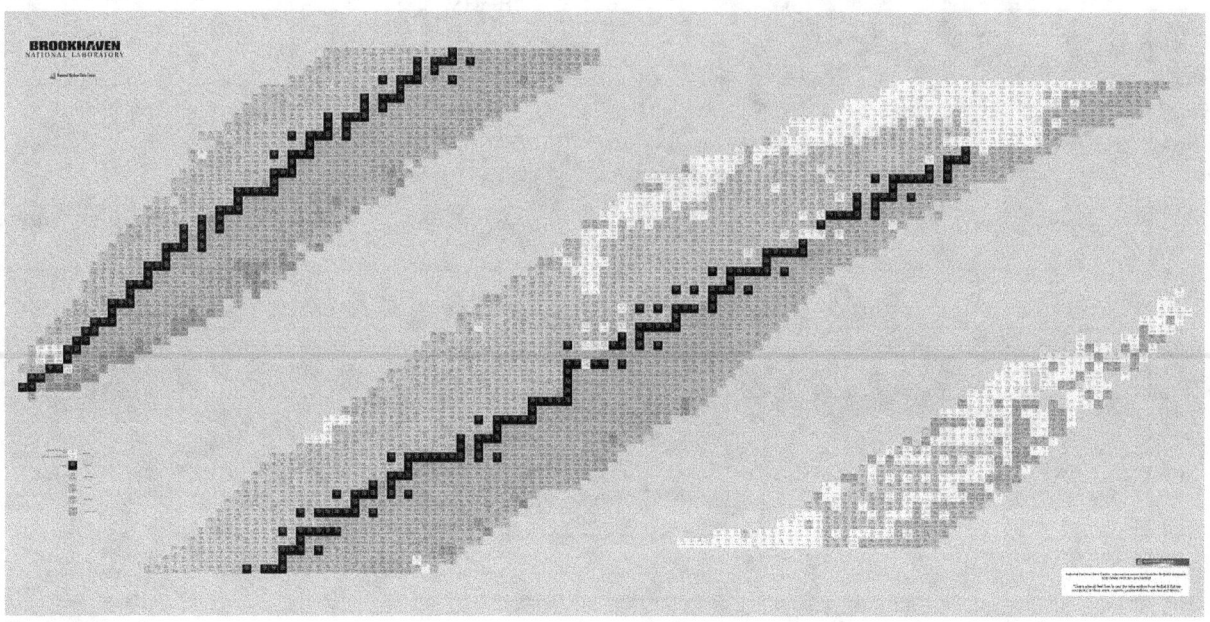

Decay modes

None (stable)		Neutron emission	
Beta decay		Alpha decay	
Proton emission		Spontaneous fission	

- Positron emission or Electron capture

Stable isotopes

https://upload.wikimedia.org/wikipedia/commons/8/80/Isotopes_and_half-life.svg

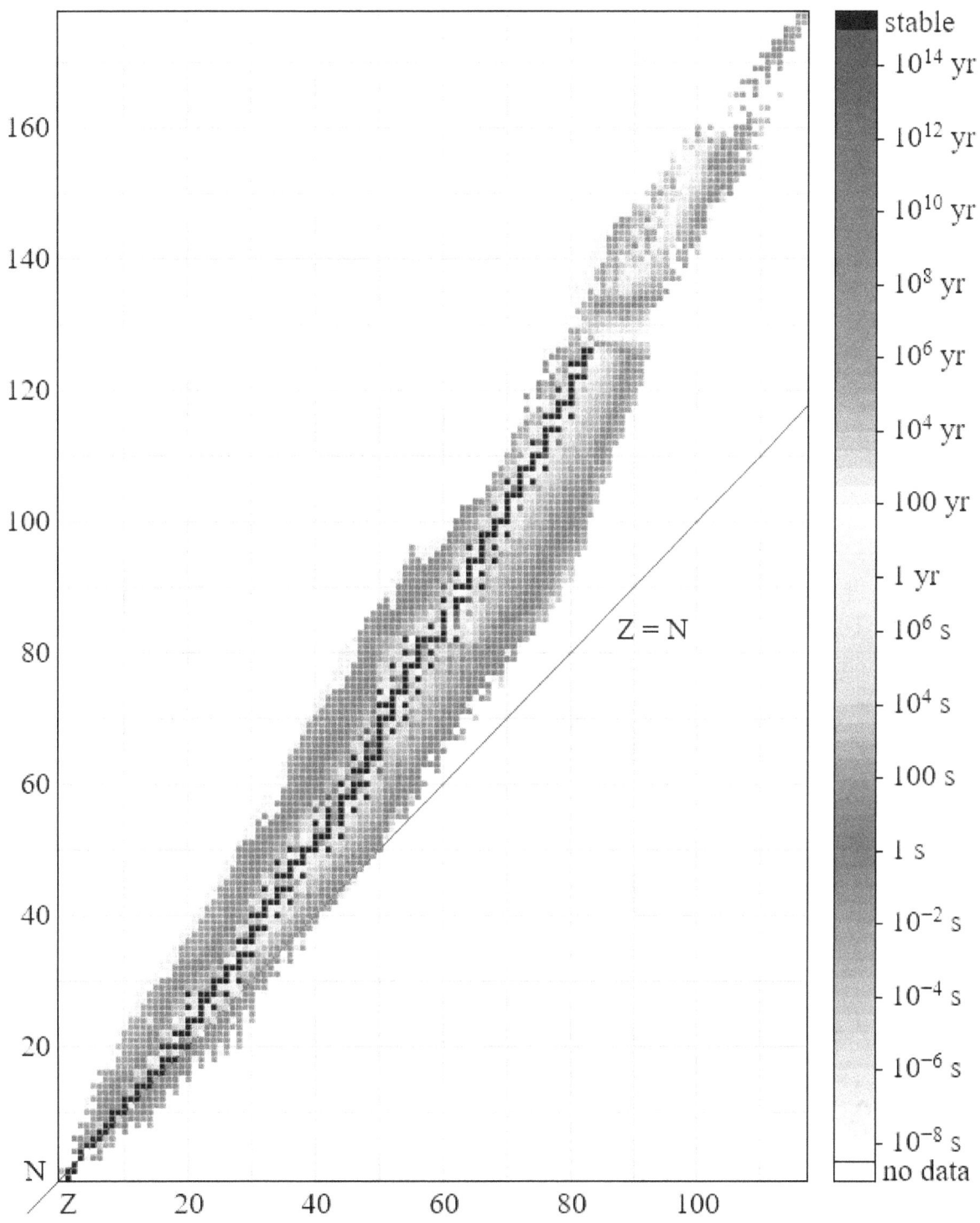

====

180

Type of decay

https://en.wikipedia.org/wiki/Radioactive_decay#Decay_modes_in_table_form
https://www-nds.iaea.org/relnsd/vcharthtml/VChartHTML.html
nds.contact-point@iaea.org <nds.contact-point@iaea.org>;

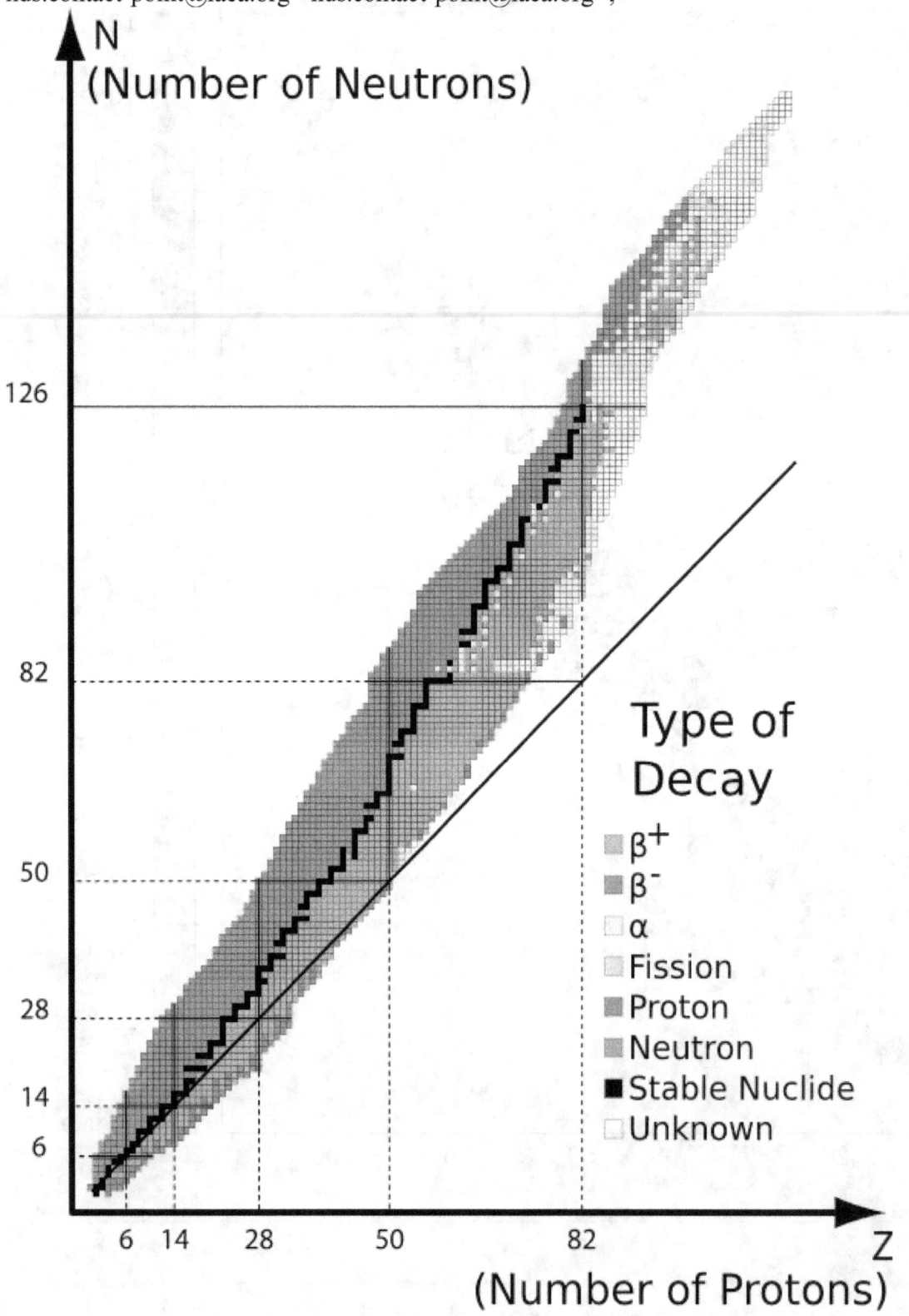

Tables of Nuclides (in Russian). Григорьев И.С. Физический Справочник, Москва

Т а б л и ц а 37.1. **Таблица нуклидов [1, 2]**

Эле-мент	A	Период полураспада	Тип распада или относительная распространенность стабильного изотопа, %	Энергия, МэВ (относительная интенсивность, %)	
				групп частиц	γ-излучения
$_1$H	1	} Стабилен	99,985 (1)	—	—
	2		0,015 (1)	—	—
	3	12,33 (6) года	β⁻	0,0186	—
$_2$He	3	} Стабилен	1,38 (3)·10⁻⁴	—	—
	4		99,999862 (3)	—	—
	6	0,8081 (20) с	β⁻	3,508	—
	8	0,122 (2) с	β⁻, β⁻n (12)	—	0,98 (88)
$_3$Li]	6	} Стабилен	7,5 (2)	—	—
	7		92,5 (2)	—	—
	8	0,842 (6) с	β⁻ / 2α	13 / 1,6	—
	9	0,176 (2) с	β⁻ / n	13,61 / 0,76	—
	11	0,0085 с	β⁻, β⁻n (61)	—	—
$_4$Be	7	53,44 (9) сут	э. з.	—	0,477 (10,3)
	9	Стабилен	100	—	—
	10	1,6 (2) 10⁶ лет	β⁻	0,555	—
	11	13,81 (8) с	β⁻ / β⁻α	11,5 / 0,77	} 2,14 (32); 4,67 (2,1); 5,85 (2,4); 6,79 (4,4); 7,99 (1,7)
	12	0,0114 (5) с	β⁻	11,7	—
$_5$B	8	0,769 (4) с	β⁺ / 2α	14,1 / 8,3; 1,6	—
	10	} Стабилен	19,9 (2)	—	—
	11		80,1 (2)	—	—
	12	0,02041 (6) с	β⁻ (~100) / β⁻3α (1,5)	13,37 / 0,192 (1,5)	} 4,43 (1,3)
	13	0,01736 (16) с	β⁻, β⁻n (0,28)	13,44	3,68 (7)
	14	0,0161 (12) с	β⁻	14,0	6,09; 6,73
$_6$C	9	0,1265 (9) с	p	8,2 (60); 1,1 (40)	—
	10	19,42 (6) с	β⁺	1,87	0,511 (200, ан); 0,717 (100); 1,023 (1,7)
	11	20,40 (4) мин	β⁺ (> 99) / э. з. (0,19)	0,97	} 0,511 (200, ан.)
	12	} Стабилен	98,90 (3)	—	—
	13		1,10 (3)	—	—
	14	5730 (40) лет	β⁻	0,156	—
	15	2,449 (4) с	β⁻	9,82 (32); 4,51 (68)	5,299 (68)
	16	0,747 (8) с	β⁻n (> 98,8)	0,79; 1,72	—
$_7$N	12	0,01097 (4) с	β⁺ (~100) / 3α (~3)	16,4 / 0,195	0,511 (200, ан.); 4,43 (2,4)
	13	9,961 (4) мин	β⁺	1,20	0,511 (200, ан.)
	14	} Стабилен	99,634 (9)	—	—
	15		0,366 (9)	—	—
	16	7,13 (2) с	β⁻ / α (0,0006)	10,40 (26); 4,27 (74) / 1,7	2,75 (1); 6,13 (69); 7,11 (5)
	17	4,169 (8) с	β⁻ / n	8,68(1,6); 7,81(2,6) / 4,1(95)	} 0,87(3); 2,19 (0,5)
	18	0,63 (3) с	β⁻	0,40(45); 1,21(45); 1,81(5) 9,4	0,82 (59); 1,65 (59); 1,98 (100); 2,47 (41)
$_8$O	13	0,0089 (2) с	p	6,40/100/; 6,97/24/	—
	14	70,599 (22) с	β⁺	4,12 (0,6); 1,811 (99)	0,511 (200, ан.); 2,312 (99)
	15	122,24 (16) с	β⁺	1,74	—
	16	} Стабилен	99,762 (15)	—	—
	17		0,038 (3)	—	—
	18		0,200 (12)	—	—
	19	26,91 (8) с	β⁻	4,60; 3,3	0,197 (97); 1,37 (59)
	20	13,57 (10) с	β⁻	2,8	1,06 (100)
	21 [2]	3,4 с	β⁻	6,4	1,73; 1,79; 2,80; 3,52

Table of Nuclides (in Russian). Григорьев И.С. Физический Справочник, Москва

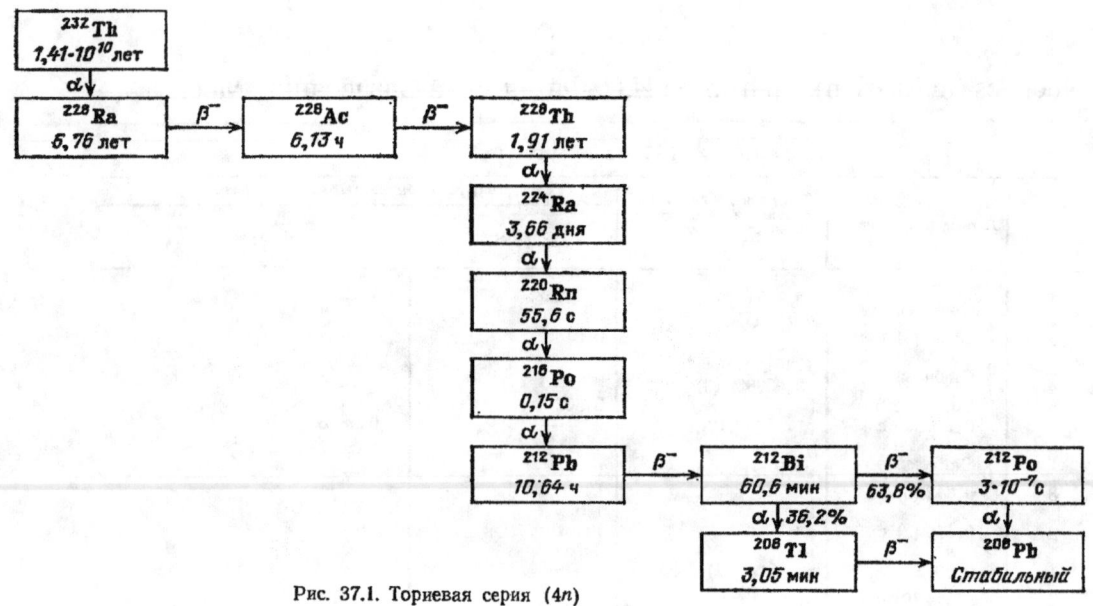

Рис. 37.1. Ториевая серия (4n)

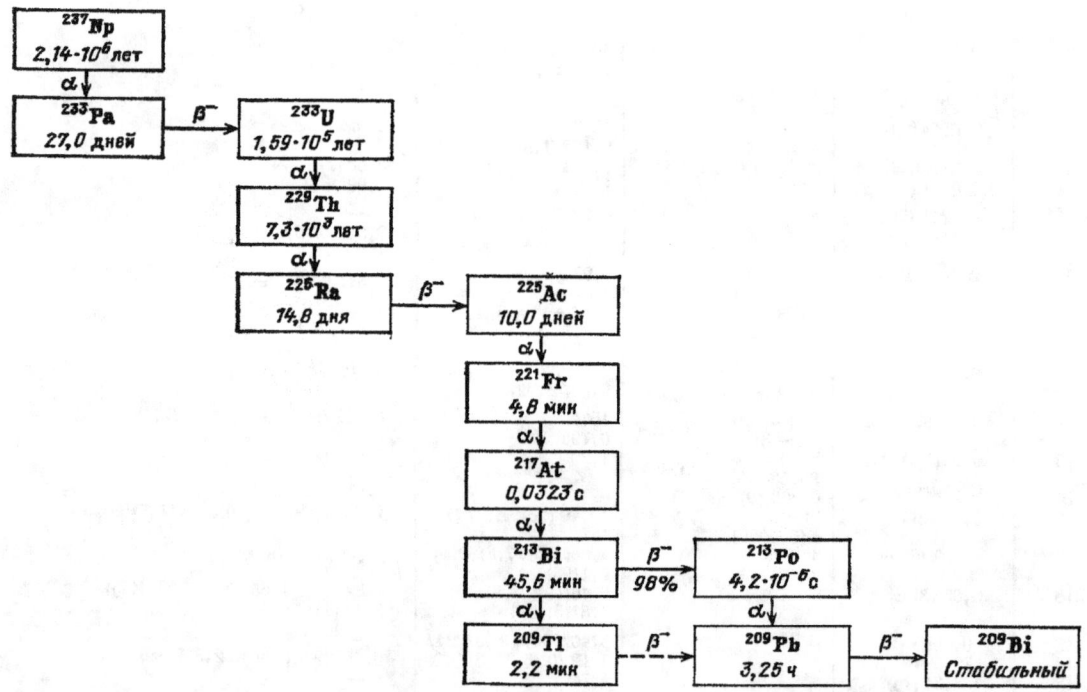

Рис. 37.2. Нептуниевая серия (4n+1)

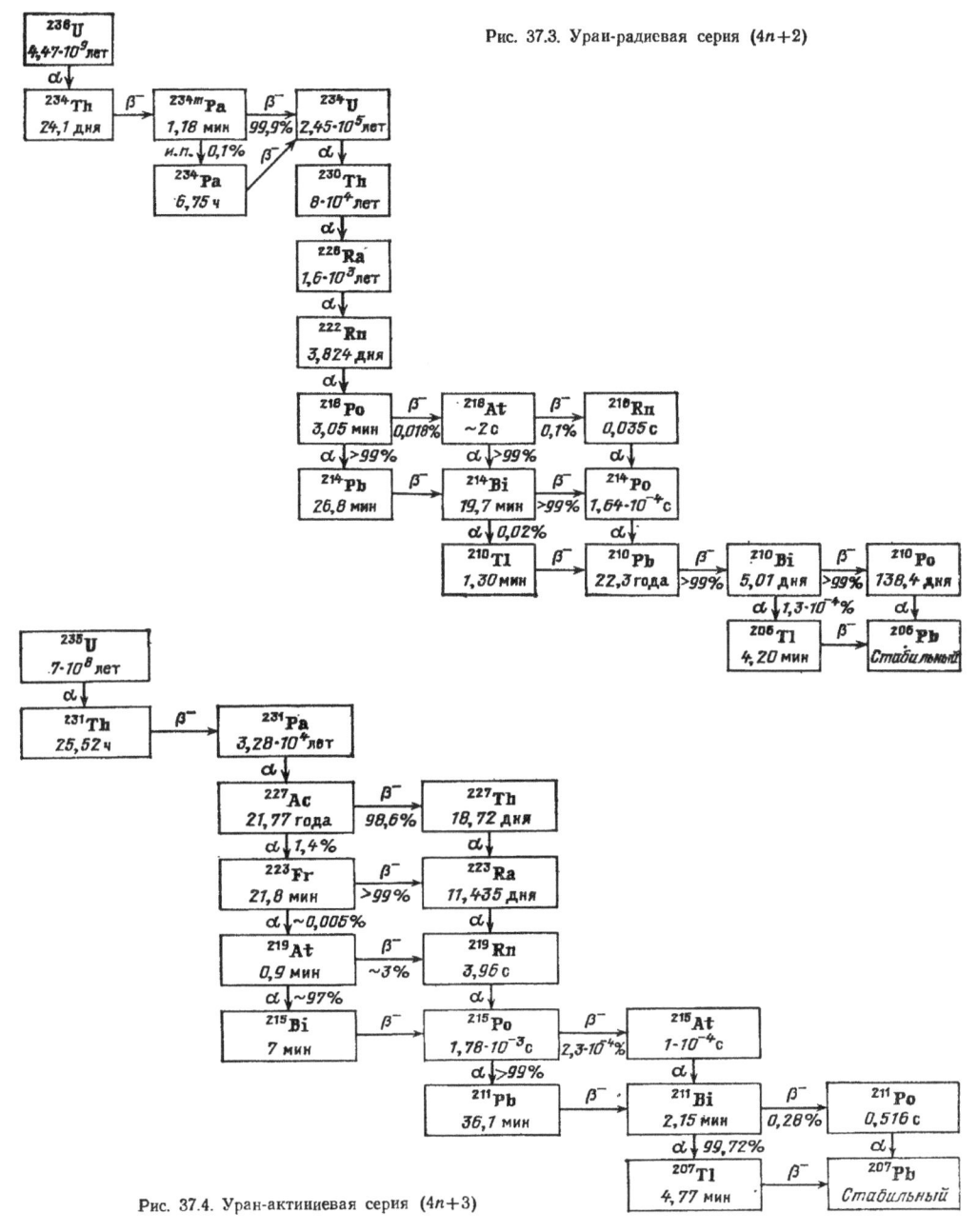

Рис. 37.3. Уран-радиевая серия (4n+2)

Рис. 37.4. Уран-актиниевая серия (4n+3)

Таблица 39.1. **Дефекты масс** $\Delta M = M - A$, кэВ [1]
(N — число нейтроиов; Z — число протонов; $A = N + Z$ — массовое число; с — масса получеиа в результате интерполяции или экстраполяции на основе имеющейся систематики)

N	Z	A	Элемент	Дефект массы, кэВ
1	0	1	n	8071,431 (39)
0	1		H	7289,034 (23)
1	1	2	H	13 135,84 (4)
2	1	3	H	14 949,94 (5)
1	2		He	14 931,32 (5)
3	1	4	H	25 920 (500)
2	2		He	2424,94 (4)
1	3		Li	25 130 (300)
4	1	5	H	33 790 (800)
3	2		He	11 390 (50)
2	3		Li	11 680 (50)
4	2	6	He	17 597,0 (35)
3	3		Li	14 087,3 (8)
2	4		Be	18 375 (6)
5	2	7	He	26 111 (30)
4	3		Li	14 908,2 (9)
3	4		Be	15 770,1 (9)
2	5		B	29 940 (100)
6	2	8	He	31 609 (12)
5	3		Li	20 946,9 (9)
4	4		Be	4941,76 (10)
3	5		B	22 921,9 (13)
2	6		C	35 085 (25)
6	3	9	Li	24 954,8 (20)
5	4		Be	11 348,0 (4)
4	5		B	12 416,1 (10)
3	6		C	28 912,1 (39)
7	3	10	Li	33 830 (250)
6	4		Be	12 607,6 (6)
5	5		B	12 051,7 (5)
4	6		C	15 702,9 (7)
3	7		N	39 500 (c)
8	3	11	Li	40 940 (120)
7	4		Be	20 176 (6)
6	5		B	8667,9 (4)
5	6		C	10 650,0 (11)
4	7		N	25 230 (100)
8	4	12	Be	25 030 (40)
7	5		B	13 369,5 (13)
6	6		C	0,0 (0,0)
5	7		N	17 338 (1)
4	8		O	32 070 (260)
9	4	13	Be	34 900 (c)
8	5		B	16 562 (4)
7	6		C	3125,038 (18)
6	7		N	5345,6 (9)
5	8		O	23 105 (10)
10	4	14	Be	40 970 (c)
9	5		B	23 657 (30)
8	6		C	3019,922 (24)
7	7		N	9863,444 (23)
6	8		O	8008,3 (5)
5	9		F	33 610 (c)
10	5	15	B	29 530 (c)
9	6		C	9873,2 (8)
8	7		N	101,514 (36)
7	8		O	2855,4 (7)
6	9		F	17 660 (c)
11	5	16	B	38 000 (c)
10	6		C	13 693 (16)

N	Z	A	Элемент	Дефект массы, кэВ
9	7		N	5681,6 (23)
8	8		O	—4737,02 (4)
7	9		F	10 692 (14)
6	10		Ne	24 110 (140)
12	5	17	B	45 270 (c)
11	6		C	21 060 (c)
10	7		N	7870 (15)
9	8		O	—809,9 (8)
8	9		F	1951,66 (18)
7	10		Ne	16 478 (26)
12	6	18	C	25 370 (c)
11	7		N	13 274 (30)
10	8		O	—783,03 (30)
9	9		F	872,5 (7)
8	10		Ne	5319 (5)
7	11		Na	25 320 (c)
13	6	19	C	34 430 (c)
12	7		N	15 600 (300)
11	8		O	3331,4 (27)
10	9		F	—1487,33 (13)
9	10		Ne	1750,9 (6)
8	11		Na	12 930 (12)
13	7	20	N	22 200 (c)
12	8		O	3799 (8)
11	9		F	—17,1 (6)
10	10		Ne	—7043,0 (5)
9	11		Na	6844 (7)
8	12		Mg	17 568 (27)
14	7	21	N	26 950 (c)
13	8		O	8120 (80)
12	9		F	—47 (7)
11	10		Ne	—5733,1 (11)
10	11		Na	—2185,8 (7)
9	12		Mg	10 912 (16)
14	8	22	O	9490 (220)
13	9		F	2826 (30)
12	10		Ne	—8026,1 (5)
11	11		Na	—5184,0 (7)
10	12		Mg	—394,1 (19)
9	13		Al	18 210 (c)
15	8	23	O	17 950 (c)
14	9		F	3350 (170)
13	10		Ne	—5155,1 (21)
12	11		Na	—9529,6 (8)
11	12		Mg	—5470,6 (15)
10	13		Al	6768 (25)
15	9	24	F	8650 (c)
14	10		Ne	—5949 (10)
13	11		Na	—8417,5 (8)
12	12		Mg	—13 930,6 (7)
11	13		Al	—52 (4)
10	14		Si	10 740 (120)
16	9	25	F	12 840 (c)
15	10		Ne	—2150 (90)
14	11		Na	—9357 (7)
13	12		Mg	—13 190,8 (11)
12	13		Al	—8912,9 (11)
11	14		Si	3824 (10)
16	10	26	Ne	—190 (c)
15	11		Na	—6888 (23)
14	12		Mg	—16 212,4 (9)
13	13		Al	—12 207,6 (10)
12	14		Si	—7143,1 (31)
11	15		P	11 260 (c)

N	Z	A	Элемент	Дефект массы, кэВ
17	10	27	Ne	6670 (c)
16	11		Na	—5630 (80)
15	12		Mg	—14 585,0 (14)
14	13		Al	—17 194,3 (7)
13	14		Si	—12 385,3 (15)
12	15		P	—590 (c)
17	11	28	Na	—1130 (120)
16	12		Mg	—15 016,4 (22)
15	13		Al	—16 848,2 (8)
14	14		Si	—21 491,2 (6)
13	15		P	—7159,5 (38)
12	16		S	4190 (120)
18	11	29	Na	2660 (150)
17	12		Mg	—10 750 (50)
16	13		Al	—18 212 (5)
15	14		Si	—21 893,7 (8)
14	15		P	—16 949,3 (18)
13	16		S	—3160 (50)
19	11	30	Na	8380 (300)
18	12		Mg	—9790 (c)
17	13		Al	—15 890 (40)
16	14		Si	—24 431,7 (9)
15	15		P	—20 204,5 (27)
14	16		S	—14 062,0 (31)
13	17		Cl	4840 (c)
20	11	31	Na	10 610 (c)
19	12		Mg	—3900 (c)
18	13		Al	—15 100 (100)
17	14		Si	—22 948,7 (10)
16	15		P	—24 439,5 (6)
15	16		S	—19 044,1 (16)
14	17		Cl	—7070 (50)
21	11	32	Na	16 410 (c)
20	12		Mg	●2890 (c)
19	13		Al	—11 290 (c)
18	14		Si	—24 092 (7)
17	15		P	—24 304,7 (8)
16	16		S	—26 015,1 (6)
15	17		Cl	—13 329 (8)
14	18		Ar	—2210 (130)
21	12	33	Mg	4130 (c)
20	13		Al	—9370 (c)
19	14		Si	—20 570 (50)
18	15		P	—26 336,9 (21)
17	16		S	—26 585,9 (8)
16	17		Cl	—21 003,0 (9)
15	18		Ar	—9385 (30)
21	13	34	Al	—4150 (c)
20	14		Si	—20 250 (c)
19	15		P	—24 550 (50)
18	16		S	—29 931,25 (28)
17	17		Cl	—24 438,3 (8)
16	18		Ar	—18 379,2 (30)
15	19		K	—1480 (c)
22	13	35	Al	—840 (c)
21	14		Si	—15 040 (c)
20	15		P	—24 940 (80)
19	16		S	—28 846,27 (21)
18	17		Cl	—29 013,73 (10)
17	18		Ar	—23 048,9 (16)
16	19		K	—11 169 (20)
22	14	36	Si	—12 670 (c)
21	15		P	—20 770 (c)

Таблица 42.2. Пробег R, мг/см², ионов водорода ${}^1_1H^+$ в различных веществах [1]

E_m — энергия на единицу массы падающей частицы; E — полная энергия

E_m, МэВ/а.е.м.	Be	C	Al	Ti	Ni	Ge	Zr	Ag	Eu	Ta	Au	U	E, МэВ
0,0125	0,059	0,070	0,102	0,158	0,193	0,216	0,231	0,246	0,373	0,421	0,449	0,491	0,0126
0,0160	0,068	0,082	0,118	0,181	0,222	0,247	0,265	0,283	0,430	0,488	0,521	0,575	0,0161
0,0200	0,078	0,093	0,134	0,205	0,251	0,279	0,300	0,320	0,489	0,556	0,595	0,661	0,0202
0,0250	0,089	0,106	0,151	0,231	0,283	0,315	0,338	0,362	0,554	0,632	0,678	0,757	0,0252
0,0320	0,102	0,122	0,173	0,264	0,323	0,360	0,386	0,414	0,635	0,727	0,781	0,877	0,0322
0,0400	0,116	0,139	0,196	0,298	0,365	0,406	0,436	0,468	0,719	0,824	0,886	0,999	0,0403
0,0500	0,132	0,158	0,222	0,337	0,413	0,458	0,493	0,529	0,815	0,935	1,006	1,139	0,0504
0,0600	0,147	0,175	0,246	0,374	0,458	0,507	0,546	0,586	0,905	1,038	1,117	1,268	0,0605
0,0700	0,161	0,193	0,270	0,409	0,500	0,554	0,597	0,641	0,990	1,136	1,223	1,391	0,0705
0,0800	0,175	0,209	0,293	0,443	0,541	0,600	0,646	0,694	1,072	1,230	1,324	1,509	0,0806
0,0900	0,190	0,226	0,316	0,476	0,582	0,644	0,694	0,746	1,152	1,321	1,423	1,623	0,0907
0,1000	0,204	0,242	0,339	0,510	0,622	0,688	0,742	0,797	1,230	1,410	1,519	1,734	0,1008
0,1250	0,240	0,281	0,397	0,592	0,721	0,798	0,861	0,924	1,421	1,629	1,754	2,006	0,1260
0,1600	0,293	0,337	0,480	0,709	0,861	0,952	1,029	1,103	1,686	1,931	2,077	2,378	0,1612
0,2000	0,357	0,402	0,580	0,848	1,025	1,133	1,227	1,314	1,993	2,280	2,448	2,804	0,2016
0,2500	0,444	0,486	0,714	1,031	1,241	1,371	1,486	1,591	2,389	2,728	2,926	3,351	0,2519
0,3200	0,580	0,614	0,919	1,306	1,564	1,726	1,874	2,006	2,973	3,387	3,628	4,151	0,3225
0,4000	0,756	0,774	1,179	1,649	1,964	2,164	2,355	2,520	3,685	4,188	4,481	5,122	0,4031
0,5000	1,007	0,998	1,542	2,122	2,511	2,761	3,010	3,222	4,642	5,262	5,624	6,422	0,5039
0,6000	1,293	1,253	1,947	2,642	3,110	3,412	3,727	3,992	5,676	6,421	6,855	7,814	0,6047
0,7000	1,615	1,539	2,394	3,210	3,761	4,119	4,506	4,828	6,789	7,666	8,174	9,301	0,7055
0,8000	1,971	1,857	2,881	3,824	4,461	4,877	5,344	5,726	7,976	8,990	9,577	10,881	0,8062
0,9000	2,359	2,207	3,405	4,480	5,207	5,682	6,235	6,684	9,230	10,385	11,054	12,541	0,9070
1,0000	2,780	2,589	3,967	5,177	5,997	6,533	7,178	7,697	10,546	11,848	12,602	14,275	1,0078
1,2500	3,967	3,685	5,524	7,088	8,153	8,851	9,746	10,461	14,095	15,789	16,769	18,935	1,2597
1,6000	5,944	5,552	8,056	10,148	11,584	12,529	13,827	14,869	19,647	21,933	23,269	26,189	1,6125
2,0000	8,643	8,132	11,435	14,165	16,068	17,317	19,155	20,629	26,796	29,798	31,589	35,443	2,0156
2,5000	12,661	11,975	16,361	19,938	22,471	24,137	26,772	28,843	36,865	40,850	43,256	48,349	2,5195
3,2000	19,456	18,433	24,518	29,373	32,867	35,196	39,120	42,169	52,948	58,480	61,830	68,810	3,2250
4,0000	28,827	27,268	35,569	42,008	46,704	49,898	55,475	59,875	74,010	81,494	86,028	95,428	4,0312
5,0000	42,846	40,387	51,855	60,461	66,794	71,210	79,124	85,476	104,075	114,385	120,447	133,199	5,0390
6,0000	59,307	55,720	70,767	81,732	89,870	95,609	106,176	114,724	138,207	151,568	159,272	175,591	6,0468
7,0000	78,125	73,180	92,199	105,704	115,830	122,961	136,444	147,441	176,200	192,816	202,300	222,377	7,0546
8,0000	99,234	92,701	116,062	132,276	144,562	153,185	169,769	183,485	217,769	237,963	249,315	273,414	8,0624
9,0000	122,542	114,237	142,282	161,361	175,962	186,185	206,058	222,734	262,856	286,831	300,174	328,493	9,0702
10,0000	147,989	137,754	170,795	192,866	209,966	221,870	245,277	265,037	311,387	339,290	354,688	387,461	10,078
11,0000	175,544	163,200	201,546	226,713	246,486	260,164	287,342	310,256	363,240	395,146	412,594	450,213	11,086
12,0000	205,152	190,534	234,483	262,868	285,442	300,978	332,155	358,304	418,272	454,277	473,872	516,684	12,094

1143

E_m, МэВ/а. е. м	H	He	N	O	Ne	Ar	Kr	Xe	Rn	Майлар	$(CH_2)_n$	Вода	E, МэВ
0,0125	0,022	0,056	0,089	0,095	0,109	0,174	0,292	0,403	0,562	0,068	0,052	0,067	0,0126
0,0160	0,026	0,065	0,104	0,111	0,126	0,200	0,334	0,463	0,651	0,080	0,061	0,079	0,0161
0,0200	0,030	0,076	0,119	0,127	0,144	0,227	0,377	0,522	0,739	0,092	0,070	0,091	0,0202
0,0250	0,035	0,088	0,136	0,146	0,165	0,256	0,423	0,585	0,829	0,105	0,081	0,105	0,0252
0,0320	0,041	0,104	0,159	0,169	0,191	0,292	0,478	0,659	0,933	0,122	0,094	0,123	0,0322
0,0400	0,048	0,122	0,182	0,194	0,218	0,327	0,533	0,730	1,033	0,139	0,107	0,142	0,0403
0,0500	0,056	0,144	0,210	0,223	0,249	0,367	0,592	0,807	1,139	0,159	0,123	0,164	0,0504
0,0600	0,063	0,166	0,237	0,251	0,279	0,403	0,645	0,875	1,232	0,178	0,138	0,185	0,0605
0,0700	0,070	0,186	0,263	0,277	0,306	0,436	0,694	0,937	1,316	0,196	0,152	0,206	0,0705
0,0800	0,077	0,206	0,287	0,302	0,333	0,468	0,740	0,995	1,394	0,214	0,166	0,225	0,0806
0,0900	0,083	0,226	0,311	0,326	0,358	0,498	0,784	1,050	1,469	0,231	0,179	0,244	0,0907
0,1000	0,090	0,245	0,333	0,350	0,383	0,527	0,827	1,103	1,541	0,248	0,193	0,262	0,1008
0,1250	0,105	0,290	0,388	0,406	0,442	0,597	0,928	1,231	1,712	0,289	0,225	0,305	0,1260
0,1600	0,125	0,349	0,460	0,481	0,521	0,691	1,065	1,402	1,943	0,345	0,269	0,362	0,1612
0,2000	0,146	0,414	0,539	0,563	0,607	0,796	1,219	1,596	2,204	0,409	0,319	0,424	0,2016
0,2500	0,172	0,492	0,635	0,663	0,714	0,929	1,414	1,844	2,541	0,490	0,383	0,500	0,2519
0,3200	0,208	0,602	0,770	0,804	0,867	1,125	1,707	2,218	3,051	0,609	0,477	0,607	0,3225
0,4000	0,251	0,732	0,932	0,974	1,053	1,372	2,079	2,695	3,706	0,755	0,592	0,735	0,4031
0,5000	0,308	0,906	1,153	1,206	1,311	1,721	2,606	3,372	4,635	0,957	0,750	0,908	0,5039
0,6000	0,372	1,096	1,402	1,468	1,604	2,120	3,209	4,144	5,692	1,185	0,928	1,103	0,6047
0,7000	0,444	1,307	1,684	1,765	1,936	2,575	3,892	5,021	6,885	1,442	1,129	1,323	0,7055
0,8000	0,525	1,540	2,004	2,101	2,312	3,089	4,657	6,002	8,209	1,730	1,353	1,572	0,8062
0,9000	0,616	1,799	2,364	2,479	2,736	3,665	5,505	7,082	9,662	2,050	1,602	1,852	0,9070
1,0000	0,718	2,084	2,762	2,897	3,206	4,301	6,433	8,251	11,237	2,403	1,877	2,164	1,0078
1,2500	1,019	2,902	3,909	4,102	4,560	6,119	9,045	11,508	15,585	3,418	2,673	3,067	1,2597
1,6000	1,554	4,298	5,856	6,140	6,845	9,138	13,298	16,756	22,446	5,153	4,050	4,620	1,6125
2,0000	2,336	6,248	8,522	8,934	9,948	13,167	18,863	23,567	31,173	7,563	5,992	6,792	2,0156
2,5000	3,582	9,216	12,465	13,075	14,479	18,947	26,709	33,053	43,166	11,178	8,953	10,083	2,5195
3,2000	5,827	14,351	19,060	19,995	21,953	28,283	39,129	47,972	61,844	17,293	14,045	15,702	3,2250
4,0000	9,081	21,609	28,071	29,413	32,030	40,662	55,340	67,335	85,883	25,701	21,140	23,481	4,0312
5,0000	14,110	32,706	41,483	43,369	46,857	58,666	78,649	94,977	119,798	38,231	31,808	35,137	5,0390
6,0000	20,125	45,939	57,172	59,674	64,112	79,413	105,264	126,386	157,996	52,904	44,359	48,838	6,0468
7,0000	27,062	61,206	75,032	78,221	83,712	102,808	135,109	161,429	200,471	69,628	58,705	64,477	7,0546
8,0000	34,874	78,444	94,976	88,900	105,595	128,788	168,137	200,041	247,074	88,323	74,781	81,953	8,0624
9,0000	43,542	97,563	116,955	121,671	129,683	157,304	204,352	242,228	297,738	108,953	92,552	101,233	9,0702
10,0000	53,047	118,429	140,896	146,465	155,891	188,297	243,626	287,921	352,412	131,476	111,983	122,269	10,078
11,0000	63,367	140,982	166,715	173,205	184,154	221,721	285,865	337,042	411,094	155,833	133,023	144,996	11,086
12,0000	74,495	165,192	194,371	201,846	214,428	257,523	330,985	389,490	473,713	181,985	155,646	169,377	12,094

Т а б л и ц а 42.3. Массовая тормозная способность веществ, МэВ/(мг·см$^{-2}$), для ионов гелия $_2^4$He$^+$ [1]

E_m — энергия на единицу массы падающей частицы; E — полная энергия

E_m, МэВ/а. е. м	Be	C	Al	Ti	Ni	Ge	Zr	Ag	Eu	Ta	Au	U	E, МэВ
0,0125	0,877	0,726	0,532	0,354	0,287	0,261	0,241	0,223	0,142	0,121	0,112	0,096	0,0500
0,0160	0,993	0,821	0,602	0,400	0,325	0,295	0,273	0,253	0,161	0,137	0,126	0,108	0,0640
0,0200	1,110	0,918	0,673	0,447	0,363	0,330	0,305	0,282	0,180	0,154	0,141	0,121	0,0801
0,0250	1,241	1,026	0,752	0,500	0,406	0,368	0,341	0,316	0,201	0,172	0,158	0,135	0,1001
0,0320	1,403	1,163	0,852	0,566	0,461	0,419	0,387	0,358	0,227	0,195	0,180	0,154	0,1281
0,0400	1,554	1,294	0,945	0,631	0,514	0,469	0,432	0,400	0,254	0,219	0,201	0,173	0,1601
0,0500	1,704	1,428	1,040	0,697	0,571	0,520	0,479	0,444	0,282	0,245	0,226	0,194	0,2001
0,0600	1,819	1,539	1,115	0,751	0,617	0,561	0,518	0,480	0,308	0,267	0,247	0,212	0,2402
0,0700	1,906	1,631	1,173	0,795	0,657	0,596	0,551	0,511	0,328	0,287	0,266	0,229	0,2802
0,0800	1,972	1,709	1,219	0,832	0,689	0,624	0,576	0,536	0,347	0,305	0,282	0,243	0,3202
0,0900	2,020	1,773	1,254	0,863	0,715	0,649	0,598	0,557	0,365	0,319	0,296	0,255	0,3602
0,1000	2,054	1,827	1,280	0,887	0,737	0,668	0,614	0,575	0,379	0,332	0,307	0,265	0,4003
0,1250	2,094	1,932	1,317	0,927	0,774	0,702	0,645	0,604	0,405	0,354	0,331	0,286	0,5003
0,1600	2,077	2,018	1,323	0,948	0,797	0,723	0,663	0,622	0,424	0,372	0,349	0,304	0,6404
0,2000	2,010	2,040	1,299	0,944	0,799	0,723	0,664	0,622	0,431	0,380	0,357	0,312	0,8005
0,2500	1,901	1,999	1,248	0,922	0,784	0,711	0,651	0,610	0,431	0,381	0,357	0,312	1,0007

Т а б л и ц а 42.4. Пробег R, мг/см², ионов гелия $^{4}_{2}He^{+}$ в различных веществах [1]

E_m —энергия на единицу массы падающей частицы; E— полная энергия

E_m, МэВ/а.е.м.	Be	C	Al	Ti	Ni	Ge	Zr	Ag	Eu	Ta	Au	U	E, МэВ
0,0125	0,085	0,099	0,142	0,219	0,267	0,307	0,337	0,371	0,583	0,685	0,746	0,874	0,0500
0,0160	0,100	0,117	0,166	0,255	0,310	0,356	0,390	0,428	0,673	0,790	0,860	1,007	0,0640
0,0200	0,114	0,135	0,190	0,292	0,356	0,406	0,444	0,487	0,764	0,897	0,976	1,143	0,0801
0,0250	0,131	0,155	0,218	0,333	0,406	0,462	0,505	0,552	0,867	1,017	1,107	1,296	0,1001
0,0320	0,152	0,180	0,252	0,385	0,470	0,532	0,580	0,634	0,996	1,167	1,269	1,486	0,1281
0,0400	0,174	0,206	0,288	0,438	0,534	0,603	0,658	0,718	1,127	1,320	1,435	1,679	0,1601
0,0500	0,198	0,235	0,327	0,498	0,607	0,684	0,745	0,812	1,275	1,490	1,620	1,895	0,2001
0,0600	0,221	0,262	0,365	0,553	0,674	0,758	0,825	0,898	1,410	1,645	1,788	2,090	0,2402
0,0700	0,243	0,287	0,400	0,604	0,737	0,827	0,900	0,979	1,535	1,789	1,944	2,271	0,2802
0,0800	0,263	0,311	0,433	0,654	0,796	0,892	0,971	1,055	1,654	1,924	2,090	2,441	0,3202
0,0900	0,283	0,334	0,465	0,701	0,853	0,955	1,039	1,128	1,766	2,053	2,229	2,602	0,3602
0,1000	0,303	0,356	0,497	0,747	0,908	1,016	1,105	1,199	1,874	2,176	2,362	2,756	0,4003
0,1250	0,351	0,409	0,574	0,857	1,041	1,162	1,263	1,369	2,129	2,467	2,675	3,118	0,5003
0,1600	0,418	0,480	0,680	1,006	1,218	1,358	1,477	1,597	2,466	2,852	3,086	3,591	0,6404
0,2000	0,496	0,559	0,802	1,175	1,419	1,579	1,718	1,854	2,840	3,277	3,539	4,110	0,8005
0,2500	0,599	0,658	0,959	1,389	1,672	1,858	2,022	2,178	3,304	3,802	4,099	4,751	1,0007
0,3200	0,752	0,802	1,191	1,700	2,036	2,259	2,461	2,647	3,964	4,547	4,892	5,656	1,2808
0,4000	0,945	0,977	1,475	2,075	2,474	2,738	2,986	3,209	4,743	5,423	5,825	6,718	1,6010
0,5000	1,211	1,216	1,860	2,577	3,055	3,372	3,682	3,955	5,759	6,564	7,039	8,099	2,0013
0,6000	1,510	1,481	2,283	3,120	3,680	4,052	4,430	4,758	6,838	7,773	8,323	9,551	2,4016
0,7000	1,842	1,777	2,745	3,708	4,353	4,782	5,236	5,622	7,989	9,060	9,687	11,088	2,8018
0,8000	2,207	2,103	3,244	4,337	5,072	5,560	6,095	6,544	9,207	10,418	11,126	12,709	3,2021
0,9000	2,604	2,460	3,779	5,006	5,833	6,381	7,004	7,521	10,485	11,842	12,633	14,402	3,6023
1,0000	3,031	2,848	4,349	5,714	6,635	7,245	7,961	8,549	11,821	13,328	14,205	16,164	4,0026
1,2500	4,228	3,955	5,920	7,642	8,811	9,584	10,553	11,339	15,403	17,304	18,411	20,866	5,0033
1,6000	6,211	5,826	8,460	10,711	12,251	13,272	14,645	15,759	20,971	23,466	24,929	28,141	6,4042
2,0000	8,906	8,403	11,835	14,723	16,729	18,054	19,967	21,512	28,111	31,320	33,237	37,383	8,0052
2,5000	12,910	12,232	16,742	20,475	23,109	24,850	27,556	29,696	38,143	42,332	44,862	50,241	10,007
3,2000	19,669	18,656	24,857	29,860	33,450	35,850	39,838	42,951	54,141	59,869	63,338	70,594	12,808
4,0000	28,981	27,434	35,838	42,415	47,199	50,459	56,090	60,545	75,070	82,738	87,383	97,044	16,010
5,0000	42,903	40,463	52,011	60,740	67,151	71,624	79,576	85,969	104,928	115,401	121,564	134,555	20,013
6,0000	59,245	55,686	70,787	81,859	90,060	95,848	106,434	115,008	138,814	152,317	160,110	176,642	24,016
7,0000	77,926	73,019	92,063	105,656	115,831	123,000	136,481	147,486	176,530	193,264	202,825	223,087	28,018
8,0000	98,882	92,397	115,752	132,035	144,354	153,004	169,563	183,268	217,797	238,083	249,497	273,752	32,021
9,0000	122,021	113,778	141,783	160,909	175,528	185,766	205,590	222,233	262,558	286,598	299,989	328,433	36,023
10,0000	147,287	137,127	170,092	192,190	209,289	221,197	244,530	264,234	310,742	338,683	354,114	386,980	40,026
11,0000	174,648	162,394	200,627	225,799	245,552	259,221	286,299	309,135	362,231	394,145	411,613	449,291	44,029
12,0000	204,051	189,540	233,337	261,704	284,239	299,753	330,801	356,851	416,882	452,868	472,467	515,303	48,031

E_m, МэВ/а.е.м.	H	He	N	O	Ne	Ar	Kr	Xe	Rn	Майлар	(CH₂)_n	Вода	E, МэВ
0,0125	0,032	0,081	0,123	0,132	0,149	0,237	0,409	0,596	0,917	0,097	0,075	0,095	0,0500
0,0160	0,038	0,097	0,146	0,156	0,176	0,277	0,474	0,688	1,056	0,115	0,089	0,113	0,0640
0,0200	0,045	0,113	0,170	0,181	0,204	0,318	0,540	0,780	1,194	0,133	0,103	0,133	0,0801
0,0250	0,053	0,132	0,197	0,210	0,236	0,364	0,613	0,879	1,337	0,154	0,120	0,155	0,1001
0,0320	0,062	0,158	0,232	0,247	0,277	0,420	0,699	0,996	1,501	0,180	0,140	0,183	0,1281
0,0400	0,073	0,186	0,269	0,286	0,319	0,475	0,783	1,107	1,659	0,207	0,162	0,213	0,1601
0,0500	0,085	0,219	0,312	0,330	0,367	0,536	0,874	1,225	1,822	0,238	0,186	0,247	0,2001
0,0600	0,096	0,251	0,352	0,371	0,411	0,590	0,954	1,327	1,963	0,267	0,208	0,279	0,2402
0,0700	0,106	0,282	0,390	0,410	0,452	0,639	1,027	1,419	2,087	0,294	0,229	0,309	0,2802
0,0800	0,116	0,311	0,425	0,446	0,490	0,685	1,093	1,503	2,200	0,319	0,249	0,336	0,3202
0,0900	0,125	0,338	0,458	0,480	0,526	0,727	1,156	1,580	2,305	0,343	0,268	0,363	0,3602
0,1000	0,134	0,364	0,490	0,513	0,560	0,768	1,214	1,654	2,404	0,366	0,286	0,388	0,4003
0,1250	0,154	0,425	0,563	0,588	0,639	0,861	1,350	1,824	2,633	0,421	0,329	0,445	0,5003
0,1600	0,179	0,500	0,654	0,684	0,739	0,981	1,523	2,042	2,927	0,493	0,386	0,518	0,6404
0,2000	0,206	0,579	0,750	0,783	0,844	1,109	1,711	2,278	3,245	0,570	0,447	0,594	0,8005
0,2500	0,236	0,671	0,863	0,900	0,969	1,264	1,940	2,569	3,640	0,665	0,522	0,683	1,0007
0,3200	0,277	0,795	1,016	1,060	1,142	1,486	2,271	2,991	4,217	0,799	0,627	0,804	1,2808
0,4000	0,323	0,937	1,193	1,246	1,346	1,756	2,677	3,513	4,932	0,959	0,753	0,943	1,6010
0,5000	0,385	1,121	1,428	1,493	1,620	2,126	3,238	4,232	5,919	1,174	0,922	1,128	2,0013
0,6000	0,451	1,320	1,687	1,765	1,925	2,542	3,867	5,038	7,022	1,412	1,107	1,330	2,4016
0,7000	0 525	1,538	1,979	2,073	2,269	3,013	4,573	5,944	8,255	1,678	1,314	1,558	2,8018

2 4 6 8 10 12 0   Рис. 42.2. Пробег электронов в алюминии

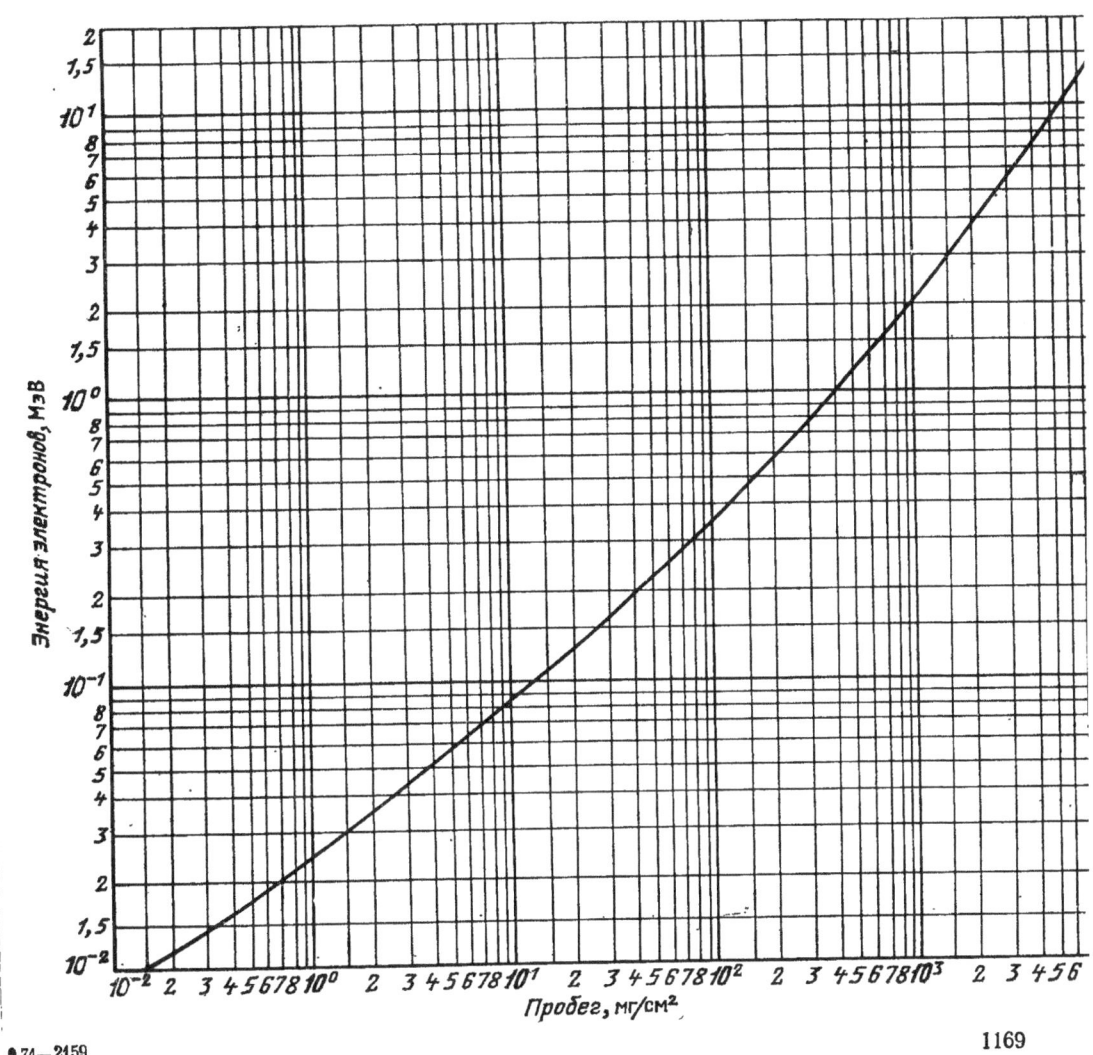

Пробег, мг/см²

74—2159   1169

Таблица 42.21. Толщина защиты из свинца, мм, в зависимости от кратности ослабления и энергии γ-излучения (широкий пучок) [8]

Кратность ослабления	Энергия γ-излучения, МэВ															
	0,1	0,2	0,3	0,4	0,5	0,6	0,7	0,8	0,9	1,0	1,5	2	3	4	6	10
2	1	2	3	4	5	7	8	10	11,5	13	17	20	21	20	16	13,5
5	2	4	6	9	11	15	19	22	25	28	38	43	46	45	38	30
10	3	5,5	9	13	16	21	26	30,5	35	38	51	59	65	64	55	42
30	3,5	7	11,5	17	23	30	36,5	43	49,5	55	73	85	93	92	80	63
100	5	10	16	23	30	38,5	47	55	63	70	96,5	113	122	121	109	87
500	6,5	14	22	31	40	51	61	72	82	92	129	150	163	161	149	119
10^3	7	15	24	33	44	57	69,5	81	92	102	141	165	160	178	165	133
$5 \cdot 10^3$	9	19	30	42	55	70	85	99	112	124	170	198	219	217	203	166
10^4	10,5	21	33	45,5	59	75	91	106	120	133	183	213	236	234	220	180
$5 \cdot 10^4$	11,5	23,5	37	52	69	87	105	123	140	156	214	247	263	272	258	215
10^5	11,5	24	38	54	72	92	111	130	148	165	227	262	289	289	275	229

1170

Decay chains

The four most common modes of radioactive decay are: alpha decay, beta decay, inverse beta decay (considered as both positron emission and electron capture), and isomeric transition. Of these decay processes, only alpha decay changes the atomic mass number (A) of the nucleus, and always decreases it by four. Because of this, almost any decay will result in a nucleus whose atomic mass number has the same residue mod 4, dividing all nuclides into four chains. The members of any possible decay chain must be drawn entirely from one of these classes. All four chains also produce helium-4 (alpha particles are helium-4 nuclei).

Three main decay chains (or families) are observed in nature, commonly called the thorium series, the radium or uranium series, and the actinium series, representing three of these four classes, and ending in three different, stable isotopes of lead. The mass number of every isotope in these chains can be represented as $A = 4n$, $A = 4n + 2$, and $A = 4n + 3$, respectively. The long-lived starting isotopes of these three isotopes, respectively thorium-232, uranium-238, and uranium-235, have existed since the formation of the earth, ignoring the artificial isotopes and their decays since the 1940s.

Due to the relatively short half-life of its starting isotope neptunium-237 (2.14 million years), the fourth chain, the neptunium series with $A = 4n + 1$, is already extinct in nature, except for the final rate-limiting step, decay of bismuth-209. The ending isotope of this chain is now known to be thallium-205. Some older sources give the final isotope as bismuth-209, but it was recently discovered that it is very slightly radioactive, with a half-life of 1.9×10^{19} years.

There are also non-transuranic decay chains of unstable isotopes of light elements, for example those of magnesium-28 and chlorine-39. On Earth, most of the starting isotopes of these chains before 1945 were generated by cosmic radiation. Since 1945, the testing and use of nuclear weapons has also released numerous radioactive fission products. Almost all such isotopes decay by either β^- or β^+ decay modes, changing from one element to another without changing atomic mass. These later daughter products, being closer to stability, generally have longer half-lives until they finally decay into stability.

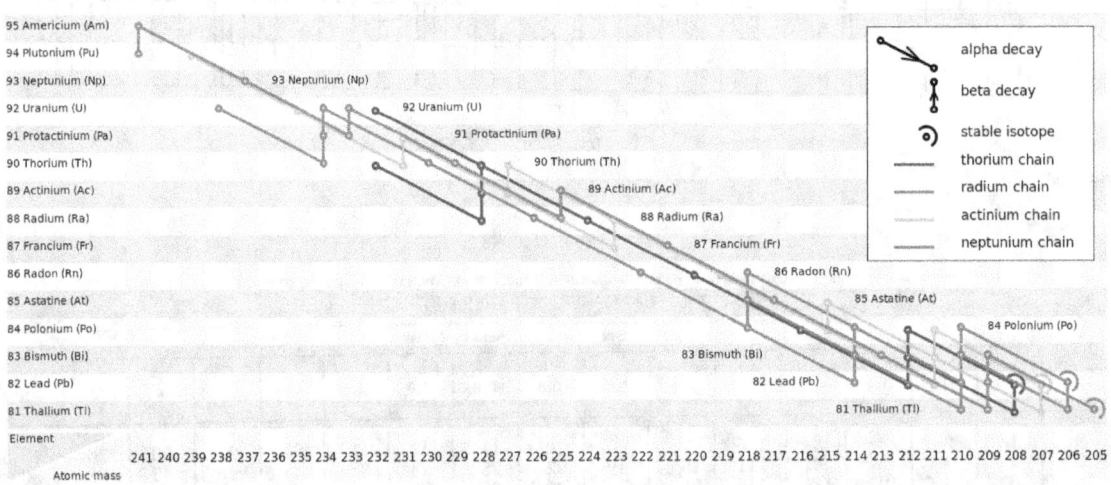

This diagram illustrates the four decay chains discussed in the text: thorium (4n, in blue), neptunium (4n+1, in purple), radium (4n+2, in red) and actinium (4n+3, in green).

Thorium series.

The 4n chain of Th-232 is commonly called the "thorium series" or "thorium cascade". Beginning with naturally occurring thorium-232, this series includes the following elements: actinium, bismuth, lead, polonium, radium, radon and thallium. All are present, at least transiently, in any natural thorium-containing sample, whether metal, compound, or mineral. The series terminates with lead-208.

The total energy released from thorium-232 to lead-208, including the energy lost to neutrinos, is 42.6 MeV

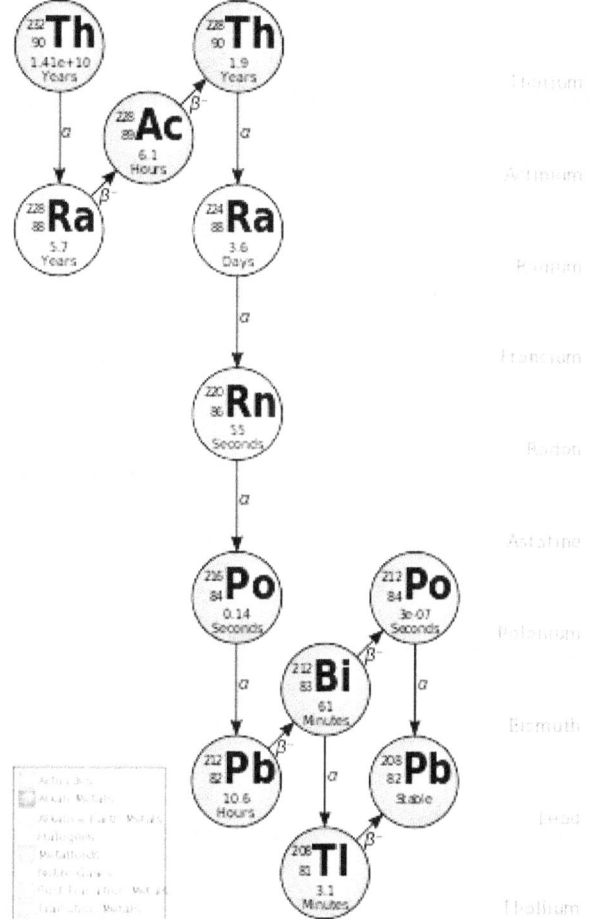

nuclide	historic name (short)	historic name (long)	decay mode	half-life (a=year)	energy released, MeV	product of decay
252Cf			α	2.645 a	6.1181	248Cm
248Cm			α	3.4×105 a	5.162	244Pu
244Pu			α	8×107 a	4.589	240U
240U			β−	14.1 h	.39	240Np
240Np			β−	1.032 h	2.2	240Pu
240Pu			α	6561 a	5.1683	236U
236U		Thoruranium[8]	α	2.3×107 a	4.494	232Th
232Th	Th	Thorium	α	1.405×1010 a	4.081	228Ra
228Ra	MsTh1	Mesothorium 1	β−	5.75 a	0.046	228Ac
228Ac	MsTh2	Mesothorium 2	β−	6.25 h	2.124	228Th
228Th	RdTh	Radiothorium	α	1.9116 a	5.520	224Ra
224Ra	ThX	Thorium X	α	3.6319 d	5.789	220Rn
220Rn	Tn	Thoron, Thorium Emanation	α	55.6 s	6.404	216Po
216Po	ThA	Thorium A	α	0.145 s	6.906	212Pb
212Pb	ThB	Thorium B	β−	10.64 h	0.570	212Bi

212Bi	ThC	Thorium C	β− 64.06% α 35.94%	60.55 min	2.252 6.208	212Po 208Tl
212Po	ThC′	Thorium C′	α	299 ns	8.955	208Pb
208Tl	ThC″	Thorium C″	β−	3.053 min	4.999	208Pb
208Pb	ThD	Thorium D	stable	.	.	.

Neptunium series

The 4n + 1 chain of Np-237 is commonly called the "neptunium series" or "neptunium cascade". In this series, only two of the isotopes involved are found naturally, namely the final two: bismuth-209 and thallium-205. A smoke detector containing an americium-241 ionization chamber accumulates a significant amount of neptunium-237 as its americium decays; the following elements are also present in it, at least transiently, as decay products of the neptunium: actinium, astatine, bismuth, francium, lead, polonium, protactinium, radium, thallium, thorium, and uranium. Since this series was only studied more recently[when?], its nuclides do not have historic names. One unique trait of this decay chain is that it does not include the noble-gas radon, and thus does not migrate through rock nearly as much as the other three decay chains.

The total energy released from californium-249 to thallium-205, including the energy lost to neutrinos, is 66.8 MeV.

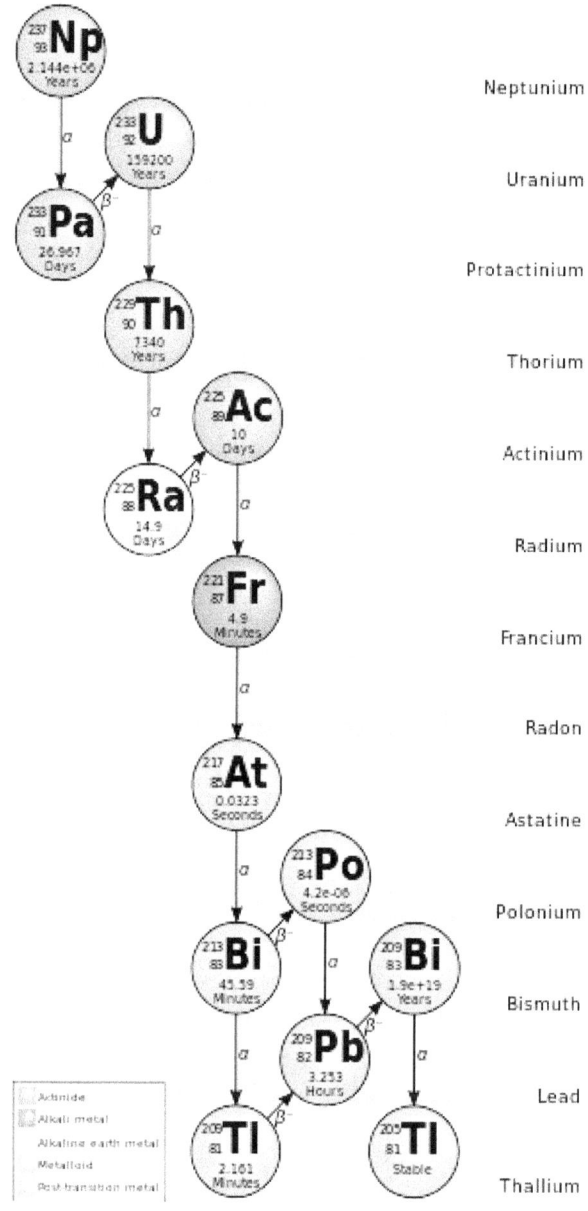

nuclide	decay mode	half-life (a=year)	energy released, MeV	product of decay
249Cf	α	351 a	5.813+.388	245Cm
245Cm	α	8500 a	5.362+.175	241Pu
241Pu	β−	14.4 a	0.021	241Am
241Am	α	432.7 a	5.638	237Np
237Np	α	2.14·106 a	4.959	233Pa
233Pa	β−	27.0 d	0.571	233U
233U	α	1.592·105 a	4.909	229Th
229Th	α	7340 a	5.168	225Ra
225Ra	β−	14.9 d	0.36	225Ac
225Ac	α	10.0 d	5.935	221Fr
221Fr	α	4.8 min	6.3	217At

217At	α	32 ms	7.0	213Bi
213Bi	β− 97.80% α 2.20%	46.5 min	1.423 5.87	213Po 209Tl
213Po	α	3.72 μs	8.536	209Pb
209Tl	β−	2.2 min	3.99	209Pb
209Pb	β−	3.25 h	0.644	209Bi
209Bi	α	1.9·1019 a	3.137	205Tl
205Tl	.	stable	.	.

Uranium series

The 4n+2 chain of U-238 is called the "uranium series" or "radium series". Beginning with naturally occurring uranium-238, this series includes the following elements: astatine, bismuth, lead, polonium, protactinium, radium, radon, thallium, and thorium. All are present, at least transiently, in any natural uranium-containing sample, whether metal, compound, or mineral. The series terminates with lead-206. The total energy released from uranium-238 to lead-206, including the energy lost to neutrinos, is 51.7 MeV.

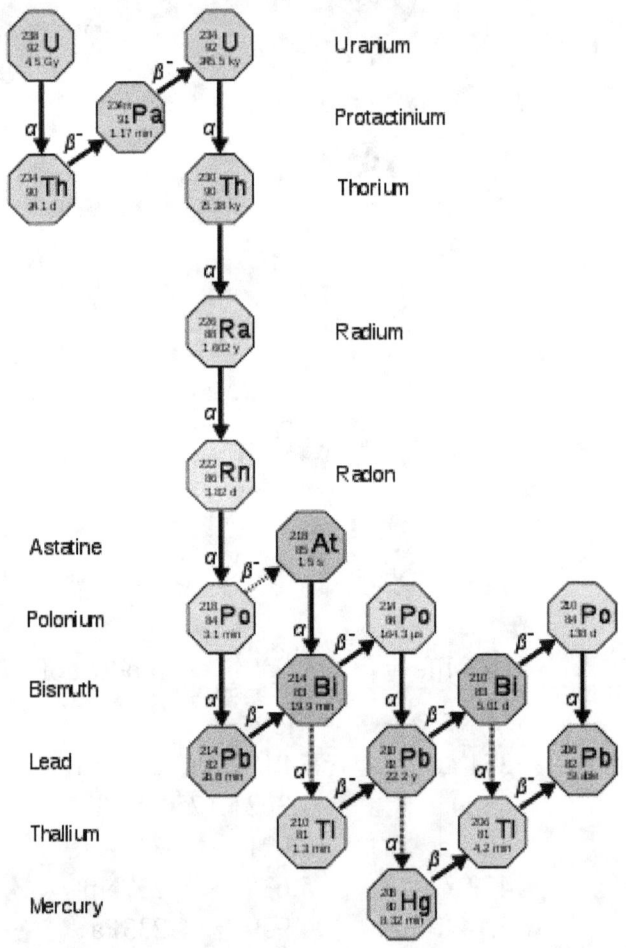

(More comprehensive graphic)

parent nuclide	historic name (short) [citation needed]	historic name (long)	atomic mass [RS 1]	decay mode [RS 2]	branch chance [RS 2]	half life [RS 2]	energy released, MeV [RS 2]	daughter nuclide [RS 2]	Subtotal MeV

238U	UI	Uranium I	238.051	α	100 %	4.468·109 a	4.26975	234Th	4.2698
234Th	UX1	Uranium X1	234.044	β−	100 %	24.10 d	0.273088	234mPa	4.5428
234mPa	UX2, Bv	Uranium X2, Brevium	234.043	IT	0.16 %	1.159 min	0.07392	234Pa	4.6168
234mPa	UX2, Bv	Uranium X2, Brevium	234.043	β−	99.84 %	1.159 min	2.268205	234U	6.8110
234Pa	UZ	Uranium Z	234.043	β−	100 %	6.70 h	2.194285	234U	6.8110
234U	UII	Uranium II	234.041	α	100 %	2.455·105 a	4.8598	230Th	11.6708
230Th	Io	Ionium	230.033	α	100 %	7.54·104 a	4.76975	226Ra	16.4406
226Ra	Ra	Radium	226.025	α	100 %	1600 a	4.87062	222Rn	21.3112
222Rn	Rn	Radon, Radium Emanation	222.018	α	100 %	3.8235 d	5.59031	218Po	26.9015
218Po	RaA	Radium A	218.009	β−	0.020 %	3.098 min	0.259913	218At	27.1614
218Po	RaA	Radium A	218.009	α	99.980 %	3.098 min	6.11468	214Pb	33.0162
218At			218.009	β−	0.1 %	1.5 s	2.881314	218Rn	30.0428
218At			218.009	α	99.9 %	1.5 s	6.874	214Bi	34.0354
218Rn			218.006	α	100 %	35 ms	7.26254	214Po	37.3053
214Pb	RaB	Radium B	214.000	β−	100 %	26.8 min	1.019237	214Bi	34.0354
214Bi	RaC	Radium C	213.999	β−	99.979 %	19.9 min	3.269857	214Po	37.3053
214Bi	RaC	Radium C	213.999	α	0.021 %	19.9 min	5.62119	210Tl	39.6566
214Po	RaC'	Radium C'	213.995	α	100 %	164.3 μs	7.83346	210Pb	45.1388
210Tl	RaC"	Radium C"	209.990	β−	100 %	1.30 min	5.48213	210Pb	45.1388
210Pb	RaD	Radium D	209.984	β−	100 %	22.20 a	0.063487	210Bi	45.2022
210Pb	RaD	Radium D	209.984	α	1.9·10−6 %	22.20 a	3.7923	206Hg	48.9311
210Bi	RaE	Radium E	209.984	β−	100 %	5.012 d	1.161234	210Po	46.3635
210Bi	RaE	Radium E	209.984	α	13.2·10−5 %	5.012 d	5.03647	206Tl	50.2387
210Po	RaF	Radium F	209.983	α	100 %	138.376 d	5.40745	206Pb	51.7709
206Hg			205.978	β−	100 %	8.32 min	1.307649	206Tl	50.2387
206Tl	RaE"	Radium E"	205.976	β−	100 %	4.202 min	1.532221	206Pb	51.7709
206Pb	RaG	Radium G	205.974	stable -	-	-	-	-	51.7709

Jump up ^ "The Risk Assessment Information System: Radionuclide Decay Chain". The University of Tennessee.

^ Jump up to: a b c d e "Evaluated Nuclear Structure Data File". National Nuclear Data Center.

Actinium series

The 4n+3 chain of uranium-235 is commonly called the "actinium series" or "plutonium cascade". Beginning with the naturally-occurring isotope U-235, this decay series includes the following elements: actinium, astatine, bismuth, francium, lead, polonium, protactinium, radium, radon, thallium, and thorium. All are present, at least transiently, in any sample containing uranium-235, whether metal, compound, ore, or mineral. This series terminates with the stable isotope lead-207.

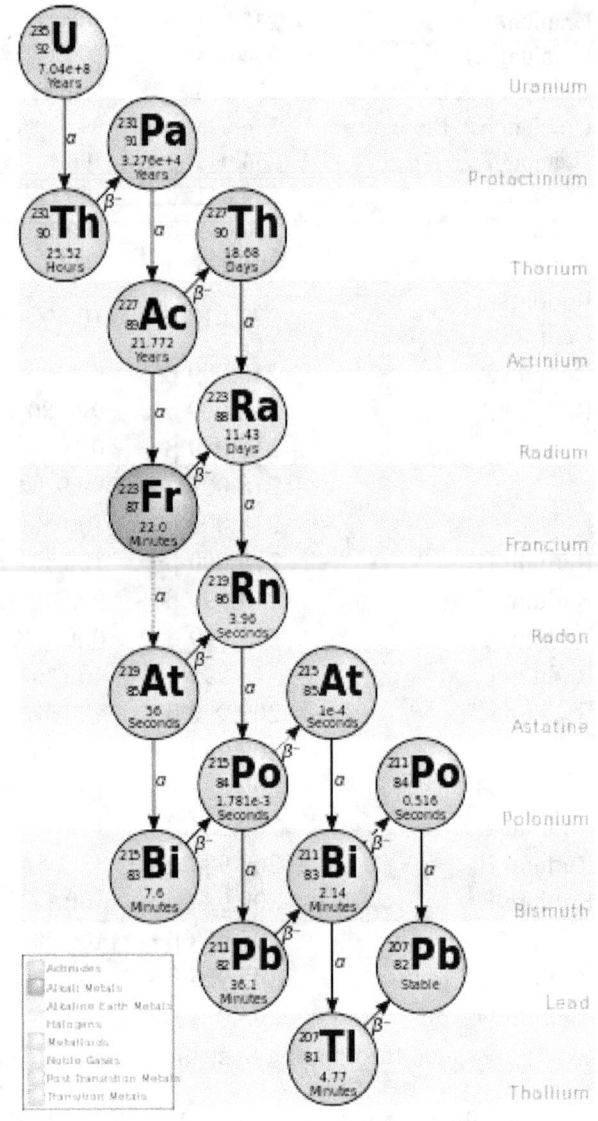

(More detailed graphic)

The total energy released from uranium-235 to lead-207, including the energy lost to neutrinos, is 46.4 MeV.

nuclide	historic name (short)	historic name (long)	decay mode	half-life (a=year)	energy released, MeV	product of decay
251Cf			α	900.6 a	6.176	247Cm
247Cm			α	1.56·107 a	5.353	243Pu
243Pu			β−	4.95556 h	0.579	243Am
243Am			α	7388 a	5.439	239Np
239Np			β−	2.3565 d	0.723	239Pu
239Pu			α	2.41·104 a	5.244	235U
235U	AcU	Actin Uranium	α	7.04·108 a	4.678	231Th
231Th	UY	Uranium Y	β−	25.52 h	0.391	231Pa
231Pa	Pa	Protactinium	α	32760 a	5.150	227Ac
227Ac	Ac	Actinium	β− 98.62% α 1.38%	21.772 a	0.045 5.042	227Th 223Fr
227Th	RdAc	Radioactinium	α	18.68 d	6.147	223Ra

223Fr	AcK	Actinium K	β− 99.994% α 0.006%	22.00 min	1.149 5.340	223Ra 219At
223Ra	AcX	Actinium X	α	11.43 d	5.979	219Rn
219At			α 97.00% β− 3.00%	56 s	6.275 1.700	215Bi 219Rn
219Rn	An	Actinon, Actinium Emanation	α	3.96 s	6.946	215Po
215Bi			β−	7.6 min	2.250	215Po
215Po	AcA	Actinium A	α 99.99977% β− 0.00023%	1.781 ms	7.527 0.715	211Pb 215At
215At			α	0.1 ms	8.178	211Bi
211Pb	AcB	Actinium B	β−	36.1 min	1.367	211Bi
211Bi	AcC	Actinium C	α 99.724% β− 0.276%	2.14 min	6.751 0.575	207Tl 211Po
211Po	AcC'	Actinium C'	α	516 ms	7.595	207Pb
207Tl	AcC"	Actinium C"	β−	4.77 min	1.418	207Pb
207Pb	AcD	Actinium D	.	stable	.	.

Periodical Table

Group →	1	2	3		4	5	6	7	8	9	10	11	12	13	14	15	16	17	18
Period																			
1	1 H																		2 He
2	3 Li	4 Be												5 B	6 C	7 N	8 O	9 F	10 Ne
3	11 Na	12 Mg												13 Al	14 Si	15 P	16 S	17 Cl	18 Ar
4	19 K	20 Ca	21 Sc		22 Ti	23 V	24 Cr	25 Mn	26 Fe	27 Co	28 Ni	29 Cu	30 Zn	31 Ga	32 Ge	33 As	34 Se	35 Br	36 Kr
5	37 Rb	38 Sr	39 Y		40 Zr	41 Nb	42 Mo	43 Tc	44 Ru	45 Rh	46 Pd	47 Ag	48 Cd	49 In	50 Sn	51 Sb	52 Te	53 I	54 Xe
6	55 Cs	56 Ba	57 La	*	72 Hf	73 Ta	74 W	75 Re	76 Os	77 Ir	78 Pt	79 Au	80 Hg	81 Tl	82 Pb	83 Bi	84 Po	85 At	86 Rn
7	87 Fr	88 Ra	89 Ac	**	104 Rf	105 Db	106 Sg	107 Bh	108 Hs	109 Mt	110 Ds	111 Rg	112 Cn	113 Nh	114 Fl	115 Mc	116 Lv	117 Ts	118 Og

*		58 Ce	59 Pr	60 Nd	61 Pm	62 Sm	63 Eu	64 Gd	65 Tb	66 Dy	67 Ho	68 Er	69 Tm	70 Yb	71 Lu
**		90 Th	91 Pa	92 U	93 Np	94 Pu	95 Am	96 Cm	97 Bk	98 Cf	99 Es	100 Fm	101 Md	102 No	103 Lr

Data of nucleus: https://www-nds.iaea.org/relnsd/vcharthtml/VChartHTML.html .

Comparison of neutronicity of reactions

Reactants	Products	Q	n/MeV
First-generation fusion fuels			
$^2D + ^2D$	$\rightarrow ^3He + ^1_0n$	3.268 MeV	0.306
$^2D + ^2D$	$\rightarrow ^3T + ^1_1p$	4.032 MeV	0
$^2D + ^3T$	$\rightarrow ^4He + ^1_0n$	17.571 MeV	0.057
Second-generation fusion fuel			
$^2D + ^3He$	$\rightarrow ^4He + ^1_1p$	18.354 MeV	0
Third-generation fusion fuels			
$^3He + ^3He$	$\rightarrow ^4He + 2^1_1p$	12.86 MeV	0
$^{11}B + ^1_1p$	$\rightarrow 3\ ^4He$	8.68 MeV	0
Net result of D burning (sum of first 4 rows)			
6 D	$\rightarrow 2(^4He + n + p)$	43.225 MeV	0.046
Current nuclear fuel			
$^{235}U + n$	$\rightarrow 2\ FP + 2.5n$	~200 MeV	0.001

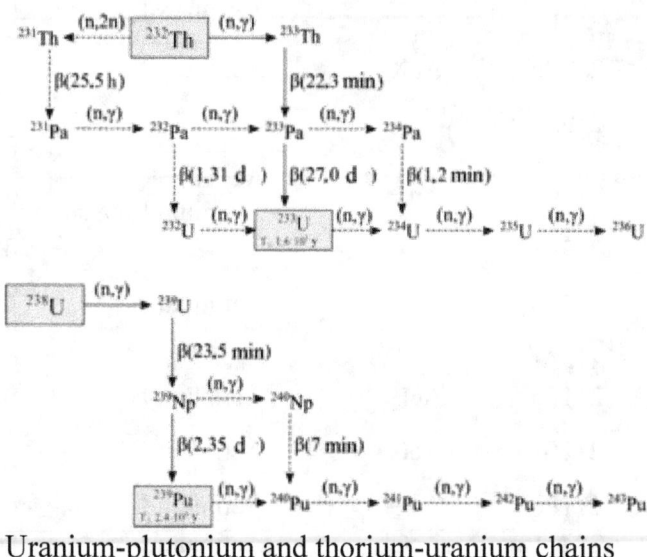

Uranium-plutonium and thorium-uranium chains

===

Short biography of Bolonkin, Alexander Alexandrovich <abolonkin@gmail.com>

Alexander A. Bolonkin was born in the former USSR. He holds doctoral degree in aviation engineering from Moscow Aviation Institute and a post-doctoral degree in aerospace engineering from Leningrad Polytechnic University. He has held the positions of senior engineer in the Antonov Aircraft Design Company and Chairman of the Reliability Department in the Clushko Rocket Design Company. He has also lectured at the Moscow Aviation Universities. Following his arrival in the United States in 1988, he lectured at the New Jersey Institute of Technology and worked as a Senior Scientist at NASA and the US Air Force Research Laboratories.
Bolonkin is the author of more than 250 scientific articles and books and has 17 inventions to his credit. His most notable books include The Development of Soviet Rocket Engines (Delphic Ass., Inc., Washington , 1991); Non-Rocket Space Launch and Flight (Elsevier, 2006); New Concepts, Ideas, Innovation in Aerospace, Technology and Human Life (USA, NOVA, 2007); Macro-Projects: Environment and Technology (NOVA, 2008); Human Immortality and Electronic Civilization, 3-rd Edition, (Lulu, 2007; Publish America, 2010):Life and Science. Lambert Academic Publishing, Germany, 2011, 205 pgs. ISBN: 978-3-8473-0839-3.
http://www.archive.org/details/Life.Science.Future.biographyNotesResearchesAndInnovations; Femto technology and Revolutionary {rojectsts, Lambert, 2011, p.530; New Methods of Optimization and their Application, Moscow High Technical University named Bauman (in Russian: Новые Методы Оптимизации и их применение. МВТУим. Баумана, 1972г., 220 стр). List and links of Bolonkin's publication: https://archive.org/details/List5.3s8518. Homepage: http://Bolonkin.narod.ru , https://en.wikipedia.org/wiki/Alexander_Bolonkin .